Introducing the Magnetic Horoscope, and Our N... Forget Dogs, Rats, and Monkeys of Chinese Horo...

"...se things are marvels, just beautiful."

2016/2004/1992/1980
The Lifter,
When people need dirty jobs done, or heavy industry done, you'll be there (to junk an iron car, or lift a nut out of a crack). Beware, not all problems require heavy lifting.

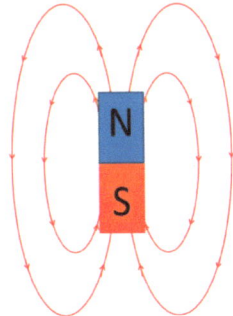

2017/2005/1993/1981
The Bar Magnet,
You are fundamental, the root reason of all good. Beware, sometimes your analytical behavior turns people off.

2018/2006/1994/1982
The Generator,
You are the self-less, un-tiring worker for the people (power the house, power the factory). Beware, younger technology may outshine you and get more press, but you are the backbone of society.

2007/1995/1983/1971
The Speaker
Yeah, you want music. You are the rebellious, artistic type, yet also common, subtle and respected. You are strong and confident, and, as you get brittle with age, you know there are strong replacements.

2008/1996/1984/1972
The Pure Electric,
You're out to save natural resources, but not yet entirely practical. You have lots of panache and flair. Beware, there are many ways to save nature.

2009/1997/1985/1973
The Games,
You are fun loving and active. You can be the muse for many practical ideas. Beware being too frivolous.

What is your magnetic personality?

2015/2003/1991/1979
Mini Quadcopter,
You've come a long way, baby. Maybe you'll fly away with people someday.

2014/2002/1990/1978
The Hybrid Electric,
You're out to save natural resources, in a practical way. You will compromise to get the job done. Beware, standard combustion engine light-car solutions also work, so don't get arrogant.

2013/2001/1989/1977
The Horseshoe Magnet,
You are willing to be re-shaped to solve problems, so you are practical. You teach society how to exploit magnetic forces.

2012/2000/1988/1976
Joe-cool flying ace,
You're out to cause a flying revolution. You'd replace jet engines, and allow vertical take offs and flying taxis where battery and plane design go together. You are the Venture Capitalists of the age. Beware blimps, helicopters, and old proven technology.

2011/1999/1987/1975
Mr. Utility Motor,
You're the strong silent type. All people know and respect you. You are satisfied to be very useful. Beware being too humble, and letting ridiculous other ideas claim to be better without challenge.

2010/1998/1986/1974
The Drill,
You are a person of action. When others need a job done, they come to you. Beware acting too fast.

Introduction to Magnets, Motors, and Generators

Motors and generators: Magnetic fields are awesome and very practical.
We take motors and generators for granted because they are so successful and reliable. Thank magnetic fields. Magnetic fields have completely changed and improved our way of life, compared to just 200 years ago. We have electric motors everywhere, in refrigerators, fans, elevators, cars, computers, and toys. Generators power our cell phones, air conditioners, and lights.

This book shows that magnet and motor experiments have real uses and applications. Motors and generators are important to modern life styles (starting your car, powering your computer and TV, playing DVDs, electric tools, dishwashers, air conditioners, elevators, household water pressure).

When my great grandfather was a kid, he had to play drums or trumpets for music, and not electric guitars.   Thank magnetic fields and speakers. When I was a kid, boom boxes would spin music cassette tapes and not DVDs, and speakers would make the sound. Thank electric motors and magnetic speakers. Without electric motors, your life would be different. For example, without electricity, you would use a hand pump every time to turn on the sink faucet, to wash dishes, or to take a bath.

I first got into showing kids the basics of motors when helping with Cub Scouts.   I was impressed that there was a basic 1-pole motor to demonstrate the concept, and that the kids could build it themselves. It is highly likely that the scout leaders themselves learn more from doing the activity.

Electric motors are possible because magnetic fields create mechanical forces, like a vacuum cleaner sucking up dirt or a kid spinning a spinning wheel at a playground. Magnetic fields are created from electric current, which allows motors and mechanical spinning. Electric motors move half the things around us, like tools, refrigerator pumps, and Blu-ray disks. The other half, mostly for cars, is the combustion engine.

Generators are also possible using magnetic fields. Generators are very similar to electric motors, just in reverse, and supply most of the electricity you use every day. Generators work because moving wires through magnetic fields go the other way and create electric current. By steam power or water power, we spin turbines and generators to create electricity that powers our cities and lives, such as light bulbs, computers, and dishwashers.

The magnet/motor/generator concepts and history are all in the main part of the book, and the details or the more advance material are placed in the appendices, so the main part of the book keeps a lighter flow of concepts.

How to read this book:
This book covers lots of things: magnetic forces, motors, generators, experiments, and sometimes even equations. Please jump around and see whatever you want. If any one page is just not working for you, just jump to the next page.

Please try a few of the experiments listed on the right side of the pages, indicated with a star. Some are just 'go-see-them' activities, or 'search-and-find', and others involve getting magnets, compasses, and toy motors, toy generators, and volt meters.

Symbols of level or difficulty of math (with a nod to ski slopes):

Easy :
4th grade

Algebra :
8th grade

Algebra and trigonometry :
10th grade

Would you rather live like the people who struggled in the year 1900?

To use tools, there are labor intensive options that change your lifestyle:
- Live near flowing water, and use belt driven tools.
  - Example: the textile mills near water were dominating industry, with water powered looms.
- Get horses, donkeys, or people to turn tools
- Use steam engines, powered with coal or wood.

Do you like the older lifestyle? Would people ride a bicycle to power an air conditioner?

PHOTO © PAUL STARKEY

Manual water pump to irrigate crops

Without magnetic fields, electric motors, and generators, your life would be different and less convenient.
- **The whole national electrical power grid is powered by generators, leading to your 120 volt power outlet.**
- **We plug motors, computers, cell phones, and ovens into wall outlets and just expect convenience.**

Publish date: 3/14/2022, Revision date: 7/26/2024

Outline

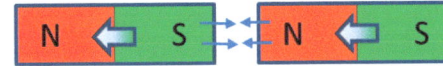

Opposite poles attract and their fields are in same direction or aligned

Chapter 1: Introduction to Magnets, Motors, and Generators, a Quick Summary of the Entire Book:

In this book on magnets and their uses, we first look at magnetic forces from permanent magnets and electro-magnets, which attract each other and magnetize and attract iron. Magnetic attraction enables motors and generators, which we use every day. Generators power our houses and cities.

Chapter 2: DC (static) Magnetic Forces and Lift Forces, Electro-magnets, and Speakers:

Get familiar with magnetic forces between permanent magnets. This is called DC magnetic forces, or Static Field magnetic forces. These forces are easy to demonstrate around the house. Just get some permanent magnets and lift some iron spoons, or attach the magnets to the refrigerator.

Chapter 3: Electric Motor Concepts:

Look at what a simple toy motor can do, to create mechanical rotation as a motor.

We open up the simple DC toy motor and look at the parts. The stator is a permanent magnetic field, and the spinning rotor has a shaft and coils. The coils on the rotors have a controllable electro-magnetic field, with current switching between coils using the commutator.

We compare DC and AC motors, and brush and brushless motors.

Chapter 4: 1-Pole and 2-Pole Demonstration Motors:

Build different demonstration motors. Motors are a conversion of electrical to mechanical power; that is, conversion between a chemical battery to a spinning shaft of the motor. For simplicity, these demonstration motors are 1-pole and 2-pole and have little power. We explore the reversal of spin direction by reversing the stator permanent magnetic field, or reversing the current in the rotor coil.

Chapter 5: Electric Motors Around the House:

Let's explore where electric motors are obviously contributing to people's lives. Find all the motors in a house, like computer hard drives, fans, pumps in refrigerators, clothes washers and water well. Motors are also in a car, train, or boat. Find future uses of electric motors, like robots, EV cars, and airplanes.

Chapter 6: Generators:

A simple toy motor can also be a generator, just spin the shaft by hand, by hot steam, or flowing water. Various toy generators introduce all the power plants around the country. Coal, natural gas, nuclear heat all can heat water to form steam and turn a turbine and generator. 'Green' alternatives also exist like hydroelectric, geothermal, wind mills, solar thermal and solar cells.

Toy motor

Ceiling fan

Refrigerator

One-pole motor demonstration

Two-pole motor demonstration

Emergency generator

Two-pole generator demonstration

Industrial generators in hydroelectric power dam

Table of Contents: Magnets, Motors, and Generators

Chapter 1: Manual Labor in the Past without Magnetic Motors and Generators...   7

Chapter 2: Static Magnetic Forces 22
A: Some Uses of Magnetic Forces
B: Lift Using Iron and Electro-Magnets with Iron Cores
C: People's Use of Static Attraction: Games, Speakers, and Gripping

Chapter 3: Motors and Magnets Flipping or Getting Pulled, enabling life's conveniences as we know it 53
A: Motor Introduction: Magnets Rotate to Align, and Move in Closer
B: Motors, the Basic DC Motor Using Electro-Magnets
C: Direct Current, Alternating Current Motors: Not That Different
D: DC Brush and DC Brushless Motors

Chapter 4: 1-Pole and 2-Pole Demonstration Motor Experiments ... 90
A: 1-Pole and 2-Pole Demonstration Motor Experiments

Chapter 5: How Many Electric Motors in Your House or Cities?... 103
A: How Many Electric Motors in Your House?
B: Imagine Life Without Electric Motors and Magnetic Fields
C: People Movers and Cars
D: People Movers and Boats, Trains, Elevators, Airplanes.
E: Electric Motors and Robots

Chapter 6: Generators, Magnets flipping near coils, enabling quality of life as we know it 148
A: Generators, Motors can Work Both Ways
B: Examples of Generators for Power Plants
C: Imagine Life Without Magnetic Fields for Electric Generators and Power Plants
D: History of Power Plants to Spin Generators

Chapter 7: Summary of Magnets, Motors, and Generators ... 187

Glossary of Magnets, Motors, and Generators ... 194
Appendices:
A: Main Concept: Magnet Force and Energy 195
B: Motors and Energy, the Buddha Way 203
C: 1 and 2 Pole Commutator: Current versus Time ... 208
D: Observations: Rotor Coil's Magnetic Field ... 210
E: Simple Circuit for 1-pole and 2-pole Motors ... 215
F: Fundamental Magnetic Forces 221
G: Equations for Magnetic Fields, Forces, and Torques.... 227
H: Subtle Coil Behavior of 1-pole Motor 229

Lift or Attraction Forces:
- Alignment (same direction) and Gradient of magnetic fields

"Hard to argue with the complete every-where use of magnetic fields, from speakers, to motors, to generators, to computer hard drives, to lifting cars."

Motors:
- Magnetic Forces
- Turn / Twist for Motors:
  - Alignment, no Gradient, of magnetic fields
- Turn / Twist for Generators

Magnets twisting to align

Industrial motor

Build 1-pole and 2-pole motor:
- See spin, and currents

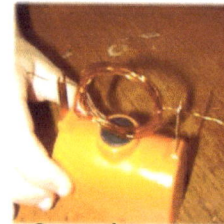

One-pole motor demonstration

Two-pole motor demonstration

Generators:

Voltage generation by spinning shaft of motor manually

Emergency generator by spinning shaft of motor using combustion engine

Industrial generator / power plant by spinning shaft of generator using hot steam or flowing water

Feel free to skip around the book, if there is something you most want to know.

Recommended Hands-on Kits to Try Experiments

This book is designed to have simple demonstrations of magnetic forces, magnetic motors, and magnetic generators. You'd have a good-old hands-on fun time, if you gather these items.

- 4 disk magnets
- 2 bar magnets
- Wire/nail/battery holder
- Iron filings (in a plastic pouch / gel)
- Compass
- 2 toy motors

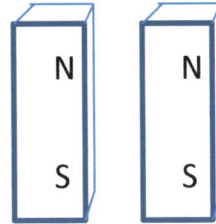

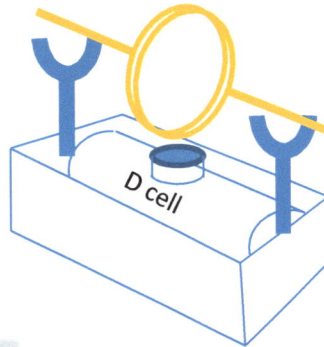

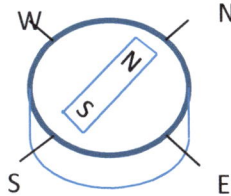

Optional motors/generators

Motor demonstration: 2-pole EUDAX STEM DIY Simple Electric Motor DC Motors

Motor demonstration: EUDAX School DIY Dynamo Lantern Educational STEM Building

Generator demonstration from air piston: Sunnytech Hot Air Stirling Engine

Generator demonstration from steam piston: Sunnytech Mini Hot Live Steam Engine

Toy DC motor: Adafruit DC Toy/Hobby Motor

Permanent magnets: DIYMAG Powerful N52 Neodymium Disc Magnets

Orienteering Compass

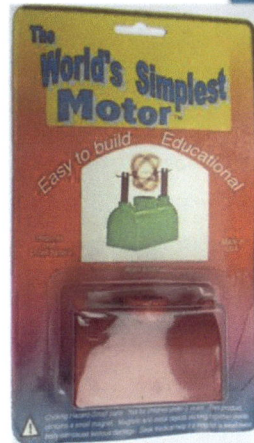

Motor demonstrations: TEDCO Simplest Motor A 1-pole motor

Bar magnets

D cell

Get a few, or all, of these toy magnets, electric motors, and generators, and try out some of the experiments in the book.

Will You Take Away These Iron Nuggets?

Here are the main points about magnets, electromagnets, forces and torques, motors, generators, and lifestyle, after you finish reading the book.

"Repeat these takeaways until you can eagerly tell a friend about electric motors, generators, and modern lifestyle!"

1. **Forces**: Magnets are attracted together, with attraction between opposite poles and alignment of magnetic fields in same direction.

2. **Electro-magnets**: Electricity running through a coil behaves just like a permanent magnet, and also allows the magnetic field to be turned off or reversed.

3. **Iron**: Iron does not have a strong net magnetic field until a bias magnetic field is applied, but then has a strong magnetic field. Iron in the electromagnet makes the coil magnetic field much larger and the mechanical torque much larger in motors.

4. **Twist of rotor**: Motors turn because the magnetic field of the middle rotor magnet wants to rotate to align itself with the outer magnet. This middle rotor needs to be an electro magnet so its field can also be switched on and off at the right rotation angle so that the rotor magnetic field stays 90 degrees sideways to the outer magnet field (from stator). The rotor coil is only excited when it is sideways to the stator.

5. **Common motors**: Here are 6 electric motors you probably have in your home: pump for refrigerant in refrigerator, motors for spinning washer and dryer, motor for electric drill, motor for fuel pump for heat, motor to spin Blu-ray/DVDs, starter motor for car engine.

6. **Electricity generation**: Generators create electrical power because a coil (the rotor) is spinning inside a magnetic field. The coil is spinning from mechanical forces, like a steam engine, or combustion engine, or water flow in a turbine. The outer stator magnetic field exerts a force along the wire on the moving charges in the moving rotor wires, which means a voltage is created.

7. **Common generators**: If you have a car, then you have a generator (alternator) in the car. In your garage, you also might have an emergency portable generator when power goes out. The biggest generators are and should be at big power plants, because that is the most efficient.

8. **Lifestyle**: You depend on electricity for your modern lifestyle, with lights, refrigerators, cars, computers, and cell phones.

Opposite poles attract

Iron core increases the electromagnet coil's magnetic field

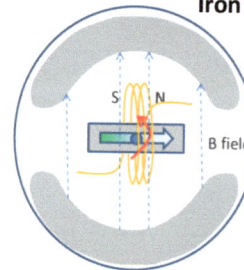

Rotors in motors twist to get their magnetic fields in same direction

Magnets waiting to align

Magnets after alignment of fields, strongly attracted

Refrigerator with motor to push coolant

Generator demonstration using flame, piston, and motor

Convenient electric tools using motors

Car with both starter motor and generator

Turbine shaft with blades to spin generator using steam

Here is trivial pursuit for motor engineers. The harmony between motors and magnetic fields are in these main points.

This chapter describes all the good that magnetic fields do for us, and is a summary of the uses of magnetic fields described through-out the book.

The magnetic fields are a fundamental field and force of nature. Electric fields and magnetic fields are both so fundamental. Any electronics can't function without either.

In the last 100 years, people have learned how to use magnetic fields for a lot of good things – motors for home dish washers, refrigerators and water pumps, generators at power plants to keep the wall outlet powered, and generators in cars to keep the spark plugs sparking.

"Life was harder back in the day, before the age of electric motors and refrigerators (yes, they need a motor)."

Chapter 1: Introduction to Magnets, Motors, and Generators, a Quick Summary of the Entire Book:

In this book on magnets and their uses, we first look at magnetic forces from permanent magnets and electro-magnets, which attract each other and magnetize and attract iron. Magnetic attraction enables motors and generators, which we use every day.

Here are some of the many applications of magnetic fields and motors and generators, all to make life easier.

- Factories: electric motors
- Computers: magnetic hard drives and all the electronic circuits
- Power plants: spinning coils in magnetic fields to get electric power
- Kitchen: dish washer, blenders, refrigerators

Why have people gone from colonial living, with hand washing and manual harvesting of crops, to driving cars, buying food at the grocery store, and talking to people on the other side of the Earth? Because people want more convenience, more knowledge, and more opportunity. Magnetic fields, being a fundamental part of nature, have something to do with that.

The various ways an electric motors can be designed is amazing. Some motors take a direct current (DC) from a battery, and some motors take an oscillating or alternating current (AC) from the power lines. Either way works.

Generators are needed to power all motors, and all electronics. Generators just do the reverse of an electric motor. Some mechanical force, like hot steam, spins a coil in a magnetic field and electric power is produced.

Without electric power and motors

Basic modern life with electric power

Hand wash clothes

Tilling the soil without a tractor

Ice cellar to freeze your deer meat.

Lights after dark

Drill, spinning the drill bit

Compressor in the refrigerator

Thank electricity and power plants / electric generators / electric motors to move your car and lift you up buildings.

Manual Labor in the Past Without Magnetic Motors and Generators: Who's Your Daddy?

This chapter describes all the good that magnetic fields do for us, and is a summary of the uses of magnetic fields described through-out the book.

For example, life used to be more physical. People back then probably would have laughed at exercise bicycles to stay in shape, when the same people had just picked crops to eat, walked miles to the store, taken care of horses, and hauled wood to the furnace.

"Life was harder back in the day, before the age of electric motors and refrigerators (yes, refrigerators need a motor)."

Experiment 1.1: Start a lawnmower small engine by pulling a pull cord. Imagine the strength to start a car with a large engine with a manual hand crank instead of a turn-key starter motor.

Electric Motor and Elevators: building height and urban density

1890

Stairs keep you healthy, but the rich owned the first floor before elevators, or just had their own mansions, without hiking up stairs.

Then, in the 1920s with elevators and cheaper steel, mansions in Manhattan are increasingly torn down in favor of luxury high-rises.

Elevators also allowed much higher buildings, so the density of people can be much higher in the city now. The rich were given options to live in the penthouse on top, with special elevators, sometimes in exchange for breaking down their mansions.

Electric Motor and Starting Cars and Headlights

1950

Hand crank to start a car   **Use starter electric motor**

Colonial America — 'Titans of Industry' America — **1890** — Titanic sinks — **1910** — **1920** — WW2 — Baby Boomers born — **1950**

We built a lot of character in the old days, and there were less heart attacks, and stronger legs and backs.

Tilling the soil without a tractor

Do you want to go back to the old life, when over half the people are farmers on small farms?

1920

Electricity and Cooking: used electricity from far away power plants, instead of indoor natural gas or wood.

Fire from Wood / Coal Stove, for bread, eggs, hot coffee, and heat. Cotton clothes were good because cotton doesn't burn as easily.

Experiments: Get sweaty and start a small engine lawnmower with a pull cord, and walk up a 10 story building without an elevator.

Thank electricity and power plants / electric generators / electric motors to move your car and lift you up buildings.

Manual Labor in the Past Without Magnetic Motors

Back in the day, to preserve food you needed to dig your own ice cellar, and buy or get ice from your local lake in the winter. Now, gratefully, we have electric refrigerators. Refrigerators need electric motors to push the refrigerant around, and keep your food cold.

Back in the day, people used to only eat what was in season (apples, grapes, oranges in the summer), and what was local (beef from the butcher, or fish). Now refrigerated storage trucks can bring food from all around the world, at any time of year.

"Life was harder back in the day, and you could not get your apples, bananas, pineapples all year round, because refrigerators did not keep them fresh, or transport them across the country."

Experiment 1.2: Go back 150 years (or to your backyard in the winter) and see how long ice stays frozen in a camping cooler during the summer. Back then, they would have used ice, a pit in the ground, and a covering of hay for better insulation, to get a few months of summer refrigeration.

Cooling Food and Electric Motors

electricity

Using an electric motor, a high pressure gas is pushed out a small hole, where it evaporates and cools. This cools the coils in the refrigerator.

Ice cellar to freeze your deer meat.
Ice can keep a cellar cool all summer long. 'Back in the day' before refrigerators, there was a whole industry cutting ice blocks from the lakes.

Ice box in homes.
Ice can keep food cold for a few days

Refrigerators became common
1930

1880   Refrigerators invented with pumped refrigerant gas

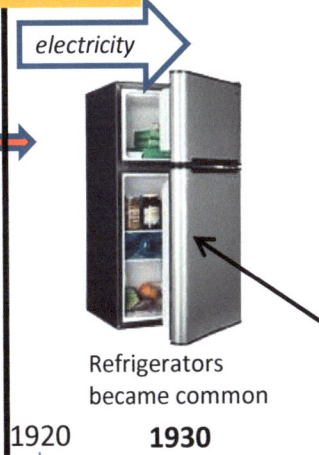

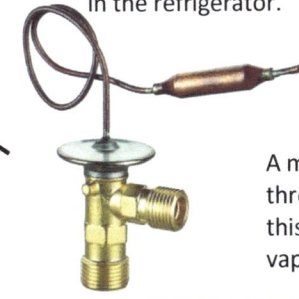

1920

A motor pushes liquid through a narrow hole in this cavity, where it vaporizes and cools.

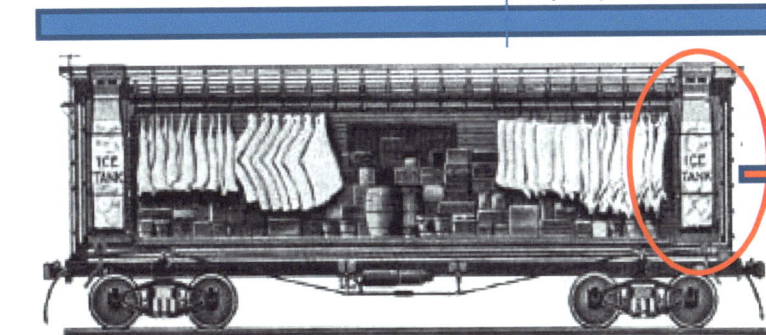

Florida and California can ship fruit around the country.

Ice

Air Conditioner

An image from the 1870s shows an early refrigerator car design. At this point, a car could travel a little over 250 miles before it would need to be "re-iced."

Refrigerated train and truck for long distance food delivery (bananas, eggs, milk, meat), using an electric air conditioner. Travel distance is not limited because the food is kept cold.

Experiments: See how long ice stays frozen in a camping cooler. That's how long a truck or train had to drive to bring you food.

Thank electricity, from power plants and generators, to power electric motors to keep food cold and help grow food.

People's First Knowledge: Earth's Magnetic Field

Earth's magnetic field is created by moving molten iron and rock near the core of the planet. Electric charges are carried along with the molten rock, creating an electric current and magnetic field.

Experiment 1.3: Find north using a compass. If you are in the northern hemisphere, north should line up with the North Star at night.

"Bring a compass on a trip, so you at least know you aren't going totally opposite to where you think you're going.

Here's a riddle:
'If you get lost somewhere
There's no need to raise a rumpus
First of all, just stay calm
And pull out your trusty _ _ _ _ _ _ _ ' "

Earth's North geographic pole, but magnetic south pole

- **Iron Needle Rotation: Compass needle lines up with Earth's magnetic field, where opposite poles attract**
- Iron Needle Forces: There is no tearing of the compass needle out of your hand. Instead, the needle just rotates to align. This is because Earth's magnetic field is mostly constant or uniform compared to the small size of a compass, discussed later.

Earth's South geographic pole, but magnetic North pole (opposite poles attract)

Compass needle rotation:

A compass needle will rotate to line up with the Earth's magnetic field. The compass needle is a permanent magnet and 'opposite poles attract'. North wants to kiss South.

In terms of physics lingo, the compass magnetic needle will have lower energy when the Earth's field and compass needle's field are aligned in same direction, which leads to the rule 'opposite poles attract'.

What is a magnetic field?
A magnetic field is a force field that is created by moving electric charges (electric currents) and magnetic dipoles (iron). Magnetic fields exert a force on other nearby moving charges (electric currents) and magnetic dipoles.

Quick facts about Earth's magnetic field:
- Earth's magnetic poles switch about every 10 thousand years.
- The Earth's magnetic field is caused by the huge currents in the Earth's iron core.
- When a planet cools down, it loses its molten core. Without moving charges, the magnetic field gets much weaker.
- The moon has little iron, and no molten core, so no magnetic field. Evidence shows that the moon was made from Earth's outer surface, of oxygen, silicon, and aluminum, possibly the splatter from a smaller planet hitting and merging with Earth during the formation of the solar system. None of those elements are magnetic.

Experiment: Get a magnetic compass and find the North direction.

Everyone should know where North is, if they are lost in the city, desert, or ocean. Just get a magnetic compass, or learn to find the North star or know the direction of sunrise in the east.

Lodestone Magnetic Compass, in Pre-Historic Times

A modern pocket compass is just an iron needle that has been magnetized using a strong permanent magnet. These modern iron needles, pivoting on a pin, are convenient but not historical.

"A thousand years ago people could still make a compass ... good for seafaring lads in middle of ocean without landmarks on a cloudy day."

Compass needles can be magnetite 'Lodestone' which is just rock with lots of iron, the only naturally occurring magnet. If a magnet is allowed to rotate freely, for example by floating the magnetic rock on a dish on water, the stone or needle will rotate to align its magnetic field in the same direction with the Earth's magnetic field.

Raw magnetic material

Magnetite or Lodestone, attracting iron nails

Naturally magnetized Magnetite attracts iron paper clip.

These old fashion compasses below allow a bar magnet to pivot without friction, to point North.

Compass 1

Experiment 1.4: Float a permanent bar magnet on a floating dish to make a compass.

Magnet floating on dish in water, points North.

Compass 2

Magnetite Lodestone bullet, pivots North-South

Compass 3

Why float, when you can pivot, to make a compass.

Ancient Chinese magnetic compass. Apparently, the Vikings did not know about magnetic objects aligning North.

Experiments: Float a bar magnet in a floating dish to measure North

Make your own magnetic compass, using a magnet that is free to spin, such as flotation on water

Magnetic Fields Allow Electric Motors You Use

Here is a conglomeration of what's to come in this book, regarding motors. Look around your house. Electricity is very useful, with refrigerators, computers, anything that moves with motors. There are uses of electric motors everywhere, even if motors are not a thing of beauty. If imitation is the best form of flattery, then electricity and electric motors have some honest respect.

"You depend on me."

DC motors: wherever you have a battery

Permanent magnets, iron cores, and wire current loops, or no permanent magnet and stator is powered by the DC current.

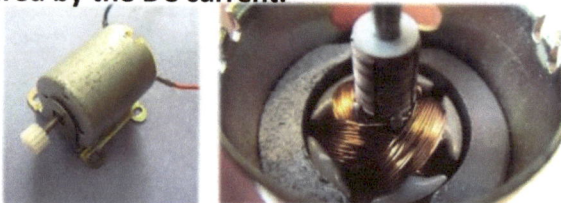

Toy DC motor

DC motor applications

Alternator / generator in car

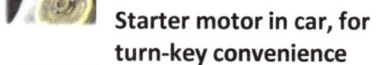

Starter motor in car, for turn-key convenience

Remote Controlled airplane, with a high energy density rechargeable battery (lithium ion)

Starter motor on snow blower, if the engine is heavy enough to make pull cord too exhausting.

AC motors: wherever you have a wall outlet

Iron cores, and wire current loops (no permanent magnet). Both the stator and the rotor are powered by the AC current.

Industrial AC motor

AC motor applications

Compressor inside the air conditioner

Washer / Dryer, spinning

Fan on your ceiling

Drill, spinning the drill bit

Compressor in the refrigerator

Treadmill, rotating motor pulls belt

Garage door opener, pulling up door

Oil pump for oil furnace

Water pump for a well

...And many other applications of motors ...they are everywhere

(see Chapter 5A to identify electric motors in your house)

Magnetic Fields Allow Motors and Generators: Motors Can Work Both Ways, to Make Electricity or Mechanical Motion

Welcome to the national power grid. There is a huge network of power generation built into everyday cities and across the countryside from power plants.
Here is the circle of electricity: that is, the conversion between electrical energy and mechanical energy, both ways.
Typically, in power plants, heated water or steam spins a shaft and a magnetic field or coils to create electricity, and electricity then does useful things, like spin drills and turn on computers.

'Dude, this just blows my mind. How simple and elegant. Motors, Electricity, you're everywhere and I'm so sorry that I have taken you for granted all my life. Besides, your electricity from huge traditional power plants is cheap and plentiful, compared to other options like gasoline driven portable generators.'

Here we have the beautiful two-way nature of motors. Motors can be used to generate mechanical motion, like spinning, when current runs through their rotor wires in a magnetic stator field. Conversely, by spinning the shaft manually and forcing the wires to move in a stator magnetic field, motors can be used to generate electricity, like at power plants, car alternators, or the portable emergency generator in your house when a tree falls on a power line.

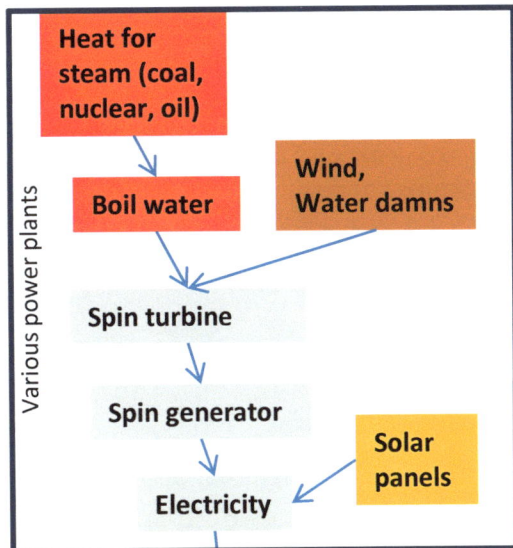

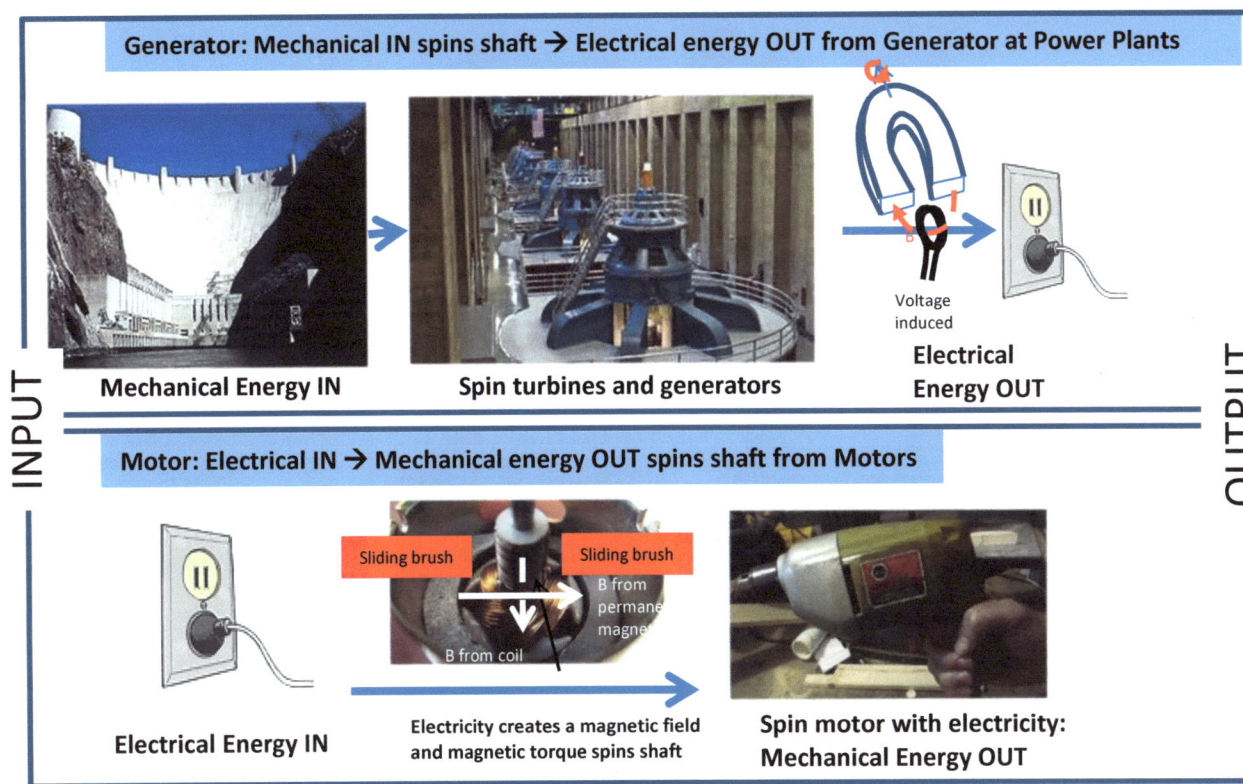

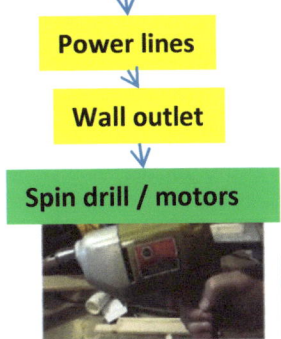

Various power plants

Heat for steam (coal, nuclear, oil) → Boil water → Spin turbine → Spin generator → Electricity

Wind, Water damns

Solar panels

Electricity → Power lines → Wall outlet → Spin drill / motors

INPUT

Generator: Mechanical IN spins shaft → Electrical energy OUT from Generator at Power Plants

Mechanical Energy IN   Spin turbines and generators   Voltage induced   **Electrical Energy OUT**

Motor: Electrical IN → Mechanical energy OUT spins shaft from Motors

Sliding brush   Sliding brush   B from permanent magnet   B from coil

Electrical Energy IN   Electricity creates a magnetic field and magnetic torque spins shaft   **Spin motor with electricity: Mechanical Energy OUT**

OUTPUT

The whole country has a power grid helping you use regular things, like computers to write, drills to build, ovens to cook food, and furnaces to heat your house.

(see Chapters 3 and 6)

Making Spin, or Making Voltage and Electric Power

Coils and a spinning magnetic field can do both functions: create mechanical spin when electrically power the coils, or generate a voltage when mechanically power or turn the shaft.

"This really makes my head spin...I'm getting recharged (and getting dizzy)."

Motors: Apply electric voltage to get mechanical spin

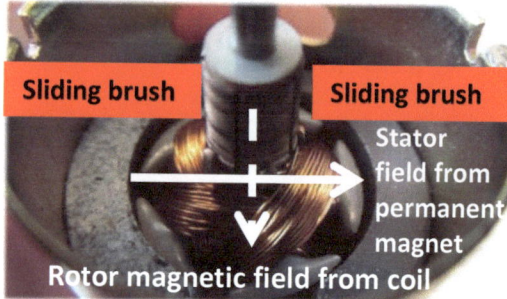

Sliding brush | Sliding brush
Stator field from permanent magnet
Rotor magnetic field from coil

Sideways magnetic orientation of magnet has most torque.

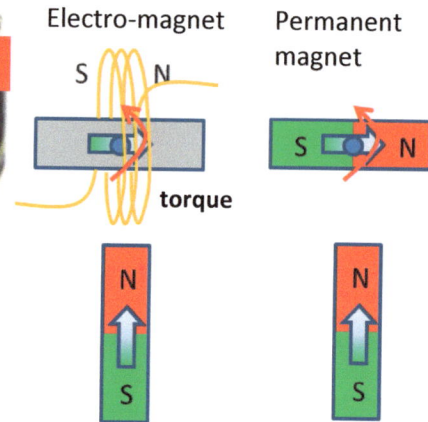

Electro-magnet | Permanent magnet

S N

torque

S N

Sequential coils, or electro-magnets:
Always excite the sideways coil as spinning so field from coil is sideways to permanent magnet and torque keeps rotor spinning in same direction. That is the purpose of the 'commutator'.

Battery powered DC motor:
Electro- magnets are required for motors because the rotor can reverse current in the coil for half the cycle, and keep torque in same direction.

Magnets want to align in same direction:
The electromagnet and permanent magnets above demonstrate that magnetic fields cause a twist. Permanent magnets can't work for a rotor. They'll just line up and stop spinning. Permanent magnets can't turn off magnet when spinning back down.
To spin, instead, use an electro-magnet to keep re-setting the rotor magnetic field sideways.

(see Chapter 3A for tilting permanent magnets and electro-magnets)

Generators: Apply mechanical spin to generate electric voltage

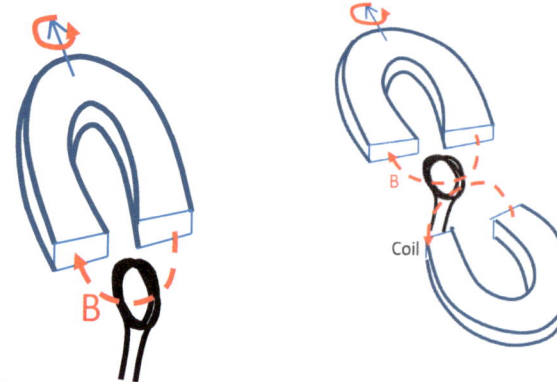

B

Coil

B

Alternating magnetic field through coil, the concept:
Alternating voltage is induced in coil when either magnet or coil spins
This spinning horseshoe magnet and coil generate a voltage, but this loose coupling of a horseshoe magnet and coil is not yet a compact generator.

136.6

A coil moving through a changing magnetic field will generate a voltage

Spinning rotor magnet to get voltage, in a DC motor:
Spin rotor or shaft mechanically to convert from mechanical energy to electrical energy.

coil

Spinning magnet

Spinning magnet near coil to get voltage:
Spinning drill chuck to spin a magnet (taped to drill) over a coil, connected to an AC voltmeter

(see Chapter 6 for generators)

Generators just need to use water or combustion engine to spin a coil in the magnetic field, which gives us all our voltage on a house walls. A magnet spinning around a coil, or a coil spinning around a magnet, will create a voltage in the coil.

Ways to Make a Shaft Spin Through History

Electric motors dominate, but there are other indispensable engines or motors, like combustion engines which make cars move, or water turbines and steam engines that spin the shaft of generators to generate electricity. When you get a shaft to spin, you can either create motion (a drill, or car, or train engine), or you can create electricity (a generator) by spinning a coil in a magnetic field, like below.

Start of the Industrial Revolution, with water and coal power

Post Industrial Revolution, when electrical power can be anywhere using transmission lines

1800-	1880-	1880-	1900-
Steam	**Water Turbine**	**Electric Motor**	**Combustion engine**

Steam

- Power plants originated after 1880
- Old-fashion steam locomotive for direct motion

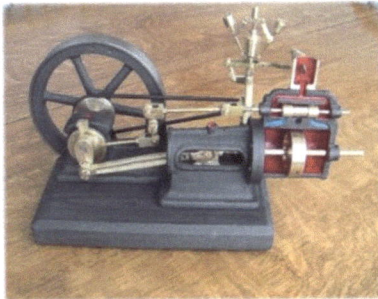

Old-time piston for powering generators or locomotives.

Modern fans inside of a steam turbine

Water Turbine

- Power plants after 1880

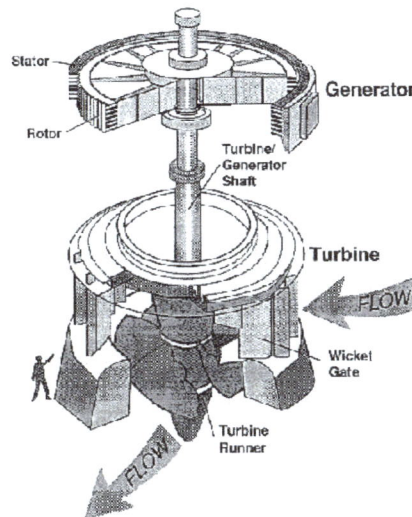

Stator
Rotor
Generator
Turbine/Generator Shaft
Turbine
FLOW
Wickel Gate
Turbine Runner
FLOW

Modern water turbines in hydroelectric power plant.

Electric Motor

- Starters in cars
- Electric cars
- Tools
- Most everything

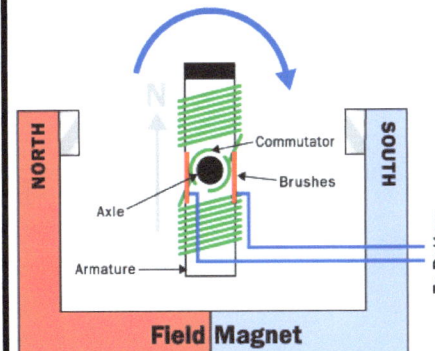

NORTH
SOUTH
N
Commutator
Brushes
Axle
Armature
To Battery
Field Magnet
©2001 HowStuffWorks

Electric motor concept:
Magnetic field in rotor coil or armature wants to rotate and align with the permanent stator magnets.

Combustion engine

- Car
- Lawnmowers
- Portable generators
- Propeller Airplanes (switch to turbofans for faster planes)

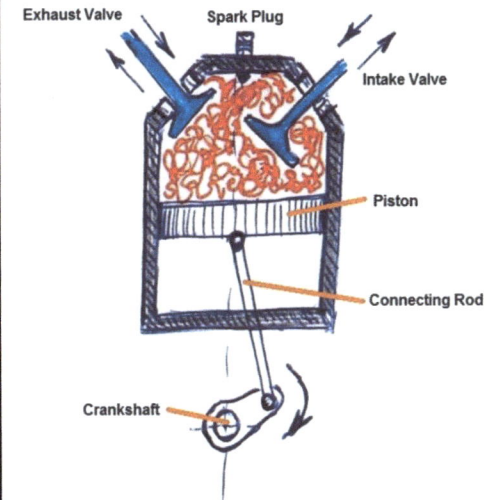

Exhaust Valve
Spark Plug
Intake Valve
Piston
Connecting Rod
Crankshaft

Combustion engine using exploding gas to push piston:
The huge energy in gasoline makes combustion engines a highly successful way to power cars.

All these ways to make a shaft spin are very common. We've been depending on electricity from spinning shafts for over 100 years.

Production and Spinning Shafts in Colonial America

Colonial America used water to power our factories. Big factories for sawing lumber or sewing fabric needed to be alongside large rivers.

Nowadays, we have a lot more convenience spinning a shaft using combustion engines and electric motors. We don't need to be near a stream or river or lake. We can have as big a factory as we want by bringing in more fuel or electricity, instead of being limited by the water flow.

Before electricity

Saw Mills

Water wheel to spin the circular saw or bandsaw, and to feed the wood:

Example of a large circular saw to cut timber into boards:

Here is a saw mill in the early 1800s, the saw mill at Old Sturbridge Village.

Flour Mills

Stone grinding wheel for wheat to make flour:

Water wheel to spin the grinding wheel:

Gearing up the grinding wheel:

Here is a grist mill in the early 1800s.

The whole purpose of the water wheel is to spin a grinding rock. The raw wheat or oats are poured down the center and slowly crunches its way out past the edges, now ground into flour.

So to have flour, there needed to be a grist mill in each village.

The shaft is gear up to make the shaft spin faster.

Now we use electric grinders or combustion engines to spin the grinder.

After electricity

Modern saw mill. We don't depend on water. Just use fuel or electricity:

Modern flour mill grinder is a lot more compact. Just bring in fuel for the combustion engine to spin the grinder:

People have used water to spin shafts for centuries. In the old days we need a body of water. Now we can be anywhere.

Magnetic Fields in Transformers, Compasses, and Computer Memory, Besides Motors and Generators

Magnetic fields have snuck their way into the basics of modern life. Below are further examples, like spark plugs for car engines, charging car batteries, computer memory, and protecting life on this planet from solar radiation.
Nothing says practicality like overwhelming use. Applications of magnetic fields have been welcomed and adapted.

"My magnetic field has done so many things, I should be King!"

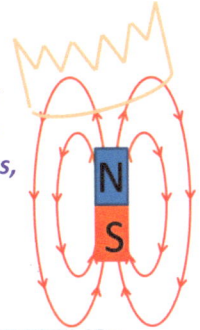

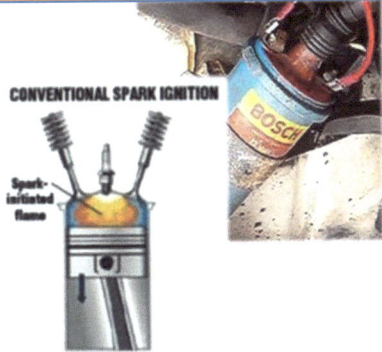

CONVENTIONAL SPARK IGNITION
Spark-initiated flame

Transformers to create high voltage spark for spark glugs

Generator or Alternator for self-contained car electrical power

Northern lights from sun's solar radiation (electrons, protons, alpha particles)

Spark Plugs:

High voltages for spark plugs (>1000 V) across a narrow gap cause lightning in the gas and combustion. The high voltages are created by quick current changes and magnetic field changes in a coil (like a generator).

- Without magnetic fields generating huge voltage, we could also use solid state transformers or piezoelectric ceramics, exploited in making Tasers. Diesel engines, where fuel spontaneously ignites from heat of compression, actually just need heat to explode.

Transformers, to convert voltage:

For high voltage power lines, from power plant to your home. Magnetic field changes between different coils create the high voltages in a transformer.

Re-charge car battery:

Charging the battery in a car uses voltages induced by an alternator / generator.

- If no alternator, then we might need to flex a 2 foot long ceramic to create voltages to charge the battery.

Computer memory:

Magnetic storage on disks is written to and read using an electro-coil or magnetic sensor above the spinning disk.

- Little magnetized dots are the 1s and 0s, magnetized in one direction or the other.
- Of course, now memory technology has advanced and transistors and resistive state changes are storing our data as well.

Earth's magnetic field protects life:

The Earth's magnetic fields protect us from the solar radiation (fast moving electrons and proton particles), which would sometimes hit our DNA and cause mutation and cancer.

- Thank the Northern Lights (aurora borealis), which displays the deflection of solar radiation, from same forces on moving charged particles that allow generators to work. Life might not exist without this protection, if DNA mutations can still reproduce as cancer.

Earth's magnetic field for navigation:

Compasses help with navigation and driving directions, or even getting oriented stepping out of the subway in a congested city.

Light:

Light is a pulse of magnetic and electric fields traveling at the speed of light.

Magnetic fields are used in most everything in nature and technology.

Transformers for low loss power distribution, enabling power transmission across the country

Magnetic disks for memory on computers

Compass for navigation on a ship, to know which way is North when the weather is cloudy and dark out, with no stars visible.

Computer Memory, Hard Drives and Magnetic Tape

Besides lifting forces for magnets, and besides twisting forces for motors and compasses, the magnetic field can be used for memory storage.

'Magnetic fields are used for itty bitty tiny 1s and 0s for magnetic memory, as well as practical refrigerator magnets, or huge electro magnets lifting cars.'

'What does Magneto from X-Men do when his computer gets dirty?
He wipes the hard drive.'

Computer Information Age

Information storage
1. Memory magnetic disk hard drives
2. Video tapes

Digital 1/0 memory on hard drive disks

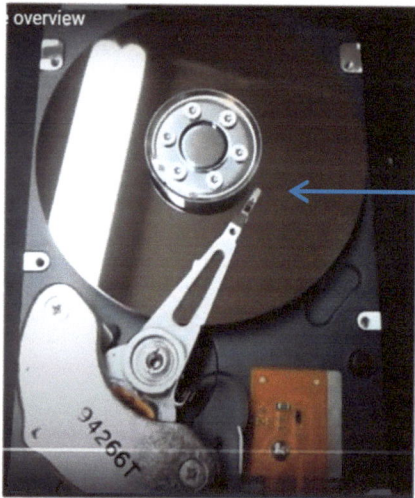

Computer hard drive disk

Magnetic memory is read by a pickup coil above the tape or film, or by a material which changes its resistance in a magnetic field (Giant Magneto Resistance GMR). Fundamentally, there is a magnetic force on electrons on the reader head due to up or down magnetic fields, so forces still are required to read the memory.

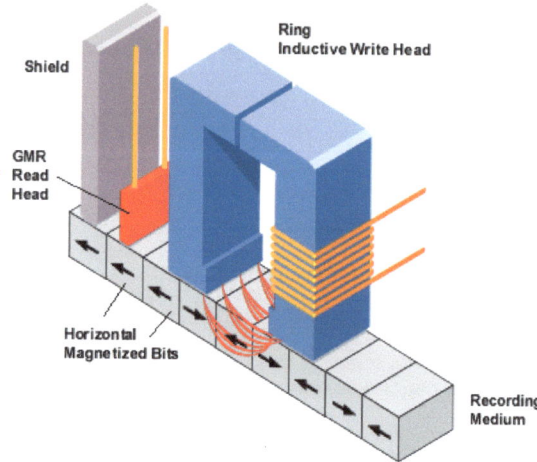

Shield
Ring Inductive Write Head
GMR Read Head
Horizontal Magnetized Bits
Recording Medium

Floating read/write magnetic head over spinning magnetic film.

A current through the reader head creates a localized magnetic field which magnetizes a spot (a '1') on the disk. The reader head also has a sensor to read the magnetic field as the disk is spinning underneath.

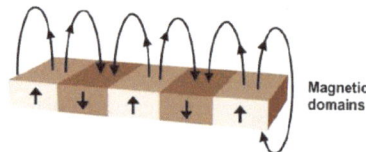

Magnetic domains

Magnetic memory on tape

Magnetic tapes for data storage: 145 Terra-Byte , much larger than disks.

Still in use:
- Magnet tape is still used for large computer data backups.

TOTAL AWESOME MIX

Video and audio cassette magnetic tapes

Outdated:
- Outdated magnetic tape cassette for music or movies, replaced by digital memory in computers, thumb drives, and by optical CDs.

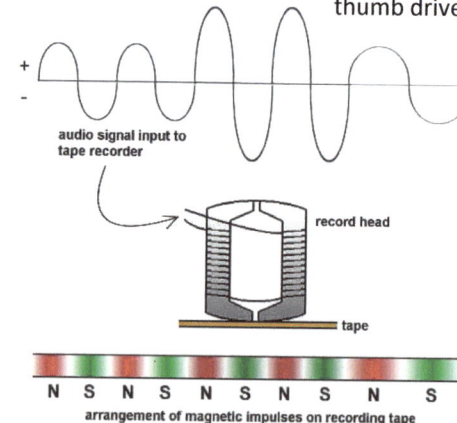

audio signal input to tape recorder

record head

tape

N S N S N S N S N S

arrangement of magnetic impulses on recording tape

Magnetized tape moving under the read head creates a voltage

Magnetic memory enabled computers and the digital age. Magnets are not just brute force grippers.

Pre and Post Electricity Through the Generations

Electricity, generated by moving magnetic fields, has helped bring us out of the cave man age, or age of physical labor farming in the field on a small farm, and into the lap of luxury. Do you picture your future as using a computer and watching TV for entertainment, or do you picture your future as endless toil in a field?

"Electricity has freed generations of people from endless drudgery. 'Toil, let my people go!' "

Whole long sweaty history without electricity and electric motors

None: 10,000 BC

Cavemen and fire: no electricity. Hard labor chopping wood with rocks.

None: 0 AD

Romans and hammers, spears, horses: no electricity. Hard labor growing crops and building boats and huts, conquering people and taking slaves.

None: 1000 AD

Knights and peasants: no electricity. Hard labor growing crops and building boats and huts.

None: 1800 AD

Hand labor cleaning clothes: no electricity. Hard labor growing crops, building boats, houses, and working in factory powered by water driven belts.

Water wheel for grinding wheat

Hand wash clothes

electricity →

1850 AD

Steam train: No electricity, just steam.

2020 AD (now)

Piles of coal for power plants and electricity.

Nuclear fuel rods

Large wind turbines

2050 AD: Will this be your standard future in 30 years?

Flying taxis with vertical take-off

Wind turbine floating in a balloon

Electric car powered by batteries or fuel cells using hydrogen.

Other power sources:
- Blimp generator above house
- Bio Mass Power Plant
- Geo thermal heat below house

Pre-history | 1800 CE | Now | Future? →

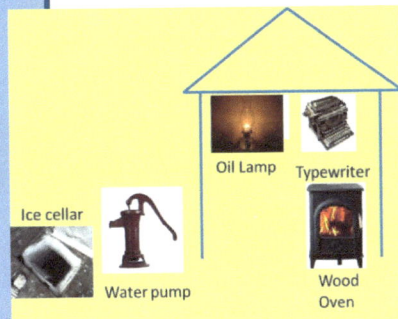

1800 house without electricity
- Oil Lamp
- Typewriter
- Ice cellar
- Water pump
- Wood Oven

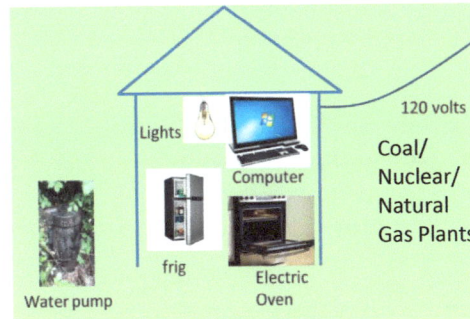

Current house with electricity, with computer, refrigerator, and oven
- Lights
- Computer
- 120 volts
- Coal/Nuclear/Natural Gas Plants
- frig
- Water pump
- Electric Oven

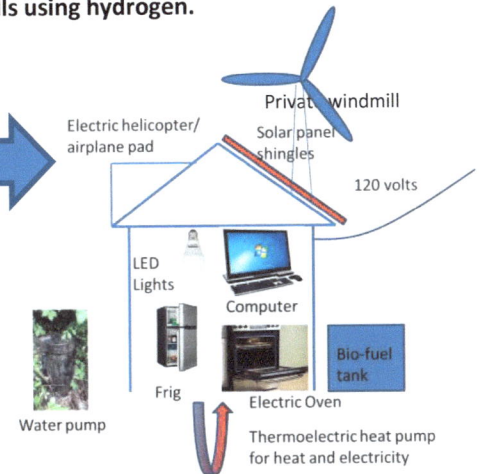

Future house with own generators, electric cars and electric helicopter.
- Electric helicopter/airplane pad
- Private windmill
- Solar panel shingles
- 120 volts
- LED Lights
- Computer
- Water pump
- Frig
- Electric Oven
- Bio-fuel tank
- Thermoelectric heat pump for heat and electricity

Electric car? Windmill generators? Computers? Electricity from magnetic fields can change lifestyles.

Future Electricity Generation and Supply

Magnetism is a fundamental property of matter. So to make electricity, we can upgrade and get an unlimited supply of energy to power the power plants, but the power plants will still be using magnetism to generate the electricity.

Old Days of water power

Just 150 years ago, water canals and rivers were all the rage to get a shaft to spin for factories, for weaving fabric or cutting wood.

For example, in the 1800s, mills were built along the banks of the Merrimack river in southern New Hampshire to make fabrics. Lowell Massachusetts and Providence Rhode Island had mills along the river. When diesel combustion engines started powering the mills, then the mill factory cold move anywhere there was lower cost labor, mostly south and the New Hampshire mills went into decline.

Modern Days of hard won electrical power

Nowadays, we have a diverse source of energy for power plants. We don't only depend on bodies of water, except for hydroelectricity, which can be sent anywhere using power lines.

Lack of energy is the root cause of a lot of conflict, so let's get lots of energy to make electricity.

- Coal
- Oil
- Natural gas
- Nuclear
- Solar cells
- Solar thermal
- Windmills
- Geothermal
- Ocean tides
- Ocean thermal gradient

Futuristic Days of unlimited electricity?

In science fiction stories, the future has limitless energy.
We assume fusion has been invented, and people can just take hydrogen from the ocean and combine them to get heat, to spin a generator.
Star Trek goes further than even fusion and assumes matter/anti-matter reactions to get huge energy and heat.
Or we assume there are huge solar panel power plants, or huge other heat sources that exist today.

The future in Star Trek does not have money, does not have energy limits, and does not have starvation. In truth, if we did have limitless free energy, then that idyllic scenario could be possible.

Want to make unlimited aluminum metal? Okay, then just run huge free electric currents through different ores.

Want to clean the atmosphere? Okay, then just run 100s of thousands of huge air filters using free electricity.

In a scenario with free unlimited electricity in the future, many upgrades are possible for civilization.

Motor and Magnetic Field Uses, Like Elevators

Here are some direct and un-intended effects of magnetic fields:

Motors all want to turn a shaft, but there are many different kinds of electric motors: AC, DC, and others.

Electric motors have had un-intended effects like tall city skylines, with elevators enabling construction of skyscraper buildings, and more free time away from chores like washing dishes or clothes.

Magnetic fields, besides motors, are used for compasses and magnetic memory.

"Magnetic fields and electric motors are mated at the hip. And you should be thankful!"

Mr. Wise
Greybeard
Motor

Question 1: When you are hiking or lost in a city and it is cloudy, how do you know where to walk?
- Take out your compass, because the sun and stars are not visible through the clouds for navigation. Of course, you need to know where North is.
- Imagine stepping out of the subway in an unfamiliar city with a map. How do you know where to go? A street sign does not tell you left or right. A compass does.

Compass for orienteering, using Earth's magnetic field

Question 2: Can a DC motor take AC current? No, because of the permanent magnetic for the stator.
- No, the motor needs a fixed current direction in the rotor relative to the permanent magnet stator. Current does not flow through the permanent magnet stator, so there is no way to synchronize the torque.

Question 3: Can a universal AC motor take DC current? Yes, because the stator magnetic gets excited by the dc current, just like an ac current.
- Yes, an AC motor can take a DC current, which is discussed in chapter 3.
- Both the rotor and stator magnetic field follow the current, so the torque on rotor is always in one direction, creating spin. Current flows in the stator, synchronized with the rotor current, so the right torque is preserved. Universal motors use coils around iron, and do not use a permanent magnet.

Many motor types, DC and AC

Question 4: What is easier, running up 40 flights of stairs, or taking the elevator?
- The easy answer uses an electric motor and an elevator. (Unless you think 'easy' means staying healthy and using any excuse to exercise, but that healthy attitude didn't catch on before the invention of elevators either. Poorer people before elevators got the top floors and buildings were shorter.)

Stairs, where rich people wanted the lower stories instead of the upper stories.

Question 5: Do elevators have an upper limit for stories? Ones with cables do. Long cables can be heavy and break.
- Yes, when the cables can't support their own length and weight, and the cables would snap. There is an upper limit if pulled up using thick cables, typically about 40 stories. The cables are very heavy and more stories adds length and weight for cable, so the cable can not even support its own weight.
- No, there is no upper limit if don't use cables and instead use motors on the elevator car itself. However, most elevators use cables and do not have motors on the car, probably because cables are the best choice: cables are not too heavy for shorter elevators and the elevator gets a counter weight. Using a counter weight, it takes a lot less power to raise an elevator because the weight of the cage is cancelled. Cars without cables don't have a counter weight. (Cage weight is cancelled, but the cable weight and passenger weight are still unbalanced.)

Elevator shaft, with cables pulled by electric motors and counter weights

Question 6: How is data recorded on a magnetic film? '1' and '0's for digital recordings are just different magnetization directions.
- Electro-magnets can write '1's or '0's to a magnetic disk in hard drives on computers. A 1 can be a left-ward magnetic field, and a 0 a right-ward magnetic field.

Data on magnetic disk and film

Here is trivial pursuit for motor engineers. The main differences between AC and DC motors are in these questions.

Magnets can be used for lifting, bearings, and levitation and trains.

This chapter introduces permanent magnets, magnetic forces, and electro-magnets with iron cores. There are speakers for sound using electromagnets pushing against permanent magnets. There are relay switches for large currents, like switches for car starters, that slam metal bars across batter electrodes. There are lifting applications for iron sorting using electromagnets and iron magnetization and attraction.

Chapter 2: DC (static) Magnetic Forces and Lift Forces:
Get familiar with magnetic forces between permanent magnets. This is called DC magnetic forces, or Static Field magnetic forces. These forces are easy to demonstrate around the house. Just get some permanent magnets and lift some iron spoons, or attach the magnets to the refrigerator.

Static Magnetic Forces

"We have permanent magnets, we have electromagnets, all with the help of iron."

Section A:
Some Uses of Magnetic Forces

A: There is the well known rule for magnetic forces that 'like repels like' and 'opposites attract'. That means N-N repulsion and N-S attraction. These forces are all due to the fact that magnetic fields are in a lower energy when they are aligned, or parallel in the same direction.

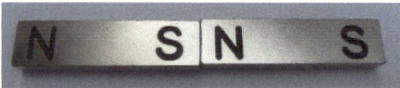

Opposites poles attract
Fields want to align, and one magnet gets attracted into the higher field of the other.

Section B:
Lift Using Iron and Electro-Magnets with Iron Cores

B: A permanent magnet can create a magnetic field, or an electric current in a coil or electro-magnet can create the same magnetic field. The fields are identical.

Iron is a great help for electro-magnets. Less power or current is required to get the same large magnetic fields because the magnetization of the iron adds to the magnetic field.

Electro-magnet with iron core
Coils also generate magnetic fields, and iron cores magnetize themselves and enhance the magnetic field.

Magnets only attract magnetic material:
Only iron is magnetic and attracted to the permanent magnet.
Copper, aluminum, and zinc are not.

Section C:
People's Use of Static Attraction and Speakers

C: We use electro-magnets for controlled lifting of iron objects, for audio speakers, for writing to computer memory, and many other things.

Speakers with electro coil and permanent magnet

Electro-magnet picking up scrap iron metal
One application of electro magnets is to pick up large steel sheets. Just by turning the current on and off, on a factory floor, a steel sheet can be moved around, grabbed and released.

Understand static magnetic field forces, from either permanent magnets or electro-magnets.

Energy and Force

We want a force, either attraction or repulsion, to pull things. Let's talk about energy.
If you know in what direction the energy is lower, then the magnets will go in that direction, either by rotation or sideways movement.
You could say opposites attract, or you could say that magnetic fields like to overlap in the same direction.

Energy is all around us and it wants to settle down: heat goes to cold, storms go to blue skies, and batteries get discharged.

Practice some Zen Buddhism, and focus on your core energy, and relax and settle down.

To predict magnetic behavior and attraction, you can first just think about where the lower energy is.

"Zen and the art of attraction"

Energy rules:

- **Lower energy → Direction of force**
  - Magnets want to get sucked into higher magnetic fields for lower energy.
  - The direction of lower energy is reversed when the magnet is flipped, the fields are opposite, and the force reverses from attraction to repulsion, or vise versa.

- **'Opposite poles attract', and 'Like poles repel'**
  - This is the same as 'Lower energy' → 'Direction of force'.
  - When opposite poles are touching each other and attractive, the two magnetic fields are in the same direction at the other magnet's location.

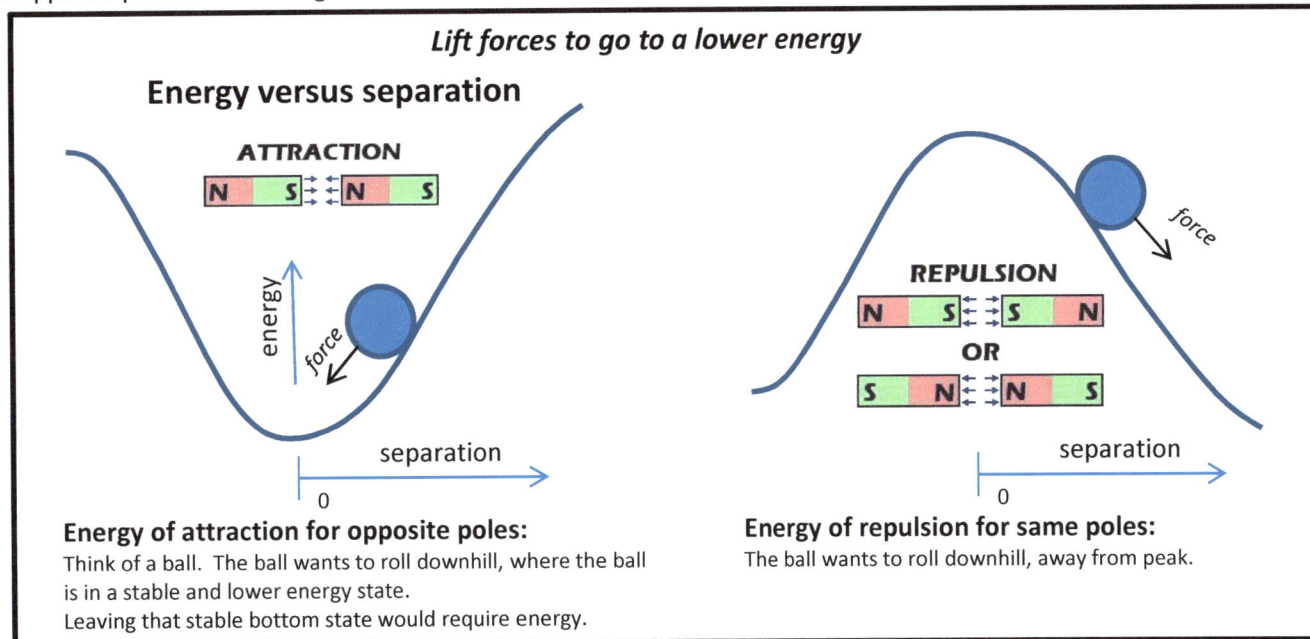

Lift forces to go to a lower energy

Energy versus separation

ATTRACTION

Energy of attraction for opposite poles:
Think of a ball. The ball wants to roll downhill, where the ball is in a stable and lower energy state.
Leaving that stable bottom state would require energy.

REPULSION

OR

Energy of repulsion for same poles:
The ball wants to roll downhill, away from peak.

See Appendix A for energy and force

Here's our rule of thumb. If the energy can be lowered by moving a magnet closer to another magnet to have parallel magnetic fields, there then is a force to do that.

Magnets can be used for lifting, bearings, and levitation and trains.

This chapter introduces permanent magnets, magnetic forces, and electro-magnets with iron cores. Some major applications of magnetic forces are speakers, relay switches, and lifting.

Permanent magnets are attracted or repulsed from each other.

"I feel attracted to you, magnetic attraction."

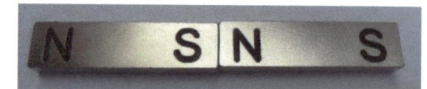

Opposite poles attract:

We all know that opposites attract, or opposite poles attract, when it comes to magnets. For example, people buy disk magnets that are all lined up, N-S, in the package.

Attraction and lifting:

Refrigerator magnets used to be made from only the weaker ferrite magnets, made from iron oxide, that can barely hold themselves up against the metal refrigerator. Now just as commonly we have rare earth magnets, that can hold up a lot more weight.

Opposites poles attract

Levitation:

When something is floating over something else, or levitating, we typically think some magnetic repulsion. Now, that magnetic levitation is not stable without some rod or electronic feedback, but that is magnetic repulsion.

In an application of levitation, people have used magnetic repulsion to make bearings. There are no mechanical surfaces rubbing against each other, so the friction is lower. Typical roller bearings with rolling balls do have some friction when going fast, so a magnetic bearing without this friction does have merit.

Simple magnetic bearing:

People dream of using magnetic levitation to make really fast trains, where the levitating train does not mechanically touch the rails. The trains hover from magnetic repulsion, and get pushed forward also from magnetic forces. There is no wheel skipping against the rail, so traction is always maintained with magnetic levitation at faster speeds. In those moments when a wheel skips off the rail going more than 200 miles per hour, there is no ability of the engine to keep pushing the train forward.

Demonstration magnetic levitating trains

Do magic tricks with magnets! Let them fly apart! Those magnetic fields are real.

Permanent Magnetic Forces, It's Not Just Fun and Games

Magnets can be used for lifting, bearings, and levitation and trains.

Permanent magnets are attracted or repulsed from each other.
The stronger magnets below are strong rare earth magnets. Close to the magnet, the magnetic field is 100 times larger compared to the Earth's magnetic field.
A compass needle is a permanent magnet too. The needle is just steel that has been magnetized in advance, to rotate and align in the same direction with the Earth's magnetic field.

"I feel attracted to you, magnetic attraction."

"Dude, I'm use to refrigerator magnets, not these high power rare earth magnets ...they fly together."

Attraction with opposite poles:
When opposite poles (North N and South S) attract, they can snap together and be all tight.

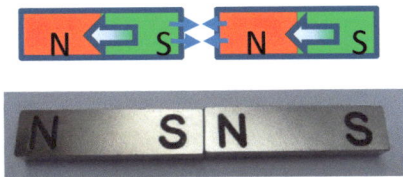

Opposites poles attract
Magnets attract when align magnetic fields in same direction and have large field gradients

Attraction holding one magnet up

Repulsion with same poles:
Like poles (North N and N) repel, they can fly apart.

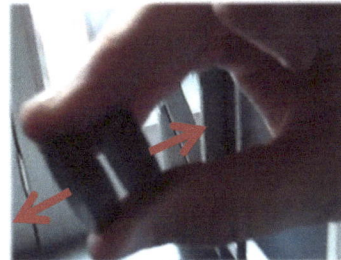

Repulsion between like poles
Magnetic fields are in opposite directions.

Attraction: squeezing finger

Attraction: Strong magnetic field controls compass needle

Attraction: side by side, alternating poles

Repulsion Game: North keeps away from North

Repulsion or Attraction Game: hidden magnet under table

Topic: This section on static magnetic forces shows that North attracts South, and North repels North. This attraction and repulsion are all because the magnets want to align their magnetic fields, to create the lowest energy state.
Practical hands-on learning: After, during, or before you read this section, get some magnets and get them to fling together or fly apart, or make strange magnetic sculptures. Strong magnets are easy to buy at an automotive store, or on the internet. Be careful when playing with two stronger ceramic rare earth magnets: they tend to snap together. They are brittle, so you could cause chips.

Do magic tricks with magnets! Let them fly apart! Those magnetic fields are real.

Rare Earth Magnets are Stronger and Smaller

Let's look at a rare earth magnet's field using a compass.
A magnetic compass is an experimenter's best friend to detect magnetic fields. We can tell there is a magnetic field just by looking for deflection of the compass needle. We use the compass as much as we can in these experiments, because compasses are so common and so relevant.

"Compass needle moves next to magnets. You control the local magnetic field."

"Pay respects to early Viking navigation using a compass, as well as jam along to music with tiny magnetic speaker headphones."

Experiment 2.1: Show compass deflection by permanent magnet

Rare earth permanent magnets overwhelm the Earth's magnetic field locally:

These strong magnets easily deflect a compass to aim toward the magnet, instead of staying aligned along the weak Earth's magnetic field.

These permanent magnetic fields are much stronger right next to these small magnets, compared to the Earth's weak magnetic field.

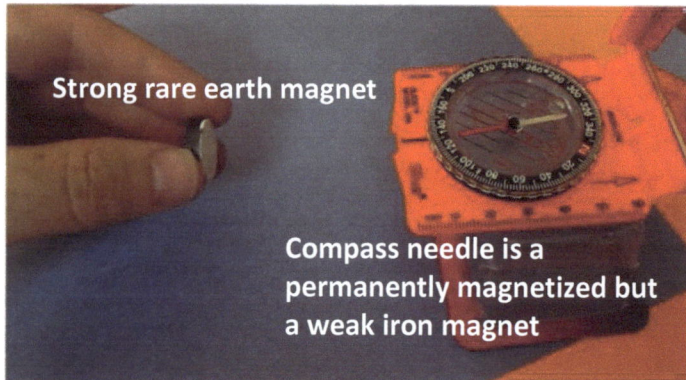

Strong rare earth magnet

Compass needle is a permanently magnetized but a weak iron magnet

The compass magnet rotates to align with field from other magnet. Rare earth permanent magnets are much stronger (1000 times) than the Earth's magnetic field.

This desire for alignment of magnetic fields in same direction is what makes motors want to rotate, discussed later.
Also, by the way, the compass 'red' side points North, but is actually the South pole of the compass needle magnet. That is because opposite poles attract, and the red side points toward the Earth's magnetic North pole.

Periodic table showing all the 'rare earth' elements in red, which are good for magnets and are not that rare in nature.

Digging for magnetic elements:
Each rare earth element is about 200 times more common than gold in the Earth's crust, so these elements are not all that rare: neodymium, samarium, cobalt, combined with iron.
'Rare' just means the elements are combined in ores, not in pure form.

Rare Earth magnet field strength:
The rare earth magnets have a very strong magnetic field. These magnets also easily snap together, or fly apart, because the fields are so large. Weaker 'ferrite' magnets (an iron oxide) are usually used for the refrigerator, which also demonstrate the same magnetic properties but just less dramatically.

Material processing:
To make magnets, we use temperature processing and materials that make a strong magnetic field. Rare earth magnets do not just naturally become good magnets. There is human engineering, and the rare earth raw material needs to be ground into small grains, pressed together tightly, aligned in same direction in a strong magnetic field (much larger than the Earth's magnetic field, by about 10,000 times, or 4 Tesla), and heated to fuse the grains together.

Experiments: Use compass to show magnetic field direction of a magnet

Rare earth magnets are helping to make electric tools stronger, and speakers more dynamic.

Attraction Between Magnets: Opposite Poles Attract

The attractive force between magnets can be strong, when the magnets are close together.

'Opposite poles attract' can be N-S kissing, or N-S side by side, as long as the opposite poles are closer together than the other 'same' pole.

Opposites attract because their magnetic fields line up.

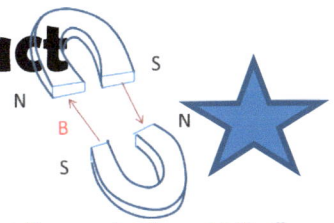

"Attraction: Don't put fingers in the middle."

Q: What did the male magnet say to the female magnet?

A: From your backside, I thought you were repulsive. However, after seeing you from the front, I find you rather attractive.

Experiment 2.2: Show permanent magnets getting attracted, end to end.

Attracting two magnets head on (fields align in same direction)

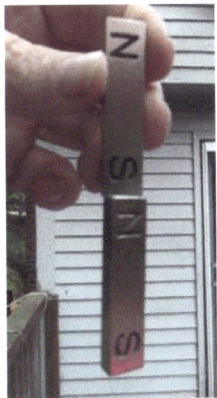

Magnetic forces are easily stronger than gravity, end to end

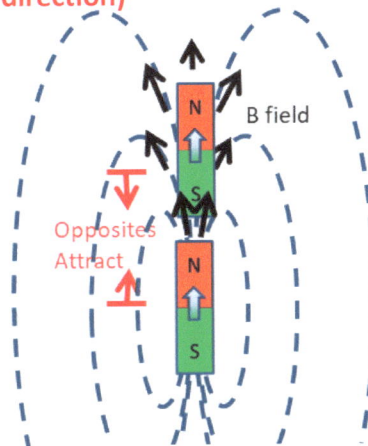

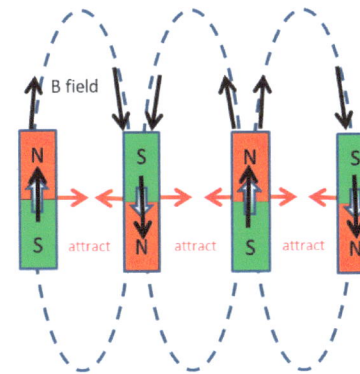

Opposite poles, or aligned magnetic fields. Attract when B field is aligned with the magnet.

Opposite poles attract. White paint side has the same pole

Magnets attract to get N and S facing each other.

Magnets on top of each other want to kiss opposite poles. Get two magnets to attract and squeeze against your finger. The finger was not crushed, but the magnetic attraction is strong enough for the magnets to stay on the finger.

Experiment 2.3: Make a chain of magnets using attraction, side by side.

Attracting two side by side magnets (fields align in same direction at magnet, within each other)

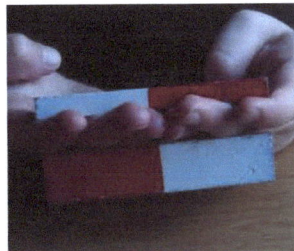

Magnetic forces are easily stronger than gravity, end to end

Opposites poles attract
Magnets are in line with magnetic B fields of neighboring magnets, so are lower energy and attractive.

Notice the poles are all opposites, so these two magnets attract and defy gravity. Magnets side by side, with opposite poles closest, have aligned magnetic fields and attract.

The same two magnets are squeezing the person's hands because they are attracted together, based on 'opposite poles attract'. But the finger tips do not look blue, so it looks like a gentle squeeze.

Magnets, if they are strong enough, like rare earth magnets, can even hurt your hands slamming together.

Magnetic fields, if they are aligned in same direction, want to get closer to each other. This is an attractive force, and this is a lower energy state with closer distance.

Make side by side chain of magnets.

Attraction sideways also holds magnets up
See Appendix G, equation 4

See Appendix G, equation 4

Experiments: Side by side attraction

Please just play with a few magnets and see the attraction, repulsion, rotation.

Like Poles of Magnets Repel: Table Hockey and More

"Why did the magnet go to the psychiatrist?
He was bipolar.

Why is the magnet on medication?
Because it's bipolar."

"Dude, have some fun."

When magnets are flipped, they can repel with a strong force. 'Like poles repel' can be N-N repulsion, or S-S repulsion.
Like poles repel because their magnetic fields are opposite, or misaligned in opposite directions.

Experiment 2.4: Show permanent magnets repulsing.

Same poles repel for magnets, or 'Like poles repel'.

Levitating a magnet: 'Like poles repel'

Play table hockey with magnets: 'Like poles repel' (fields mis-aligned)

Repulsion between magnets can allow games like table hockey or allow levitation.

For table hockey using magnets, the magnets do not need to physically touch. They are repelled by the magnetic fields not aligning, or, stated another way, because North is repelled from North.

Get yourself some strong rare earth magnets and play table hockey and have some fun

Repulsion: North keeps away from North   Repulsion: B field opposite to magnet

Forget quarters for table hockey. Instead use strong magnets. You do not need physical contact to push magnets.

These disk magnets are flat enough that they can more easily slide apart, then flip.

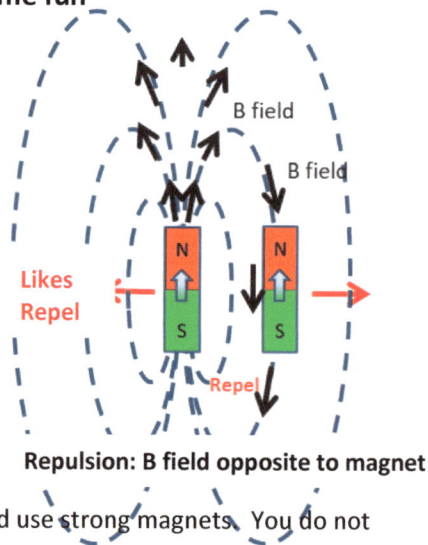

Magnetic fields passing through each other, if they are opposite directions and can not align, want to get farther from each other.

This is a repulsive force.
In terms of energy, there is a lower energy state with farther separation, which means a repulsive force.

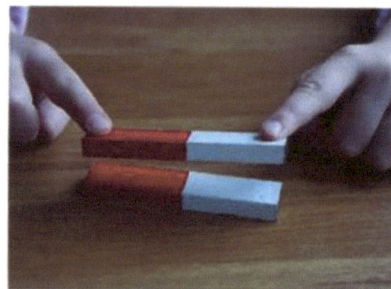

Repulsion: Like poles repel each other

Magnets can be held, levitating above other magnets, although not stably.

In the picture above, one bar magnet is hovering or suspended over a lower bar magnet, because the 'like poles repel'.

Now, one magnet is suspended, but this a not a stable condition. The person's fingers are necessary to keep the upper magnet from flying away, or rotating around and snapping onto the lower magnet.

Experiments: Magnets repelled because same poles repel

Same poles closer together repel for magnets.

Attraction Between Magnets: Hidden Magnet Tricks

Use the 'action at a distance' behavior of magnetic forces, and have a hidden magnet under the table controlling a visible magnet. These forces and fields are not changed by a non-magnetic material (wood) between the magnets.

One magnet can attract another magnet when their closest poles are opposite, or can repel another magnet when their closest poles are the same.

- Play some games with the attractive force, say, table hockey.
- Play some games with the repulsive force of the magnets. Is the repulse force less controllable, trying to flip?

Experiment 2.5: Play magic tricks by making magnet move on table, using hidden magnet underneath (opposites attract)

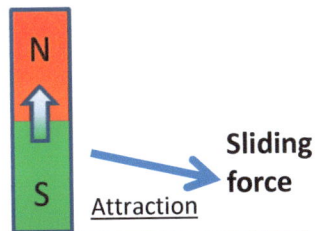

Attraction → **Sliding force**

Wood table (any non-magnetic material, such as wood, plastic, or aluminum)

Opposite poles attract for magnets.
Top magnet follows hidden magnet under table, for attraction.
Use attraction because then the top magnet won't try to flip over.

The secret magnet hidden under the table.

Impress your friends. Make a magnet appear to move by itself, albeit with a hidden magnet underneath.
...Oh no, the trick is exposed and the magician is a fraud!

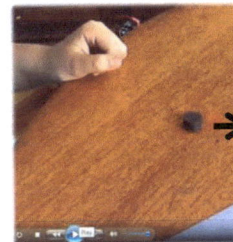

'Magic' movement by moving magnet below table.

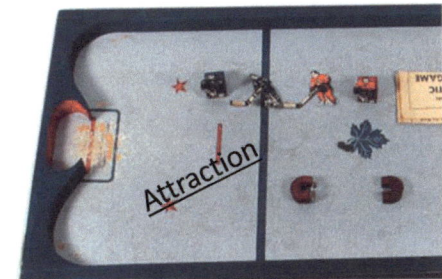

Hockey board: hockey players are attracted to magnet underneath the table.

Magnetic fish bowl cleaner

Magnetic fields, if they are aligned in same direction, want to get closer to each other.
This is an attractive force, and this is a lower energy state.
Even through a wooden table, the magnetic fields pass and experience this force.

Here is an example of simple 'magic', where the visible magnet on the table moves around, apparently by itself, but actually because the visible magnet is getting attracted to a magnet moving under the table.

Experiments: Magnets repelled because same poles repel

Static Attraction Between Magnets: Desktop Novelties and Practical Bearings

"Sometimes magnets are simply for the guy who has everything, but toys inspire new uses of magnets."

Opposite poles attract to create a very low friction magnetic bearing. There is no physical contact to the shaft so there is no wear and less friction.

Toys: People's fancy, defying gravity

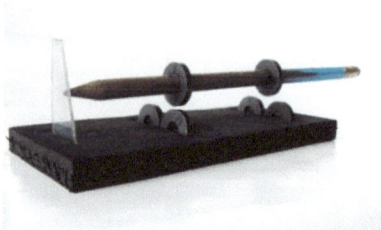

Simple magnetic bearing:
Pencil hovering because side magnets are repelling. This is stable, when stabilized by gravity and a fixed plastic wall.

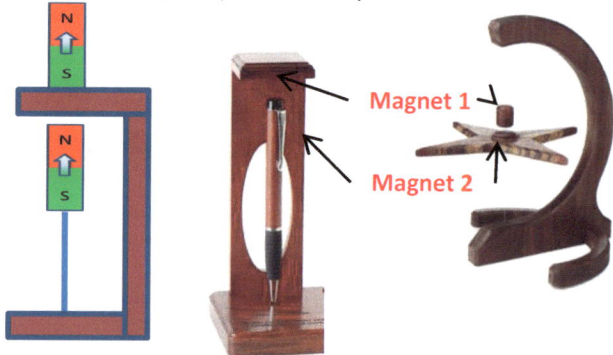

Magnet 1
Magnet 2

Attracting and balancing magnets:
Bottom magnet is fighting against string to snap into the top magnet. This padded magnet could be a punch bag, but it would hurt (and why not just hang a bag with gravity?).
Some of these magnetic toys are gadgets 'For the executive who has everything'.

Industry: People's practical use, levitating low loss bearings without two objects touching

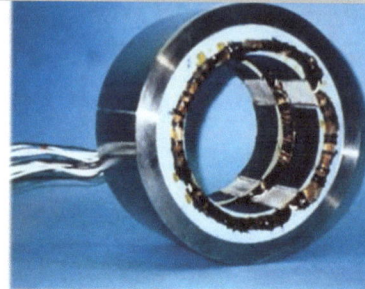

Application: Magnetic bearings with very low friction and low wear

Deltana MDH30
Receptor (Spring Cushion)
Magnet

Application: door holder
Door holders can do two things:
- Hold door open, using attractive magnets.
- Alternately, can stop the door from slamming closed, using repulsive magnets

Magnetic fields, if they are aligned, want to get closer to each other. This field alignment is an attractive force, and this is a lower energy state. The magnetic pen stays upright because it is attracted to the magnet above, and the door stays open because the magnets are aligned and attract.

Magnets are popular, for good reason, as shown by the abundance of magnetic novelty gifts (balancing pens and floating globes), and by industrial bearings with very low friction.

Some Static Uses of Magnetic Forces: Repulsion and Gee-Whiz Levitation

"You got a range of applications of magnetic forces: desktop toys for the guy who has everything, and trains floating on magnetic fields up to 200 mph."

You might think permanent magnets are good for toys (levitating pens, or pens balancing on their point), but they are used for large industrial projects too (levitating trains, lifting steel). Without people supplying any power, magnets allow forces without two objects touching.

Toys: People's fancy with levitation

Magnetic Levitation Globe

Levitation: The magnetic globe hovers and repels, but is not stable without some active feedback. The stable levitation needs some electronic control pushing the above magnet back to center.

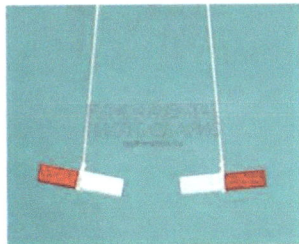

Repulsion: 'Likes repel'
These two magnets do not come closer together because 'Likes repel', which means their magnetic fields are opposite. These magnets are flush against a surface, so they can not rotate and snap together.

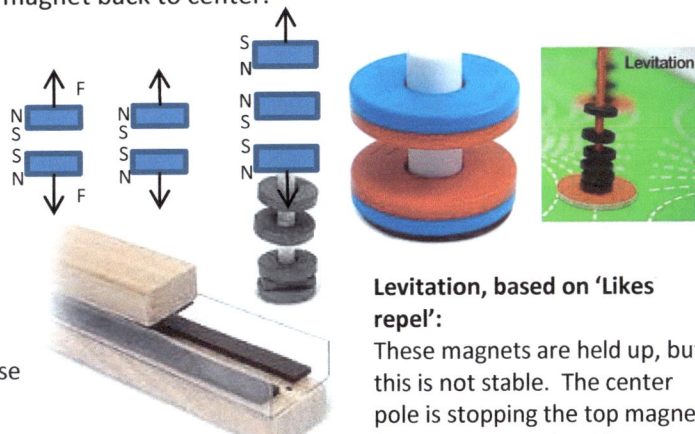

Toy train levitation concept:
Could use repel force with train magnet on top, or attract force with train magnet on bottom.

Levitation, based on 'Likes repel':
These magnets are held up, but this is not stable. The center pole is stopping the top magnet from moving to the side or flipping over. Without the pole, the magnet would flip.

Industry: People's practical use, for high speed trains

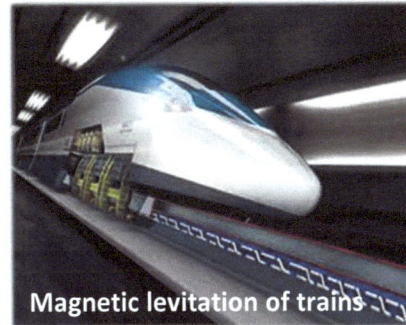

Magnetic levitation of trains
Industrial permanent magnets

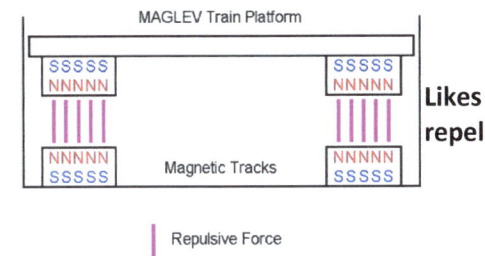

Simple diagram of magnetic support of levitating train platform

The levitating train is the dream of many train designers. Either a magnet can be on the underside of an iron train track, causing levitation, or a magnet can be over a non-magnetic metallic train track, causing eddy currents in the track and repulsion. Either case will hold up the train, above the track.

Magnetic fields can attract or repel depending on the magnet orientation.

Magnetic fields, if they are in opposite directions and can not align, want to get farther from each other. This N-N is a repulsive force, and there is a lower energy state with farther separation.

Magnets are popular, for good reason, as shown by abundant examples of motors, compasses, picking up iron, in addition to levitation. Magnets are all around us, creating motion that people take for granted.

All Electric Train Transportation Using Levitation and Propulsion from Magnetic Fields

Electric transportation is already all around us, with trolleys or subways. It is also possible to enlarge electricity's use for these more futuristic, faster technologies, like levitating trains and rail guns.

Civilian (regular everyday people) uses of magnetic fields

- Tube transport (future)
- Cars out of city (simple electric cars are good)
- Electric planes (prototypes and modern training planes due to less maintenance)

HOW MUSK'S SUPERTRAIN COULD WORK

Rail gun technology
1. Electric current flows up positive rail

2. Current flows across armature and down negative rail

3. Magnetic force is directed towards end of rails which pushes armature and train forward

Maglev technology levitates the train eradicating rail friction

Reduced air pressure in tunnel cuts wind resistance

Top speed 750mph

San Francisco · NEVADA · Los Angeles · CALIFORNIA · 380 miles · ARIZONA · San Diego

Armature · Magnets · Negative rail · Positive rail

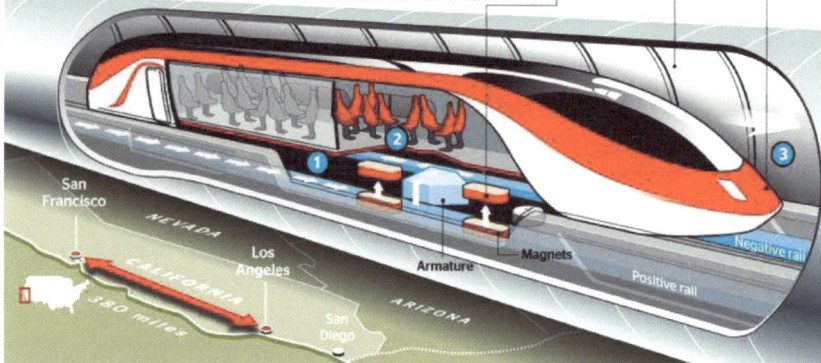

Trains: Japan's upcoming Bullet Train may be a first case application of electro-magnetic thrust and levitation with magnetic fields pulling and pushing things away.

Military uses of fields

- Electric guns (now)
  - Magnetic force on a metal beam carrying a huge electric current.
- Launch heavy fully loaded airplanes off carrier decks with short runway
  - Use magnetic forces to push planes.

Current examples of rail guns and electro-magnetic propulsion

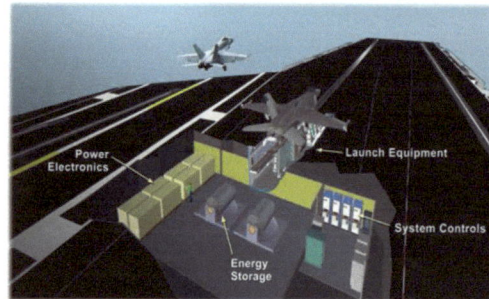

Power Electronics · Launch Equipment · System Controls · Energy Storage

Ford Aircraft Carrier:
Rail gun boosters to launch airplanes, instead of steam boosters.

Space uses magnetic fields

- Rail gun to give initial velocity to get off ground (future), instead of chemical launches.

Electro-magnetic gun: for battleship.

Us Navy Railgun - Their Most Powerful Cannon

Hyper Velocity projectiles from rail guns
Rail Guns do not use gun powder, but instead use large electric currents flowing across the projectile. The speed is not limited by any gun powder pressure.

Rail Gun propulsion of high speed levitating tube car.

No Wheels: Force without wheels, for levitation and faster travel:
- Use magnetic force on an electric current. No wheels, so no bouncing of wheels over a rough track at high speeds and losing traction.

Wheels, like today: Force between wheels and rail, but speed is limited when skipping:
- Use wheels powered by electric motors, which are powered by voltage on the rails, like current trams.
- The problem with wheels is they skip when going fast.

force

Linear motor: Force on magnets for subsonic speeds

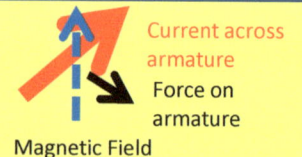

Current across armature
Force on armature
Magnetic Field

Rail gun: Force on cross bar for hypersonic speeds

Our train system, for faster trains, could be based on repulsion and attraction between permanent and electro-magnets.

See chapter 3 for linear motor
See Appendix F for fundamental force on a current

Summary of Basic Demonstrations of Permanent Magnet Lift Forces

"Engage your hands, spur your imagination."

Try these basic show-and-tell experiments for magnetic lift and alignment forces. These force experiments are great to demonstrate magnets to others (say, to scouts), and are described in more detail in this chapter. These force experiments will lead to a better understanding of motors as well, discussed in later chapters.

Go ahead, make a copy. Then check off the list.

Paint same side, or pole, of each magnet one color, based on attraction to a reference magnet.

- A consistent side (pole) of the magnet can be painted white. You can determine the consistent pole by looking at what side of the magnet gets attracted to a reference magnet.

Experiment 2.1: Show compass deflection by permanent magnet.

- A compass is an experimenter's best friend for magnets. We can tell there is a magnetic field just by looking for deflection of the compass needle. We use the compass as much as we can, because magnets are so common and so relevant.

Experiment 2.2: Make two magnets attract, end to end

- Magnets snap together. Coloring the poles really makes it easier to understand effects like the NN repulsion and the NS attraction.

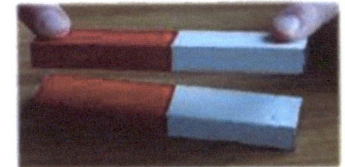

Experiment 2.3: Make magnets attract, with chains of magnets.

- Magnets can also be chained together side by side, instead of pole to pole. The poles will reverse or alternate from neighboring poles, because that way the fields are aligned in the same direction at the center of each magnet.

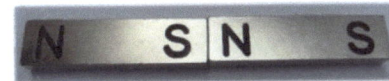

Experiment 2.4: Make two magnets repel each other

- Magnets fly apart when like poles repel.

Experiment 2.5: Play table hockey with magnets.

- Play some games with the repulsive force of the magnets, say, the game of table hockey.

Opposite poles attract, or fields align in same direction

Magnets deflect compass needles

N S N S

Opposite poles attract

Like poles repel

Experiments: Magnet field direction, magnet attraction

Permanent magnets attract each other, and repel each other. Just play with them.

Static Forces From Mysterious Magnetic Fields, Q and A

Magnets and magnetic fields create forces and torques. 'Opposite poles attract', 'Like poles repel', and opposite poles like to rotate to line up with each other, or be next to each other.

What are some applications? Magnetic forces enable relays to ring your doorbell and enable lifting iron cars. Magnets enable electric motors because magnets want to align their fields in the same direction, which causes twist or torque.

Opposite poles attract

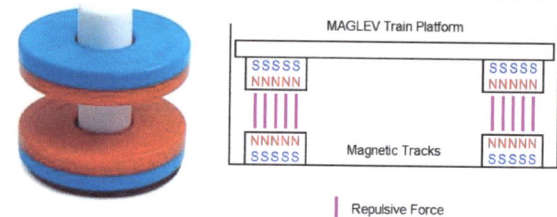

Question 1: Is N attracted to S, or N attracted to N? Opposite poles attract because that aligns their magnetic fields.

- Opposite poles attract in the magnetic field world, so N pole attracts S pole.
- This is another way of saying that the magnetic fields like to be facing in the same direction. There is lower energy when the magnetic fields are aligned in the same direction, and that means attraction.

Question 2: Can one magnet levitate or hold up another magnet? Yes, magnetic fields can be much larger than gravity, when two magnets are close together.

- Yes, repulsion (N-N or S-S) can hold up another magnet in the air. Unfortunately, it is not stable, and the hovering magnet will want to fall to the side or flip over.

Magnetic forces holding up another magnet

Question 3: What is a magnetic field? Fields, both magnetic and electric, are a way to describe forces between magnets or between charges.

- You could say that a magnetic field is a way to represent forces between electric currents. Magnetic fields represent forces between moving charges. The magnetic force happens even when those electric currents are electrically neutral, but the current must be moving.
  - Electric fields represent forces between static charges, and magnetic forces are between moving charges.
- If you want to get deeper, you could say that a magnetic field is the residual electric field of a charge that has changed location during the time the electric field travels to the other current location at the speed of light.

Magnetic fields from a magnet can be represented as a current flowing around the magnet.

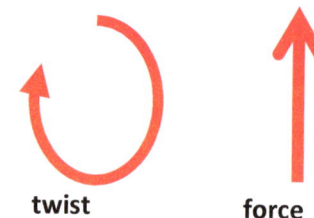

Question 4: Do permanent magnets have an electric current flowing around them?

- No, but their external magnetic field can be described by a current flowing on the surface. This effective current is not a real electric current, but it duplicates the magnetic field of the atoms in the magnet.

Question 5: What is the difference between a force and a twist (a torque)?

- A force wants to move something to the side or away from its location, and a twist wants to flip or turn something but keeps it in the same location.

twist   force

A twist just wants to rotate about the center of mass or pivot point. A twist does not move the center of mass.
A force wants to move the center of mass.

Let's talk about lifting steel objects using electro-magnets with iron cores. Magnetic fields can be generated by electric currents. All those magnetic fields from each turn of the coil add together when the electric currents are run through a coil.

A magnetic field is a magnetic field. So magnetic fields made by electric currents are the same as magnetic fields made by permanent magnets. A permanent magnet will magnetize iron (steel is mostly iron), and the magnet will get attracted to the iron.

"Electromagnet coils and iron cores are everywhere...relay switches, electric motors."

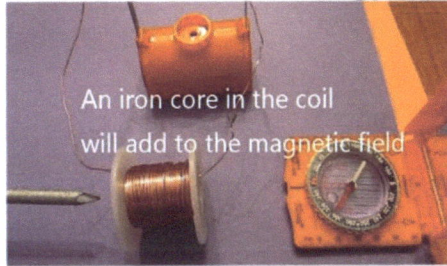

An iron core in the coil will add to the magnetic field

Electro-magnets:

Electro-magnets are extremely useful for many things. Magnetic forces can be turned on and off by turning the electric current on and off. Magnetic relays, or switches, are made with electro-coils. These relays are used to start cars, to engage the starter motors that cranks over the engine to get it started. These relays are used for door bells, where a rod is pulled with magnetic forces into a coil, which compresses a spring, and then the rod is released and hits a bell.

Iron Cores:

Iron is the dream material for magnetic fields, usually inside electro-coils. Fortunately, iron is all around us. It is 5% of the Earth's crust, and is in a lot of ores. Iron is also great for construction, because it is strong and malleable and available. Most of a car is made from iron. Most of the structure for buildings and bridges is made of iron.

Iron also has a convenient side benefit in biology of grabbing oxygen in our blood and distributing it to our organs for our survival.

Why is iron a dream magnetic material? Place iron in a magnetic field and the iron aligns with the field and enhances the magnetic field.

Well, to prove larger magnetic fields, apply electric currents in coils to create a magnetic field. Then, place an iron core in the coil and the magnetic field is made much larger. Now this electro-magnet can strongly magnetize other things like cars or scrap iron, and the resulting strong attraction allows us to pick it up. When we want to let go of whatever is attracted to the electromagnet, then we simply turn off the electric current.

Needle and magnetic field before iron core from screwdriver shaft:

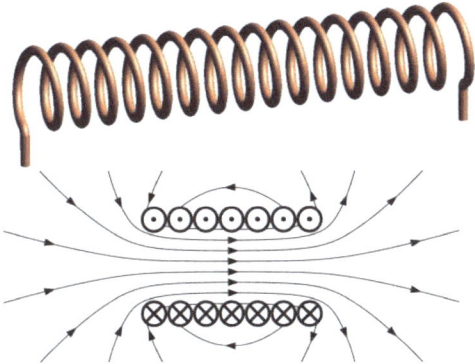

Run electric current through the coil and a strong magnetic field is created down the middle.

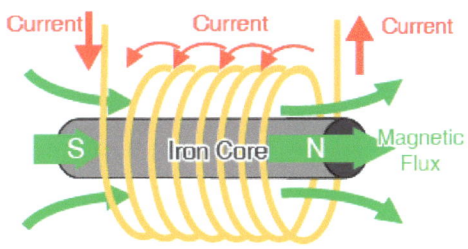

Current   Current   Current

S   Iron Core   N   Magnetic Flux

An iron core greatly enhances the magnetic field.

Electro-Magnets and Iron Cores, a Matchmaker's Dream

Let's talk about lifting steel objects using electro-magnets with iron cores.
A permanent magnet will magnetize iron (steel is mostly iron), and the magnet will get attracted to the iron. Refrigerator magnets can hold up steel forks, and bigger magnets can lift up steel cars. The horseshoe magnet shape has more lifting strength because all the high magnetic fields are focused between the two poles, where the attached iron object is stuck. Any high intensity magnetic field with lots of field gradients will cause a huge attraction to iron.

Experiment 2.6: Magnetize and lift iron stuff with permanent magnet

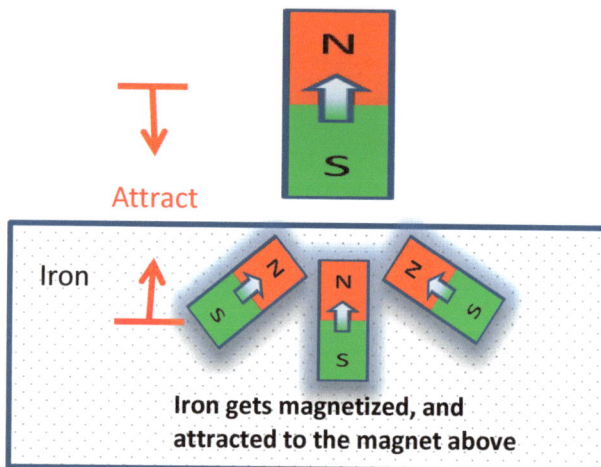

Attract

Iron

Iron gets magnetized, and attracted to the magnet above

Magnet lifting magnetized iron object, like this fork

"A magnet walks into a bar…, what does he order?
Nothing… he's still stuck to the entrance."

"Some things just got that old time magnetic attraction … especially iron"

Iron gets aligned to the external magnetic fields, and becomes it's own magnet.
Magnetic fields, when aligned, want to get closer to each other. This is an attractive force, and this is a lower energy state. Iron and a permanent magnet always have an attractive force.

Iron is magnetized by the magnet above

Iron is the main element in steel, and is a magnetic element, meaning that each atom has a magnetic moment. So when a permanent magnet gets close to iron, the iron elements like to align with the permanent magnet's field, causing attraction.

Then there is runaway attraction. As the magnet gets closer, the iron aligns more and the force increases even more.

In physics lingo, we have even lower energy when the iron gets closer to the magnet and force goes in the direction of lower energy.

See Appendix A and B for energy explanation of attraction, the Buddha way

Magnets only attract magnetic material: Only iron is magnetic and attracted to the permanent magnet.
Copper, aluminum, and zinc are not.

A horseshoe magnet has double the lifting strength of a weaker bar magnet.
because using both ends.

Permanent magnet lifts iron nails

Horseshoe shape

A horse shoe magnet has more than double the lifting strength of a bar magnet, just because of shape.

- More intense magnetic field at tips pass into whatever object you are trying to lift, with two magnet ends.
- Also, there is more field gradient, which is required to have a force pulling on the object.

The element iron, typically in steel, is king for magnetic attraction.

Magnetize a Screwdriver and Lift a Paperclip

You can permanently magnetize some types of steel (iron) by exposing it to a large magnetic field. When you magnetize iron, you get a lot more magnetic field and lift capability than simply having an electro-coil.

You can then de-magnetize the iron, say a screwdriver steel shaft, by rubbing the iron against a permanent magnet in opposite directions.

Some types of iron (hard) hold or remember the magnetic field at the magnetic field is removed, and some types of iron don't (soft).

Iron with added different elements, like carbon or nickel or zinc, holds the magnetization. 100% pure iron does not hold the magnetization after the external field is removed.

Experiment 2.7: Magnetize an iron screwdriver by rubbing against a permanent magnet in one direction.

Before applied field:
No net magnetization, with random magnetic orientation of grains or domains.

After applied field:
Get net magnetization, with parallel magnetic orientation of grains or domains.

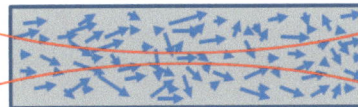

Applied field

Field from aligned magnetization of grains

No net field from random magnetization of grains

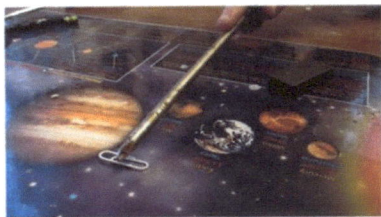

Raw Material: Ferrite magnets and rare earth magnets are made using magnetic raw material, by heating and cooling, in a magnetic field.

De-magnetizing ships: In the Earth's magnetic field, the iron hull of ships get magnetized and can set off magnetic mines. Here the submarine is getting de-magnetized using a large coil.

Iron (steel) has a memory, so when you remove the magnetic field the iron stays magnetized.

You can de-magnetize the iron screwdriver by reversing the magnet, or magnetizing the iron weakly in other direction. The extreme is to alternate the applied magnetic field to confuse the iron, gradually reducing the iron's residual magnetization. In industry, this alternating field is done using an electromagnet with alternating current.

Experiments: Magnetize iron using permanent magnet

Step 1: Un-magnetized steel
Not magnetized shaft of steel screw driver can't lift iron paper clip

The iron screwdriver starts out with no magnetization, with random moments of the grains. The screwdriver can not pick up the paper clip.

Step 2: Magnetize the steel
Rub a screwdriver against a permanent magnet.

Rest the iron screwdriver shaft on the permanent magnet, staying in one direction only.

Step 3: Magnetized steel attracts clip
Magnetized: lift iron paper clip

The iron screwdriver is now magnetized, with magnetic moments of grains lined up. The screwdriver can now pick up the paper clip, by magnetizing it.

Iron is part of most magnets. You can demonstrate how iron gets magnetized by holding an iron screwdriver over a magnet.

Electro-magnet: Make a Magnetic Field using a Battery and Coil

Hey, here is a big discovery! Both electro-magnets (loops of currents) and permanent magnets generate static magnetic fields. Look, the proof is in movement of the compass needle, when the current is applied to the coil.
Iron helps too by adding to the field, when the iron is inserted into the coil!

Experiment 2.8: Make a magnetic field with coil and battery: that is, make an electro-magnet.

The magnetic fields from the coil are just the same as from a permanent magnet. Electro-magnets take power, although an iron core helps. Here are some advantages of electro-magnets over permanent magnets:

- You can turn off the magnetic field, to turn off lifting forces. This happens for cranes lifting cars with a magnetic attachment, an electromagnet and iron core.
- With enough current, the magnetic fields can be larger than fields from permanent magnets. This happens for some medical equipment like MRI imaging machines, which use large fields to image soft tissue (organs) and blood flow in the body.

Permanent magnet attracts compass needle.

Battery

Coil

Electric current though coil attracts compass needle.
Magnetic fields are also made by electric currents.

Let's move on to the marvels of man-made magnetic fields, using electric currents. Behold, the compass moves, when current flows!

Let's celebrate this discovery! Electric currents create magnetic fields.

Here is a moment of awe, delight, and functionality: electro-magnets.

Fun facts about current measurement from its magnetic field:

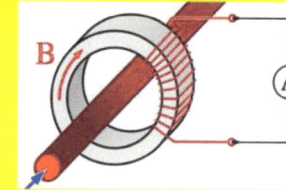

Common measurement of current using magnetic field around the current.
A magnetic field flows around a wire. A current meter measures this magnetic field, to determine the current.

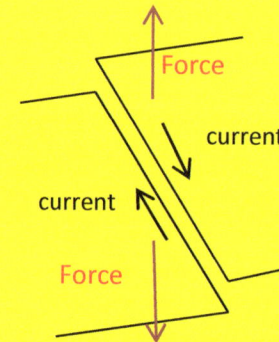

Force

current

current

Force

Precise measurement of current using force

Current feels a force in a magnetic field. Because current creates a magnetic field, two long wires carrying a current in the same direction will want to push away from each other.
A unit of current (1 Amp) can be converted to a unit of force (1 Newton) using this mechanism. Current is measured this way at the National Institute of Standards.
Thank the scientists Oersted and Ampere in the 1820s.

Experiment: Make magnetic field larger in coil using iron

A battery and a wire coil can create a magnetic field. For the same electrical power in the coil, the magnetic field can be increased by an inserted iron core in the wire coil.

Oersted and Ampere: Electricity Development

First discovery and early science had to happen, such as electrostatic forces, creation of magnetic fields, the explanations, DC and AC current, and prototype motors and generators. And then the ideas get applied by industrious engineers who see a need, like trolleys and electric lights. This was in 1820 when magnetic fields were noticed around dc currents. Science can take a while before it improves lives. Notice it took 60 to 80 years before mass applications happened, like subways in cities, car starter motors, and electric light, in the 1880s.

Early Science

Oersted: (1820, in his 40s)

- Danish
- Demonstrated a magnetic field around a dc current in a wire

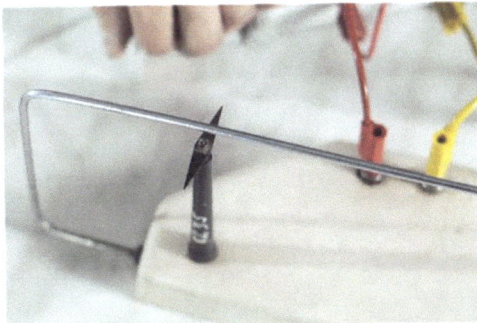

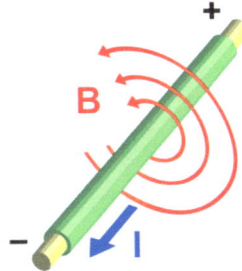

Magnetic field around a current
The compass needle aligns with the magnetic field due to the electrical current.

Education:
Oersted was home taught in Denmark in the early 1800s. He went to the University of Copenhagen and then toured the science labs around Europe for 3 years.

How did Oersted know that an electric current generates a magnetic field?
He didn't until he did the experiment in 1820. He was the first to observe the connection between a current and a magnetic field.
This magnetic field is the basis of electro-magnetics, where all the loops of wire keep adding more and more field inside the coil with each turn.

How much experimentation happened?
He didn't initially interpret the compass needle results correctly, and thought that the magnetic field was radiating like heat or light. He later showed the magnetic field was stationary and circular around the dc current.

Where were the batteries?
Oersted was able to perform the experiment because batteries were developed.

Ampere: (1820, in his 40s)

- French
- Two wires carrying current can repel with opposite currents or attract with parallel currents.
- Solenoid

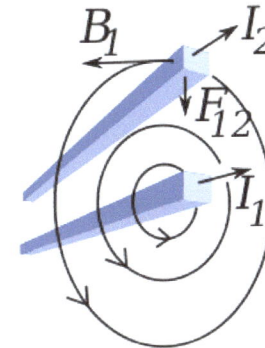

Magnetic field and force around a current

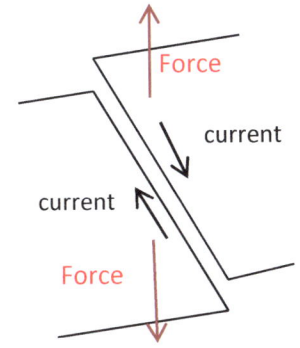

Precise measurement of current using force

Education:
Ampere was home taught on his fathers estate in France, before the French Revolution. He was free to read anything in his father's library, although at age 13 he did start to listen to lectures in mathematics and physics. Throughout his life, he became a professor despite have no formal degrees.

How did Ampere know that an electric current generates a magnetic field and a force on another current?
Ampere was fascinated by the Oersted effect of a magnetic field around a current. He repeated the experiment in the same year and made the discovery of his own about a force on another current carrying wire in the magnetic field.

Ampere's results, relating current, magnetic field, and force, are described by a law in his name and by the right hand rule.

1820 was a good year for electromagnetics. The fact that a current produces a magnetic field needed to get discovered, and the forces on other currents or current loops got discovered at the same time.

Increase Magnetic Field using Iron Core in Electro-Magnet

Iron may look dirty and rusty, but it is a beautiful thing for magnetics. Iron is like Christmas every day, or a birthday every day.

The magnetic field of a coil around an iron core can increase 10 to 1000 times, depending on the cigar shape of the core. Instead of using a coil with a hollow core of air, and wasting lots of electric current and heat to boost the magnetic field, we just let iron boost the field, for free, with less electric current.

"Really strong motors need an iron core to increase the magnetic field more than 10 fold, rather than just field from an empty coil."

Experiment 2.9: Increase magnetic field of a coil using an iron core:
Insert iron into coil, and observed compass needle point more toward coil.

Iron before applied field:
No net magnetization, with random magnetic orientation of grains.

Iron after applied field:
Get net magnetization, with parallel magnetic orientation of grains.

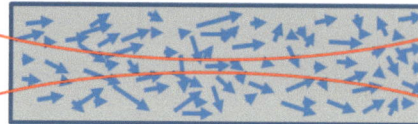

Applied field

No net field from random iron grains

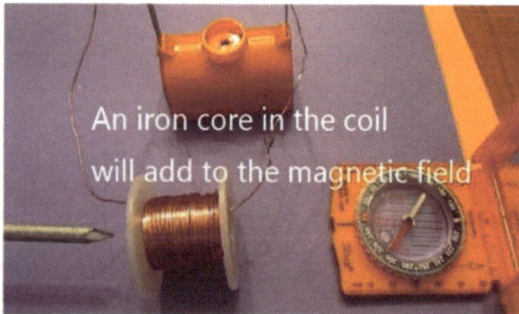

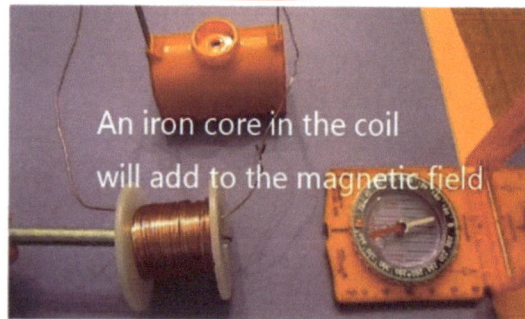

Field from aligned iron grains

Do any materials actually magnetize opposite to the applied field?
No materials magnetize opposite with any strength.

Coils are different. A metal coil will resist any changes in the magnetic field, by inducing a current in the direction to resist the magnetic field change in the coil. This is actually one way to look at or understand generators or pickup coils for guitars.

If the coil is a superconductor with no resistance, in the extreme limit of a perfect wire, then the opposite current can be huge and not allow any magnetic field inside the coil, but that is a niche case.

An iron core in the coil will add to the magnetic field

An iron core in the coil will add to the magnetic field

Needle and magnetic field before iron core from screwdriver shaft:
Some deflection of compass needle due to current from battery.

Needle and magnetic field after iron core from screwdriver shaft:
See the compass needle deflect more when the iron core is inserted into the coil.

Iron cores dramatically increase the magnetic field of electro-magnets. This increased magnetic field, for example, is used to
• Create stronger attractive forces to lift iron objects.
• Increase the torque dramatically in motors, with the same current from the battery, by just wrapping the coils around iron.

Experiments: Magnetize iron core using electromagnet, and see compass needle move by iron core

Iron can get magnetized and greatly increase the magnetic field.

Make an Electro-Magnet Pick Up Iron

Electromagnets form the same magnetic fields as permanent magnets, and pick up iron objects just the same. Here is the difference. Contrary to a permanent magnet, electromagnets can easily drop or disconnect the iron object by turning the current off, like big-boy car lifters.

"Lift iron things using electro-magnets, like me!"

Experiment 2.10: Pick up iron paper clip or nail, using coil and battery (electromagnet) and iron core.

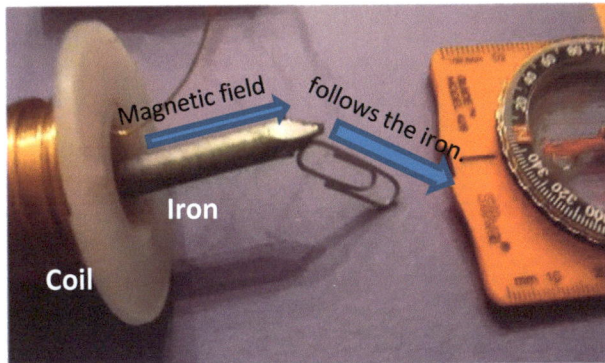

Electro-magnet magnetizes a steel paper clip, and gets attraction.

Electro-magnet with iron core (a steel screwdriver shaft):

A battery and a wire coil can create a magnetic field. For the same electrical power, the magnetic field can be increased by inserting an iron core in the wire coil, such as the steel screwdriver shaft. Iron easily magnetizes.

Choose the right iron alloy: As a point of reality or side note, soft pure iron magnetizes easily when the applied field is turned on and immediately loses the magnetization when the applied field is turned off. Not all iron alloys are the same. Some magnetize easy with no 'memory', like pure iron, but lose the magnetization after the applied magnetic field is turned off. Some magnetize hard, and keep the magnetization and we now have a permanent magnet.

Attract close iron objects:

An iron paper clip is attracted to the iron core of electro magnet because the paper clip is magnetized in the same direction.

As a general concept, things feel forces to a lower energy, and the iron object has a lower energy state when attracted and closer.

Electromagnets can pick up and drop an iron object by turning the electromagnet current on and off.

Make magnetic field with coil and battery

Wrap an insulated coil around an iron nail, and pick up other nails or paper clips. The current in the coil creates a much larger magnetic field with the help of the iron in the nail.

This coil and nail is a basic and effective creation of an electro-magnet, which you can turn on and off by disconnecting the wire.

Power hungry experiment:
- This electro-magnet drains the battery quickly. With a short length of wire, you are drawing huge currents.
- Please disconnect wire and battery after 1 second or so, when the battery heats up.

Experiments: Attract iron paper clip using magnetized iron rod

Summary of Basic Demonstrations of Iron and Electromagnet Lift Forces

"Engage your hands, spur your imagination."

Here are some household experiments to demonstrate forces from permanent magnets and electric magnets. These experiments are described in more detail in the next few pages.

Just get a battery, a spool of wire, some iron (screwdriver shaft), and a compass. A battery gives a constant current through the spool of wire, to make a magnetic field. Some iron will enhance the magnet fields and forces. A compass shows the direction of the magnetic field.

Go ahead, make a copy. Then check off the list.

Experiment 2.6: Lift iron stuff with permanent magnet (previous page)

- Iron can be magnetized and attracted to a permanent magnet.

Experiment 2.7: Magnetize an iron screwdriver by rubbing against a permanent magnet

- Iron screw drivers can be magnetized by rubbing the shaft against a permanent magnet.

Experiment 2.8: Make magnetic field with coil and battery

- Make an electro-magnet, using battery and coil.
- You could take a wire spool and use that as the coil.

Experiment 2.9: Increase magnetic field of coil using iron core

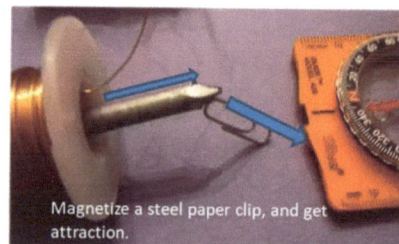

- Insert iron into coil, and observed compass needle point more toward coil. This shows a larger magnetic field.

Experiment 2.10: Attract iron paper clip or nail, using an iron core inside a coil

- Make magnetic field with coil and battery.

Magnetic force can lift magnetic objects like iron.

Magnetize a screwdriver shaft using another magnet

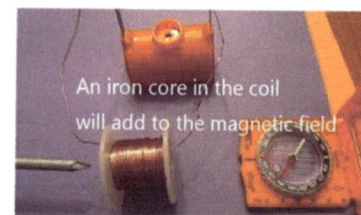

Simple electro-magnet using coil and battery

An iron core in the coil will add to the magnetic field

Increase magnetic field using iron core

Magnetize a steel paper clip, and get attraction.

Magnetized iron will magnetize and attract other iron

Experiments: Magnet field direction, magnet attraction

Iron is king, and is used as the core of electro-magnets because it can make the magnetic fields 1000 times larger.

Electro-Magnets and Lift, Q and A

Iron cores have made electro-magnets very strong and practical. The strong magnetization of the iron, due to a little bit of electric current in the coil, makes lift and motor applications possible.

Iron is everywhere in ores, dug from the ground. Iron is used for steel and tools and cars. Iron is also in your red blood cells to combine with oxygen to bring oxygen around your body.

Question 1: Why is iron so good as a core material inside a wire coil to make an electro-magnet? Iron easily magnetizes and increases magnetic fields and forces.

- The element iron has a strong magnetic moment. The applied magnetic field from the current in the coil can magnetize the iron (make the iron magnetic moments to rotate in the same direction), so the total field is much enhanced.
- Iron is used inside motors to increase the torque, and inside power line transformers to help up-convert or down-convert voltages.

Question 2: Why are some materials like aluminum so bad as a core material? Most metals are just conductive, but not magnetic. Most metals would actually reduce the magnetic field in the coil due to eddy currents around their surface that act to cancel the magnetic field.

- Aluminum is very electrically conductive, but there are no magnetic properties. The element aluminum does not have any magnetic moment. So an applied magnetic field has nothing to magnetize and rotate or align or attract.
- For AC current, this non-magnetic metal will have induced electric currents, or eddy currents, around its perimeter. These induced currents will actually reduce the magnetic field and also cause loses. Eddy currents are a negative thing for the core of an electro-magnet, due to losses, due to less magnetic field, and due to less torque.

Question 3: What advantages does an electro-magnet have over a permanent magnet? An electro magnet can be turned off to release the object.

- An electro-magnet can turn on and off, at the flip of a switch that turns the current to the coil on and off.
- An electro-magnetic can reverse the magnetic field by reversing the current.
- With extreme current, an electro-magnet can have a larger magnetic field than a permanent magnet because the field magnitude is only limited by the amplitude of the current, not by material properties of the magnet. Of course, extreme current takes a lot of power, and we would be talking about superconducting wire to avoid the resistive heating.

Question 4: How much weight can an electro-magnet lift? That depends on the current level. Large currents, more turns of the coil, and the iron cores will increase the magnetic field and the force.

- First, whatever is lifted needs to be magnetic, or able to get magnetized, like iron. Second, larger electro magnets are used to pick up large iron objects, like iron cars or industrial steel, so magnetic attraction can be strong.

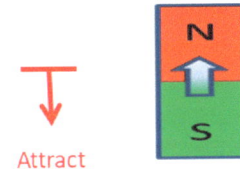

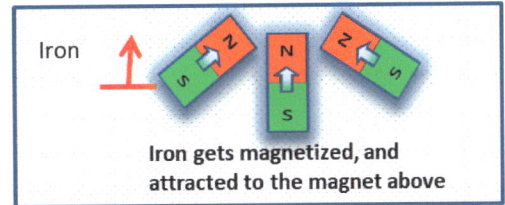

Attract

Iron

Iron gets magnetized, and attracted to the magnet above

Magnetization of magnetic materials (iron) in direction of applied magnetic field.

Magnets only attract magnetic material, like iron, not zinc, copper, or aluminum

Electro-magnet picking up scrap iron metal, and dropping it

Chapter 2C: People's Use of Static Attraction: Magnetic Games, Speakers, and Gripping Using Attraction

Let's look at the lifting or holding force of magnets.
- We can make toys like magnetic chess sets and refrigerator magnets.
- We can lift and hold iron objects, like steel sheets or cars.
- We can make a speaker for music by pushing a coil on the speaker cone against a permanent magnet.

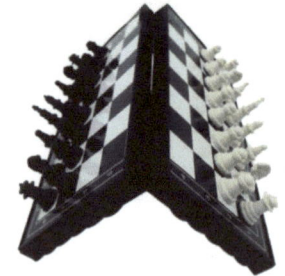
Magnetic chess game

There are many fun and practical uses of magnetic forces.

Toys:
We first get introduced to magnets for toys, like magnetic games for the car, checkers or chess. We see refrigerator magnets, which hold up favorite pictures or important telephone numbers.

Industry:
There also are some industrial uses for small items.

Small magnets: Small magnets protect your car transmission from the loose iron dust and chips from the gears over time, due to wear and tear. Magnets are used in machine shops to hold down iron objects for machining. Small magnets also can find and lift fallen steel screws.

Permanent magnetic lifter

Lifting steel: Large permanent magnets can lift up steel sheets, and then release the iron sheets by shunting the magnetic field through something else that is easier to magnetize. Large electro-magnets can also lift up steel sheets, and then release it by turning off the electric current.

Audio speakers: Electromagnets are in speakers. The current runs through a coil on the lightweight speaker cone over a permanent magnet. The electro magnet coil then gets a force toward and away from the permanent magnet, and the electrical current changes with the music. With the electromagnet speakers, the volume can be hugely amplified by increasing the current with amplifiers. The speaker can whisper or blast out over a rocket concert.

Speakers with electro coil and permanent magnet

High power switches: Mechanical switches also are created using a 'relay'. A relay has an iron core that gets pulled into a electromagnet when electric current runs through the coil. The iron core can spring out and hit a ringer for a door bell, or the iron core can push a metal bar across a gap in a wire and bring current to the starter motor if a car.

Movies and Myth:
Magnetic forces are treated like magic in the movies. The forces are greatly exaggerated to make an exciting action packed movie. Unfortunately, magnetic fields fall off very rapidly away from the magnets. There is not much force 10 feet away from an even strong magnet. Also, super huge magnetic fields would require super huge electrical currents, which is a big heating issue for the coils.

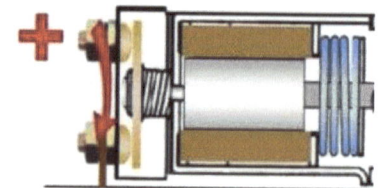
Relay switch: Metal bar against electrical contacts controlled by relay

Magnetic gripping is not just for chess sets and toys. Magnetic forces and attraction are used for heavy duty lifting of huge iron objects like cars and iron blocks, or just cleaning ground iron gear dust out of car transmissions.

People's Use of Static Attraction:
Magnetic Games and Gripping Using Attraction

Here are some fanciful gifts like magnetic games and pseudo-science heath bracelets, and here are some smaller real applications like iron filters and pick-up devices.
- Permanent magnets get attracted to iron, which magnetizes in the same direction and gets attracted.
- This magnet and iron attraction is similar to two aligned permanent magnets, with fields in the same direction.

People's toys of fancy

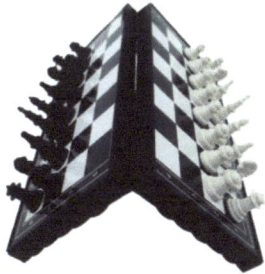

Magnetic chess game

You can play magnetic board games in the car because the pieces have magnets which get attracted to the iron sheet in the board. The pieces do not go flying everywhere when the car hits a pot hole, or when you need to tilt the board.

Refrigerator magnets

Magnetic darts game

Instead of poking holes in the wall, or endangering kids, you can use a magnetic dart board.

Unfortunately, this magnetic solution seems like a compromise for darts...magnets slip. The better traditional design, a sharp dart going into wrapped paper, is clearly more precise and won't slip, but is also sharp and dangerous.

Magnetic Bracelets

Are magnetic bracelets healthy? Do they attract the iron in your blood? ...We can neither confirm or deny.

People's practical use for smaller lifting applications

Most attraction happens where there are sharp gradients (change in magnetic field over short distances) for lifting forces.

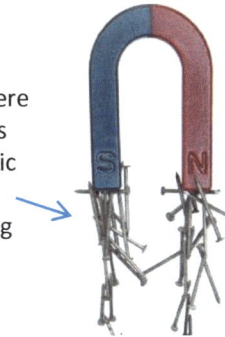

Horseshoe magnet

Telescoping magnetic pick-up tool

Magnets are mounted in a car's transmission pan to remove iron debris worn off the gears.

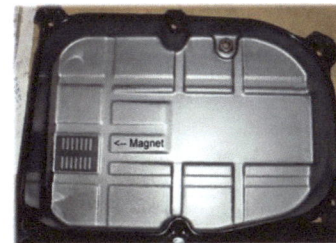

Car transmission bottom pan

Iron gear debris collected by magnet

Car transmission oil filter:

One use of magnets is to collect all the iron debris in the transmission oil of a car transmission. We want the iron pieces from the worn gears to be pulled away from the meshing of the gears, and not cause further wear.

Magnetic gripping is not just for chess sets and toys. Magnetic forces and attraction are used for heavy duty lifting of huge iron objects like cars and iron blocks, or just cleaning ground iron gear dust out of car transmissions.

Heavy Duty Lifting and Holding: Industrial Permanent Magnetic Forces

One of the first applications of magnets was simply to lift things, like steel plates and steel cars. For heavy steel utility or sewer manhole covers on the street, a magnet attachment can replace attaching a hook, with no hole or handle on the cover.

Another magnetic force application is holding things down, such as on a milling machine.

Permanent magnets are released using a parallel iron sheet between the magnet and the attached object. A lever rotates the iron sheet to steal or grab the magnetic field.

Permanent magnet lifting and holding applications for steel

Permanent magnetic lifter

Lots of neighboring alternating-reverse direction magnets will create large magnetic field gradients, and really hold on to iron objects tightly.

Halbach Array

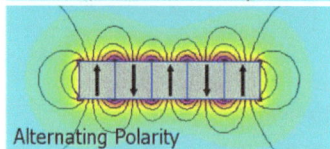

Alternating Polarity

To get the most attractive force, the magnet needs field changes and gradient

PNL0250

Lever to release magnet using iron shunt.

Handle will remove the magnetic field under the lifter, to release the object: The handle slides soft magnetic iron between the poles to steal the field inside it.

Magnetic chuck surface

Magnetic chuck locking mechanism

Holding iron down: Industrial permanent magnets on milling machine.

Hold down iron objects without needed to clamp them. This can be helpful for milling thin iron.

Large field gradients are key for larger lifting force. Alternating direction (polarity) causes more key field gradient, where there is a larger change in energy over a short distance.

Force ~ Induced Magnetic Moment * Field Gradient

where gradient is how fast the magnetic field changes with distance

See Appendix G, equation 4

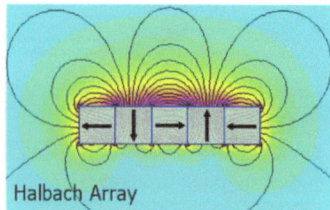

Releasing permanent magnet using iron shunt

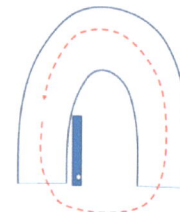

Magnetic field out bottom

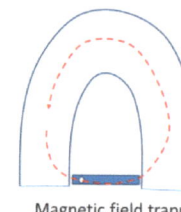

Magnetic field trapped in soft iron shunt

Field trapped in iron bridge.

Magnetic fields don't escape from soft iron short, when connecting N and S poles.

Hence force is discontinued, or un-locked.

Fun facts about magnetic lift

Lift utilities and sewer cover: Permanent magnets attract steel by magnetizing it, to lift manhole cover.

The good old days using hooks before strong magnets were invented

Old fashion manhole cover. The square knobs were for traction for horse hooves.

Here are magnets lifting steel (iron) blocks. Gripping is improved by having large changes in the magnetic field.

Heavy lifting with Industrial Electro-Magnetic Forces

As a special feature, electro-magnets can be turned on and off electrically with a switch, to pick up and release the attached object, for remote lifting on cranes.
Permanent magnets are not a good solution for lifting using cranes. Permanent magnets can not be turned on and off by an electrical switch. Still, a mechanical lever can slide in a soft iron spacer between the magnet and the iron object to steal the magnetic field and released the object.

Electro-magnet picking up scrap iron metal
An electro magnet can pick up and drop iron. One application of electro magnets is to pick up large steel sheets. Just by turning the current on and off, on a factory floor, a steel sheet can be moved around, grabbed and released.

Electro-magnetic lifter of iron objects:
- Electro-magnets lift the same as a permanent magnet but have the advantage that we can turn off the magnet (turn off the current) and drop the steel.
- Permanent magnets can not turn their magnet field and attraction on/off, unless slide in a special spacer that is very easy to magnetize. This space shunts the magnetic field away from the object getting held or lifted.
- Electro-magnet needs a powerful generator in the cab, with heavy cables leading to the electro-magnet.

Electro-magnet lifting applications: Lifting iron objects

Electro-magnet picking up scrap iron metal:
Just by turning the current on and off, on a factory floor, iron objects can be moved around, grabbed and released.

Electro magnet on a tractor arm:
Pick up random smaller iron pieces

Electro magnet on a tractor arm:
Pick up large heavy sheet of iron

More magnetic attraction: heavy duty lifting, or just grab nails.

Magnets and Coils: Speakers for TVs and Phones

Can you hear the sound from your TV or Headphones? Yeah, magnets do that. We depend on magnetic speakers in our TVs, computers, telephones, and music systems. A speaker can be as loud as we make it, louder than people yelling, with more electricity and stronger magnets.

Here's how a speaker works:

- **Coil and magnet:** The electro-magnet coil in the middle of the light speaker cone is pulled toward and pushed away from the permanent magnet as electric current changes direction, and pushes air to make sound.
  - The coil with current is alternately attracted to and repelled from the permanent magnet, depending on which way the alternating current is flowing. The coil acts like a magnet, but with reversing polarity based on current direction, moving up and down with frequency and intensity of the sound.
  - The direction of the permanent magnet does not matter. Sound waves are just air compression and expansion, so the electric current is oscillating as well at the frequency of the sound.
- **Light weight:** The light coil is wrapped around the base of the cone, because the cone is light and needs to move. The cone can oscillate at faster frequencies, above 10,000 Hz, and respond faster with less inertia. We don't put the heavy permanent magnet on the cone. A sluggish heavy speaker cone that barely moved with a little force would not generate much sound intensity. Also, a sluggish heavy speaker cone that took 1 second to ramp up from 100 Hz to 10,000 Hz would not sound very clear.

Let there be sound! Here is the long established speaker with coil and cone for high fidelity music

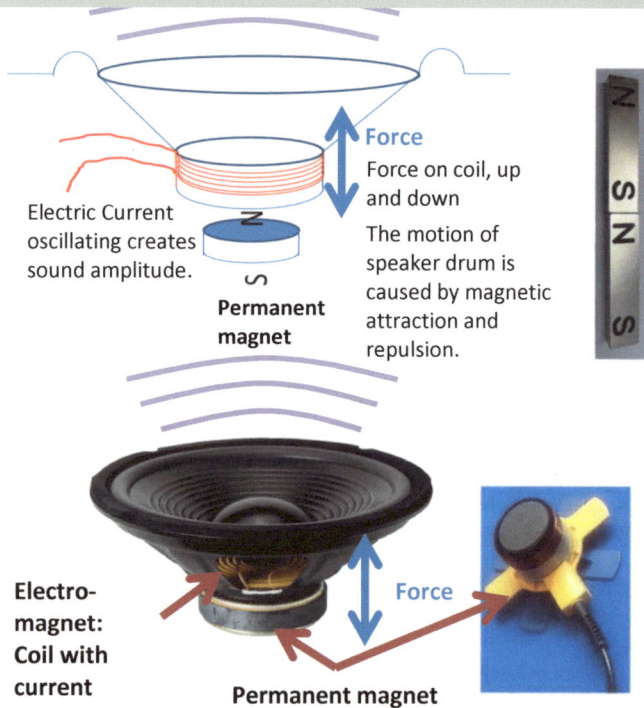

Force

Force on coil, up and down

Electric Current oscillating creates sound amplitude.

Permanent magnet

The motion of speaker drum is caused by magnetic attraction and repulsion.

Electro-magnet: Coil with current

Force

Permanent magnet

Here is one of first speakers, for low frequency voice for a telephone

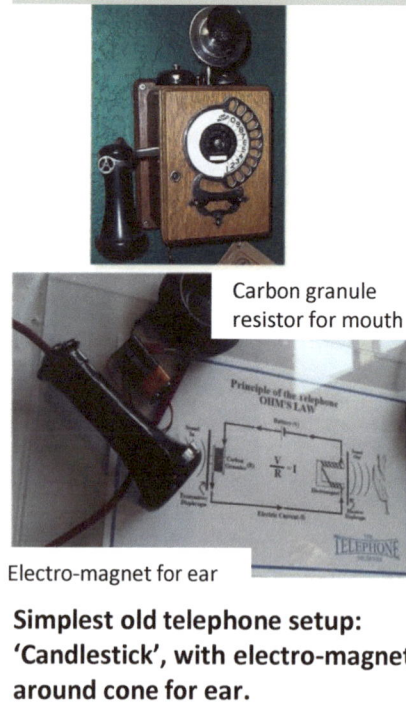

Carbon granule resistor for mouth

Electro-magnet for ear

Simplest old telephone setup: 'Candlestick', with electro-magnet around cone for ear.

The good old days before electro magnets and speakers

No amplification in first phonograph player.

Imagine no magnetic fields: Speakers could be tinny piezoelectric crystals, with distorted frequency responses, like cheap headphones.

Tinny sound from crystal (not broadband, not loud)

Public speaking before 1900.

- **Speakers are a very practical use of electro-magnet coils and permanent magnets, which are all around you (*telephones, headphones, audio speakers*).**
- **Try giving a public speech now or before 1900 without speakers: it is an act of screaming with a barrel chest and lots of diaphragm.**

Music Before and After Speakers, Oh My!

Electric speakers have changed the face and sound of music. First portable microphones just amplify the traditional pianos, horns and singers before the 60s. Then pick-up coils amplify electric instruments like electric guitars, which have no sound without electricity. Digital keyboards are all about electricity and speakers.

When the folk singer Bob Dillon first used an electric guitar in 1965 for folk music, the folk crowd there booed at Dillon for going commercial instead of staying true to original folk. Now electric guitars are the way things are, even for folk music.

Modern day music uses magnets and speakers, and has much more loud attitude

Music in the good old days with strings and horns. Hey, this was 'cool' or 'dope' in the olden days.

Evolution of music:
- Amplified music for the young and rebellious

'If it's too loud, you're too old'

Evolution of music:
- Brass bands
- String bands

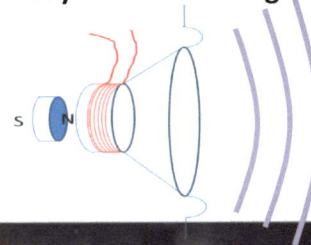

Earlier speakers:
Speakers are now more powerful because of stronger magnets, so the current day 'Wall of Sound' is a smaller size and yet still has louder sound.

Grateful Dead, 'Wall of Sound'
If your chest can feel the beat, you're probably killing your ear drums. Wear some ear plugs or go deaf in your 60s.

You can blame speakers for both the good and the bad about modern music. Electric guitars, keyboards, distortion, large venues don't happen without speakers.

Electro-Magnets: Relays to Start Car and Ring Doorbell

Relays pull in the iron core (electric current on), and release (electric current off):

Relays can wind up a hammer to hit a bell, or push a metal bar against two electrical contacts to turn on larger currents, like in a car. Relays are big coils with sliding iron or magnets. These mechanical relays came around well before transistor switches. Mechanical relays are bulky and slow and transistor switches are much faster and used everywhere in computers but overheat with large currents.

Relays and physics: Relays work because magnets are attracted to higher magnetic fields inside the coil. The iron gets magnetized in same direction as the coil and gets pulled in. Either DC or AC current can be used with an iron core, because the iron magnetization follows the current.

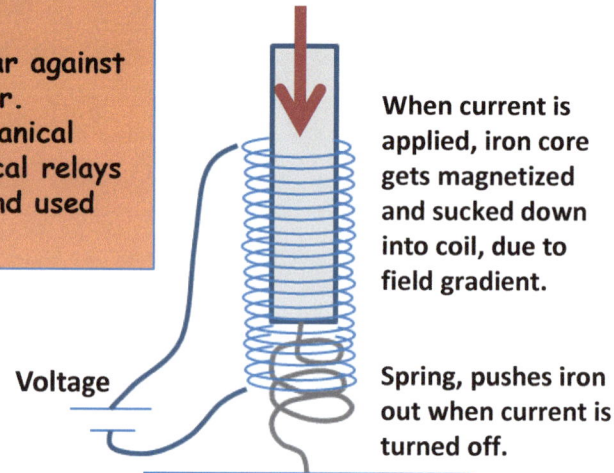

When current is applied, iron core gets magnetized and sucked down into coil, due to field gradient.

Voltage

Spring, pushes iron out when current is turned off.

The good old days in the Wild West:
Imagine no relays for door bells. In the old days, we're back to mechanical bells and pull strings, which is kind of quaint.

Servants bell pull: 'Oh Harold, please bring some tea'

Door knocker before electric door bells

Here are some applications of the simple relay, which has been around for a century:

Phone ringer with hammer hit

Electro-coil sucks in an iron core, like pulling back a hammer, when current is applied.

Magnetic iron is pulled into the electro-magnet coil, and then springs back to hit the bell when current stops.
The iron compresses a spring, so when the current is turned off, the iron shoots out again.

Door bell with hammer hit

Iron core pulled in when push door bell

Core pulling into magnetic coil

Core springs back, hits bell

Car starter switches / relays

A relay: When start a car engine, an iron core pushes metal bar across switch, to get huge currents going to starter motor.

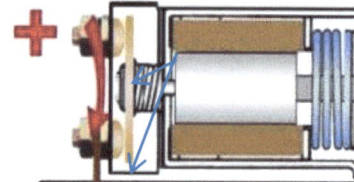

Metal bar against electrical contacts controlled by relay

Metal bar on relay held against electrical contacts to connect 12 Volts and engage starter motor

---

Relays are hardy switches for large currents: a metal bar is pushed against electric contacts using an electro-magnet.

Some Electro-Magnet Uses, Like Speakers and Relays

Before we get to motors, let's recognize that electro-magnets enable cranes for lifting, speakers for music and communications, relays to start the car, and computer memory to power the information age. These other uses are just as useful and pervasive as motors.

What did we cover? Mechanical motion ... the other applications

Workhorse of industrial Age

Lifting forces
1.   **Holding iron objects**
2.   **Refrigerator magnets**
3.   **Speakers and relays**

Twisting forces
1.   Motors creating mechanical motion, when electric voltage and current applied.
2.   Compass

Generators:
1.   Motors creating electrical energy, when shaft is spun mechanically (a generator or dynamo).

Electro-magnet picking up scrap iron metal

Question 1: What is so special about the material iron? Iron increases magnetic fields and increases magnetic forces.

- Iron can be easily magnetized. Electro-magnets can have a strong attraction to iron objects, and lift them, as iron gets magnetized and attracted.   Steel is mostly iron.
- Iron already has magnetized domains in the material, in random directions. An external magnetic field will align those domains more.

Question 2: What kind of magnets are used to amplify music? Coils are wound around light speaker cones to vibrate the cone to push air and amplify music.

- Electro-magnets are in the lightweight speaker cone of speakers, to amplify sound like music.   The electro-magnets pull the speaker cone back and forth in the field of the fixed permanent magnet, to push air and create sound.
- Cones for music need to vibrate up to 20 kHz. High end speaker cabinets will have a small 'tweeter' to vibrate at these higher frequencies, and a large base speaker to vibrate at lower frequencies.   The larger more massive cone is harder to get to vibrate at the higher frequencies, so there are two speakers.   A larger mass naturally can vibrate at lower frequencies, and a small mass naturally can vibrate at higher frequencies.
- Humans can hear the tweeter frequencies up to 20kHz when they are young, and down to the base tones of 20Hz. Larger animals like elephants and whales can hear lower frequencies. So these larger animals would need larger diameter speakers to get to lower frequencies. Small insects and bats can hear from 14kHz to 100kHz, and would need a tiny speaker cone.

Question 3: What type of magnets are used to open and close latches? Pulling an iron core into an electromagnet coil is a physical motion that can be a switch.

- Electro-magnets can release latches, called relays.   For example, in cars, these relays push a metal bar across a gap to engage the huge currents to the starter motor.
- AC motor: If the relay has a soft iron core then AC current, not just DC current, can pull the iron in.   The iron core magnetization will follow the AC current instantly, keeping the same alternating direction as the applied coil magnetic field. So the core is always an aligned and parallel magnet and feels a pull inward, no matter the direction of the AC current.
- DC motor: If the core is instead a permanent magnet, then the relay can not use AC current.   The forces would alternate between pulling in and pushing out.

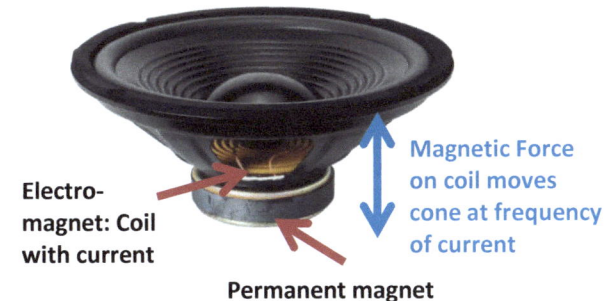

Electro-magnet: Coil with current

Magnetic Force on coil moves cone at frequency of current

Permanent magnet

Audio speakers using electro-coil and permanent magnet

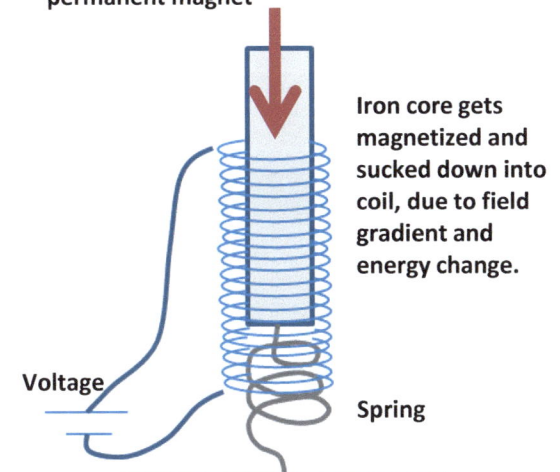

Iron core gets magnetized and sucked down into coil, due to field gradient and energy change.

Voltage

Spring

Relays to start the car starter motor, to ring the door bell, to turn on high power switch.   51

Electro-Magnets and Exaggeration in Movies

Electro-magnets and magnetic forces sometimes make an appearance in movies.   Usually there is some ridiculously large force, but the concept is there.

To get huge magnetic fields more than a few feet away from a coil, the electric currents would need to be ridiculously huge and the heat from the wire resistance would require huge cooling systems.

Magnetic miracle #1:
Grabbing cars in mid air

Grabbing cars in mid air using electro-magnet. (Fast and Furious 9)

Pulling cars in like a tracker beam

Pulling truck in and then jettisoning them away using electro-magnet. (Fast and Furious 9)

Magnetic miracle #2:
Sucking iron objects into an electro-magnet at long distances.

Terminator made from iron gets stuck to accelerator magnetic fields. (Terminator 3)

Terminator made from iron gets stuck to medical imaging MRI magnetic fields. (Terminator Genisys)

Kinda'-realistic use of magnets:
Grabbing iron objects to help sort trash

Escaping the garbage shredder using magnet sorter. (Toy Story 3)

If only magnetics fields on Earth were this strong. If we really wanted to stay close to reality, we need to go to a neutron star to get these magnetic fields for these huge forces.

Okay, magnets are used for sideways forces, like lifting iron, bearings, and magnetic levitation and trains in the future, as discussed in Chapter 2. What about rotation and turning a shaft? Remind you of anything?

Electric motors are everywhere, are based on flipping magnets, and enable life's conveniences as we known it.

This next chapter describes why magnets or rotors want to rotate, which is fundamental to a motor. Electro-magnets are essential to making motors work in order to switch currents to keep the rotor magnetic field stable in one direction even as the physical rotor and shaft rotate. This stable direction keeps a steady torque.

Chapter 3: Electric Motor Concepts:

Look at what a simple toy motor can do, to create mechanical rotation as a motor. We open up the simple DC toy motor and look at the parts. The stator is a permanent magnetic field, and the spinning rotor with shaft is a controllable electro-magnetic field, with current switching between coils using the commutator.

Motors and Magnets Flipping, enabling life's conveniences as we know it

Section A:
Motor Introduction: Magnets Rotate to Align, and Move in Closer

Magnetic fields want to be agreeable and align. This rotation can be easily demonstrated by holding a magnet near another magnet, and feeling the twist as they want to rotate and align their fields. However, that is just one flip and everything is then stationary. A motor wants to keep flipping or spinning, forever.

Middle magnet (the rotor) wants to flip and align fields

Section B:
Motors, the Basic DC Motor Using Electro-Magnets

So how is this perpetual spinning done? Does it break physics? No, at least one of the magnets must be an electric coil, and the current to the electric coil is reversed halfway through the spin cycle. Then the coil again feels the need to be aligned with the other magnetic field in the same spin direction. The whole trick is using electro-magnets and using a switch or commutator that applies current to the correct coil to keep the twist in the same direction.

Rotor, stator, and commutator in DC motor.
The middle magnet (the rotor) constantly wants to align because the current is getting switched to the proper coils.

Section C:
Direct Current and Alternating Current Motors, Not That Different

Electric motors can be powered by either a constant direct current (no oscillations) from a DC battery or an alternating current from the power grid AC wall outlet. Both work for the same reason: the rotor magnetic field is sideways to the stator magnetic field, and feels the need to align. An AC motor simply has this common magnetic field go up and down with the oscillations.

Motors can spin clockwise or counter-clockwise just by reversing the current direction to the rotor coils, without touching the field of the stator.

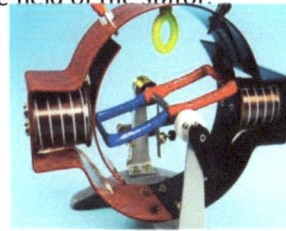

AC motor demonstration, using 2 poles, with electro-magnets for both stator and rotor:

AC drill: Clockwise or Counter-clockwise

Section D:
DC Brush and DC Brushless Motors

Transistor switches can replace mechanical brushes. Transistors don't wear and they don't cause friction.

Brushless motors are low maintenance and reliable, and are commonly used in e-bikes and EV cars.

Electric car brushless motor

Magnets rotate to align their fields, and electro-magnets enable current switches to keep this spin going.

Motor Type 1: Energy and Twist

We want twist to spin a shaft. Let's talk about energy that creates torque or twist on a magnet. This is a standard old fashion rotating magnet rotor.

This figure shows that the top magnet wants to rotate so that 'opposite poles attract', where N and S are closest and the magnetic fields are in the same direction. This desire for lower energy is just nature, so let's relax and enjoy our cars, air conditioners, ceiling fans, DVD players...

The most energy change happens when the top magnet is sideways to the bottom magnet, which means this angle has the most torque.

These motors have been around a long time and typically use brushes.

No torque, but balancing on an unstable energy plateau

Steepest change in energy, so most torque

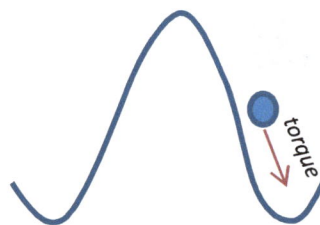

No torque, and stable energy valley

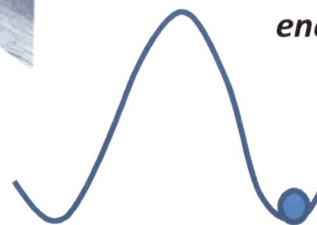

"Turn to align the magnetic fields, Padawan."

Like a ball in gravity on top of a hill, tottering to fall down.

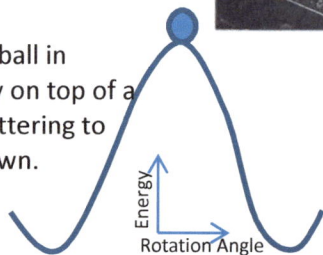

Energy / Rotation Angle

torque

Higher Energy

Lower Energy

Sideways magnetic orientation has most torque.
This is the geometry when the electro coils are turned on.

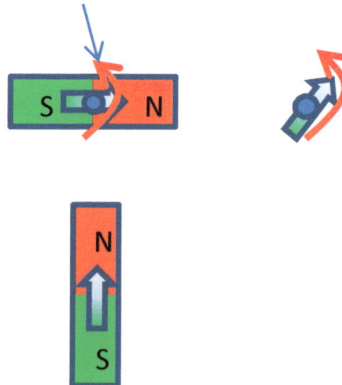

Opposite alignment and higher energy: Unstable, no torque when perfectly mis-aligned, but any tilt and will immediately want to flip.

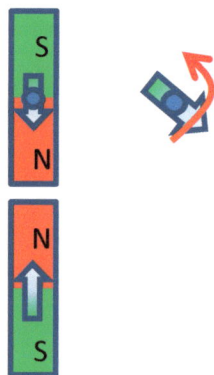

S
N

S N

N
S

Parallel alignment and lower energy: Stable, no torque and at lowest energy.

N
S

N
S

N
S

See Appendix B for energy and torque.

Here's our rule of thumb. If the energy can be lowered by rotating a magnet to have parallel magnetic fields, there then is a torque to do that. Motors design the rotor to always be in the sideways angle, to get the most torque to rotate to align.

Motor Type 2: Pull Magnets to Reduce Energy, for Linear Motor

We want twist to spin a shaft. Let's talk about energy that pulls on a magnet. This is a sideways force on magnets around a disk. This is a Linear Motor.

This figure shows that the rotor rotates because the rotor magnets wants to get pulled into the stator magnet field so that 'opposite poles attract', where N and S are closest and the magnetic fields are in the same direction. This desire for lower energy is just nature, so let's relax and enjoy our cars, air conditioners, ceiling fans, DVD players…

Linear brushless motors use electromagnets in the stator. These brushless motors use transistors for the commutator and are very reliable with no worn brushes. Without the brushes, there is less motor friction. Brushless motors are getting used in EV cars, e-bikes, and tools like drills. Linear motors are good for brushless motors, because the stator field may not be a simple uniform field, with many alternating electro coils around the stator.

This linear pulling force can also be used to pull levitating trains. The circular motor is simply rolled out on flat rails.

No pulling or torque, but balancing on an unstable energy plateau

Like a ball in gravity on top of a hill, tottering to fall down.

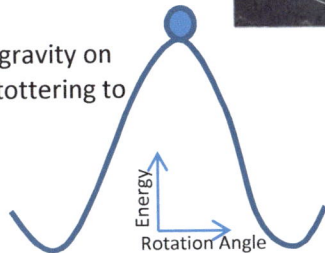

Steepest change in energy, so most torque

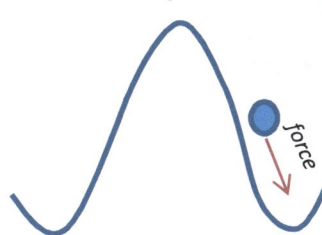

"Get pulled in to align magnetic fields, Padawan."

No pulling or torque, and stable energy valley

Energy

Rotation Angle

force

Higher Energy

Most sideways force on rotor magnet, which pulls rotor disk around.
This is the geometry when the electro coils are turned on.

Lower Energy

Opposite alignment and higher energy: Unstable, no torque when perfectly mis-aligned, but any sideways offset and will immediately want to push away.

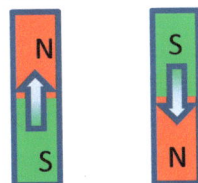

Parallel alignment and lower energy: Stable, no torque and at lowest energy. The two magnets have parallel magnetic fields and are the closest to each other.

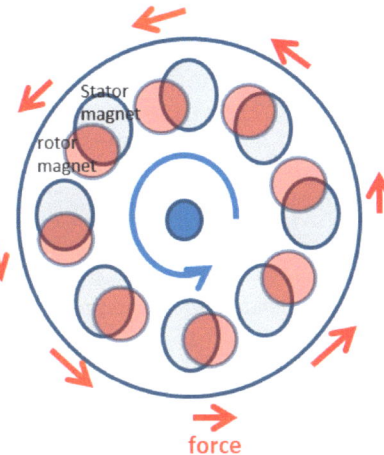

S

N

force

S

N

S

N

N

S

S

N

N

S

S

N

N

S

Stator magnet

rotor magnet

force

Motor: many magnets facing a disk. For a DC version, one set is made with commutating electromagnets.

The rotor magnet wants gets attracted to the opposite magnetic poles on the stator, and pulls the rotor disk around.

Okay, magnets are used for sideways forces, like lifting iron, bearings, and magnetic levitation and trains in the future. What about rotation and turning a shaft? Remind you of anything?

This chapter describes why magnets or rotors want to rotate, which is fundamental to a motor. Electro-magnets are essential to making motors work in order to switch currents to keep the rotor magnetic field stable in one direction even as the physical rotor and shaft rotate. This stable direction keeps a steady torque.

Magnets rotate to align their fields in the same direction. What can we do with that? Motor?
- Magnets want to rotate or flip so 'opposites attract' or fields align. Show that magnets want to rotate so that 'opposites poles attract' using two magnets. Attach a bent paper clip to the top magnet (and use tape if attraction to paper clip is not enough), and see the top magnet rotate. It can not snap down, or fly away, but it can rotate.

Magnets feel sideways forces, or a lifting force. How do we stop that for a rotor? Vibration from these sideways forces would cause wear on the bearings.
- Magnets feel forces pushing them in some direction, not just a rotation, when they are placed in a magnetic field with a field gradient. Motors don't want this force, because it causes vibration and bearing wear, so motor designs remove the field gradient using balanced top and bottom stator magnets. Motors just need a rotation or torque.

Experiment 3.1: Show magnet wanting to flip over other magnet, using suspended magnet

Force away from bottom magnet

Force toward bottom magnet after flip top magnet

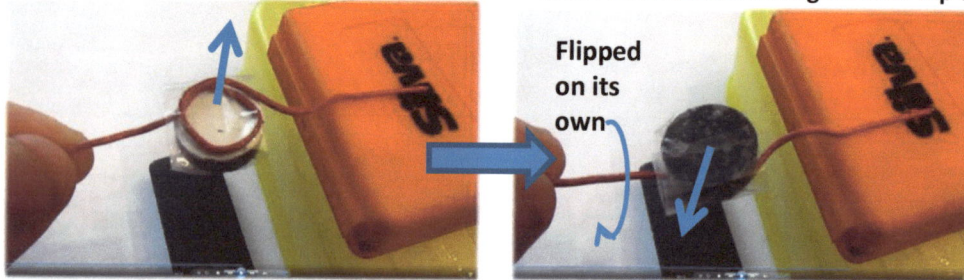

Flipped on its own

"Dude, how clever to use an iron paper clip to help the magnet stick to it. The author here is the best."

Permanent Disk Magnet, the 'rotor', rotates to align the magnetic fields with the magnet underneath, the 'stator'

Spontaneous flipping or twist, to align fields:
Here is the fundamental flipping behavior that allows electric motors to spin. The suspended magnet spontaneously flips so that opposites attract, meaning the fields or magnetic moments are in the same direction. This flip demonstrates the motor basic mechanism.

Unbalanced single sided stator field, causing force:
Note that there is also a lift or attractive force here in this magnet setup with only one bottom magnet, because there is a field gradient at the rotor magnet. In practical motors, instead, we design the DC stator fields to have no field gradient using top and bottom magnet to avoid lift forces and wear on the bearings.

Show magnet wanting to flip over other magnet, using two bar magnets in sideways orientation

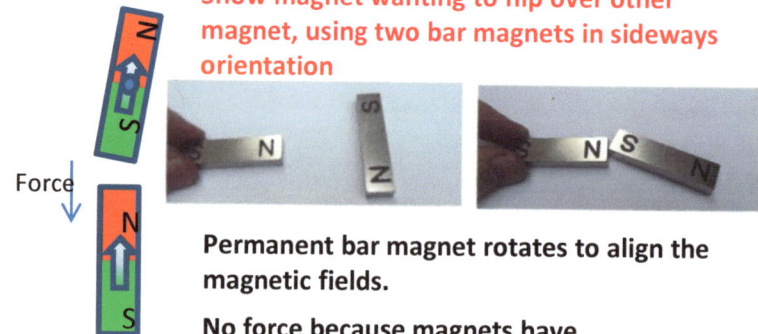

Force

Permanent bar magnet rotates to align the magnetic fields.

No force because magnets have perpendicular fields, but still have maximum torque. After the magnets start to rotate, there is a force due to unbalanced fields

This top magnet will rotate quickly, and then just stay steady, with opposites attract.

Experiments: Flipping a magnet, and difference between a force and a twist on magnet

Magnet Rotor Between Two Permanent Magnets has Twist, but No Sideways Force

"I'm strong, and I like to do flips while standing in place."

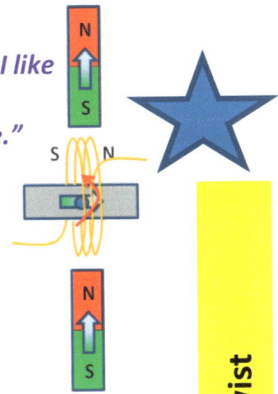

Let's 'field' a few issues with motor design.
For motors we want twist, and no sideways forces. We don't want a force against the bearings. When there are two stator magnetics, one on top and one on bottom, forces are cancelled or removed and there is much more stability, only twist but no sideways forces. This is how motors are designed.

Experiment 3.2: Show that two magnets on top and bottom will cancel the attractive or repulsive force, and will only cause a flip. The magnetic field is uniform.

Concept of balance

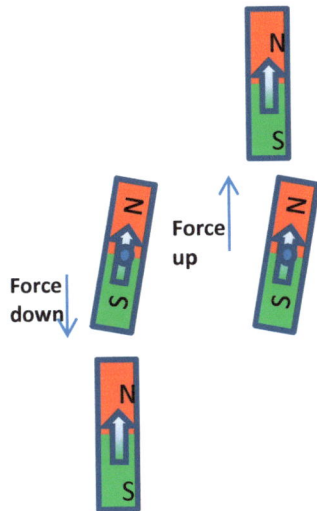

Twist or torque is strong, while forces cancel.

Twist: strength to turn, but the center of mass does not move in location, which is good for a spin-only motor.

Forces: strength to move the center of mass, or 'translate' it, which is bad for a motor.

- Use top and bottom stator magnets, to cancel forces. Design stator magnet shape to not have force, only torque, so no field gradient

Twist is strong to align magnetic fields, and forces cancel on the middle magnet (the rotor).

In this geometry, opposite forces cancel out using top and bottom permanent magnets, so the middle rotor magnet can spin without any sideways force wearing on the bearings.

Simple demonstration of torque without force

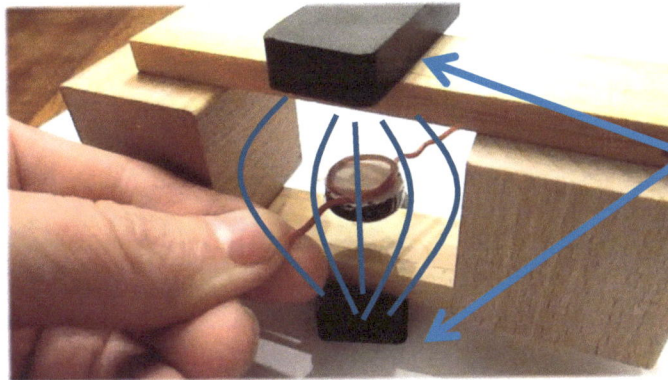

Two permanent magnets form the stator:
The permanent magnets are placed on both sides of the rotor to have forces up and down cancel each other, and to create this uniform field. The magnetic field has no slope or gradient, or change with distance, around the rotor. Hence there is only twist or torque, and no sideways or translational force.

Balanced rotor between two permanent magnets, with larger field and more torque, and no force due to cancelation.

No torque until the magnet is slightly cock-eyed, and then, boom!, a huge torque flips the magnet, like getting unbalanced and falling off a tight-rope. In this case, the middle magnet has a magnetic field starting opposite the outer magnets.

The rotor is attracted equally in opposite directions to each of the stator magnets. There is a uniform DC magnetic field between two permanent magnets, aligned. When the spinning rotor, or magnet, is placed in this uniform DC field environment, there is only twist and no pulling force. This saves wear and tear on bearings holding the spinning rotor. You'll see that stators have permanent magnets on both sides of the rotor, so forces up and down cancel each other out. These two stator magnets create this uniform magnet field around the rotor (no slope or gradient or change with distance for the middle point of the magnetic field).

For the energy point of view, energy descriptions explain why there is no up/down force on the rotor magnet. The permanent magnetic field is constant (uniform) in the middle between the two permanent magnets. The spinning magnet will see the same DC magnetic field when it moves up or down, so there is no change in energy between the rotor magnet and the constant permanent magnets. With no change in energy, there is no force. *See Appendix B for energy and torque.*

Opposite forces cancel, equal up and down. This quick, rinky-dink experiment using wooden blocks and permanent magnets shows how magnets can rotate to align, but not feel any net force.

Experiments: Create uniform magnetic field so only have twist

Rotor Needs to be Electro-magnet, not Permanent Magnet

Two permanent magnets will only align, not spin, because the goal of minimum energy is to get the fields to align. We need to play tricks with the fields, and turn them off, and run the currents through a different coil, to re-create the torque action and get the rotor to continuously want to spin in the same direction.

Try this! Regarding spin direction, if you flip the stator permanent magnet, the torque is in the other direction and the coil spins in the other direction.

"Motors twist because magnets want to align."

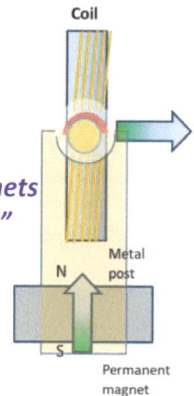

Coil
Metal post
Permanent magnet

Permanent magnets have only one-shot twist:
Single alignment is not a motor, just a static alignment

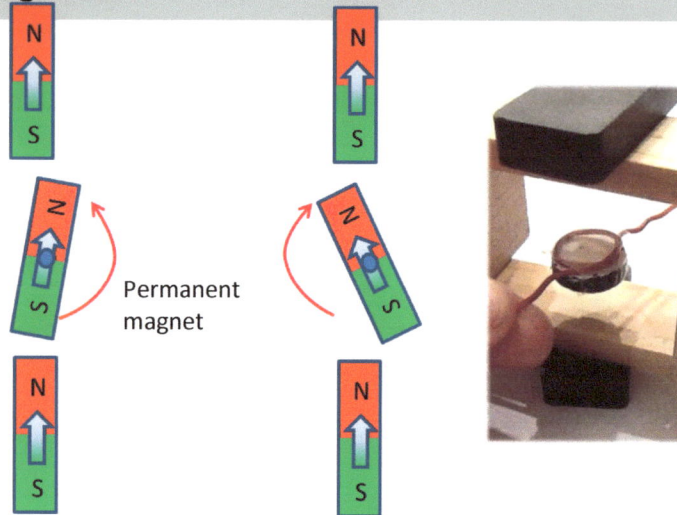

Permanent magnet

No control over permanent magnets, so they just line up without spinning.

Permanent magnets oscillate back and forth:
Permanent magnets keeps wanting to align, to the torque is reversed on the opposite side.

Electro magnets enable continuous twist:
Currents switched to sideways coil enables a motor, with switches and continuous spin

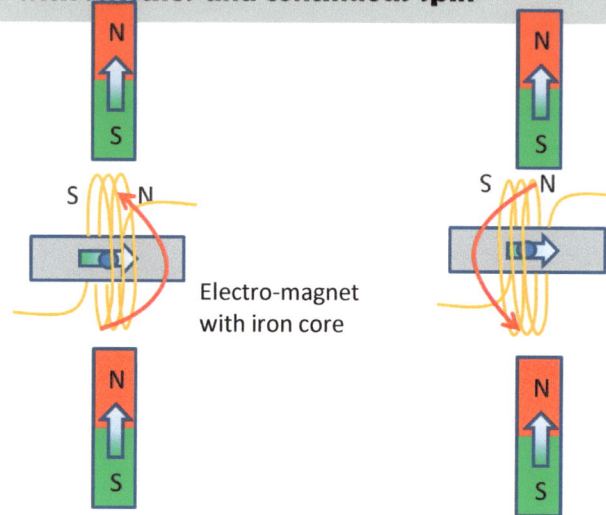

Electro-magnet with iron core

Electro-magnet rotor

Coil ½ spin later, with reversed current and reversed coil to keep rotor field and spin going in same direction

Electro-magnets have continuous spin using switched current:
Need to instead reverse twist on 'down stroke', so all permanent magnets won't keep spinning for a motor. Need electro-magnet instead to allow full 360 degree rotation.

Twist or torque on permanent magnet in a magnetic field:
When the rotor is a permanent magnet, there is no continuous rotating motion, just a single flip, or just a single attraction / repulsion. After the center magnet rotates half wave around, it wants to stay that way, with opposite poles attracting.

If the magnetic field were turned off, after the center magnet rotated so that opposites attract, then the magnet could continue rotating, without an opposite torque resisting the rotation. That is why the rotor in a motor is an electro-magnet, not a permanent magnet. The DC current in the rotor coil, which creates the same magnetic field as a permanent magnet, needs to be turned off or needs to reverse direction, to complete the turn.

Motors needs to play tricks with the rotor magnetic field, to always keep the rotor magnetic field in the same sideways direction where the torque is constant and maximum. That is why electromagnets and commutators are necessary.

Why the Rotor Needs to be an Electro-magnet

Coil

The commutator keeps or switches the current flowing only in the coil that is facing sideways, to keep the maximum torque.
When you flip the permanent magnet acting as the external field (the stator), the torque is in the other direction and so the coil spins in the other direction.

"Switching current in commutator enable continuous spinning."

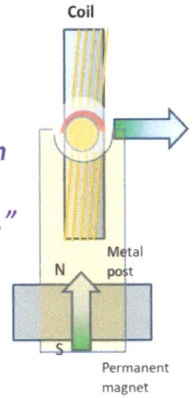

N   Metal post

S   Permanent magnet

Permanent magnet demonstration of torque:
Sideways magnetic orientation of the rotor to the stator has most torque.

Concept for torque only: Need electromagnet for rotor to reverse current when one coil on rotor is facing the other direction, which keeps a steady magnetic field from the sideways coil.

DC motor demonstration of torque:
Sideways electro-magnet has steady twisting force

- **Alignment of magnetic fields**: Coil's magnetic field wants to point in the same direction as permanent magnet, to reduce energy.
- **Coil's magnetic field**: generated from currents
- **Stator**: Permanent magnet always has a magnetic field.

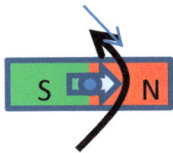

Most torque or twist at 90°:
- Sideways rotor has largest change in energy versus change in orientation, which means most torque

Least force or sideways pressure at 90°:
- Sideways rotor has no change in energy by moving away from or toward stator magnet, which means no force.

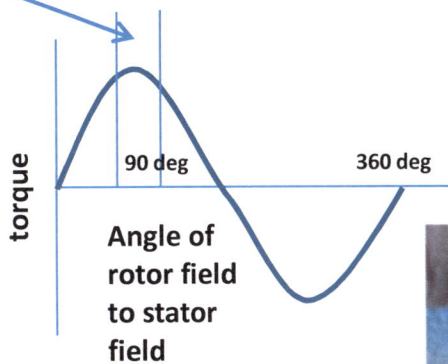

Electro-rotor always holds this optimum sideways peak torque (better using 5 pole rotor instead of 2 pole), using current switching to whichever coil is sideways.

Commutator and Brushes:
- Switching current to whichever coil that is sideways as rotor turns, to keep a steady torque to turn shaft.

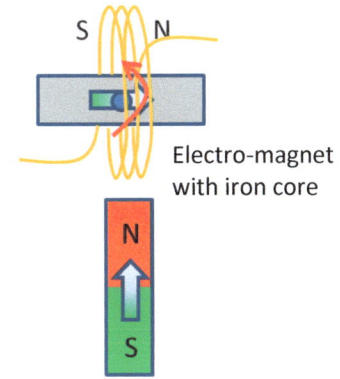

S   N

Electro-magnet with iron core

N

S

torque

90 deg   360 deg

Angle of rotor field to stator field

Voltage makes spin

Keep the spinning magnet's magnetic field oriented sideways to the permanent magnet, 90 degrees to the stator field → most torque out of the motor.

*Torque = Permanent Magnet Field (stator) * Coil Magnetic Moment (rotor) * sin(angle difference)*

Most Twist when Permanent Magnetic Field and Coil Magnetic Field are perpendicular, when
*Torque = Permanent Magnetic Field * Coil Magnetic Moment*

Other Synchronized Pushes

<u>Synchronized pushing for electric motor, two pedal bicycles, combustion engines, water wheels</u>: Bicycle riders always push when each pedal (or piston or water bucket) can create a push synchronized to a shaft going in one direction.

Are electric motors like other methods to spin a shaft? Yes. Electric motors keep switching the current to the sideways coil to keep a magnetic field sideways to the stator magnetic field. This switching is like pedaling a bicycle when you push always in the same direction, or timed explosions in a car combustion engine, or water always filling up one side of a water wheel for gravity to pull it down.

Synchronized magnetic fields

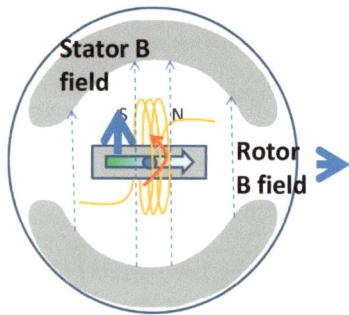

Electric current is applied to whichever coil is sideways to the stator magnetic field, as the rotor is spinning:

The commutator switches current to the right coil to get the continuous torque on the rotor.

The commutator can use brushes, which where the switching is controlled mechanically. Or a commutator can be brushless, where there are electronic switches which are controlled by the orientation of the rotor.

Synchronized feet

A bicyclist's feet only push down when bicycle pedal can push the chain in a consistent direction:

A bicycle with 2 feet is like a steady 2-pole electric motor, or like an even steadier 4 cylinder combustion engine with a 4-stroke piston process, where there is only an explosion every two strokes per piston.

Synchronized explosions

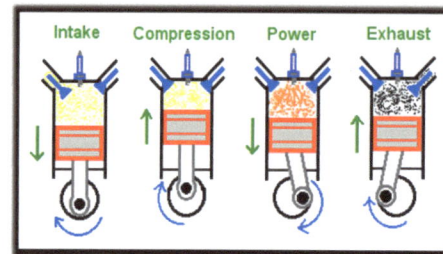

4 stages of a 4-stroke engine

The explosion of the gas in the cylinder is timed for when the piston is pushing on the crankshaft on the same side.

The timing of the explosion is very important. The explosion needs to be synchronized within a few degrees of rotation to get the most compression and efficiency from the engine. The explosion is synchronized at the top of the compression to get the most heat and pressure.

Synchronized falling water

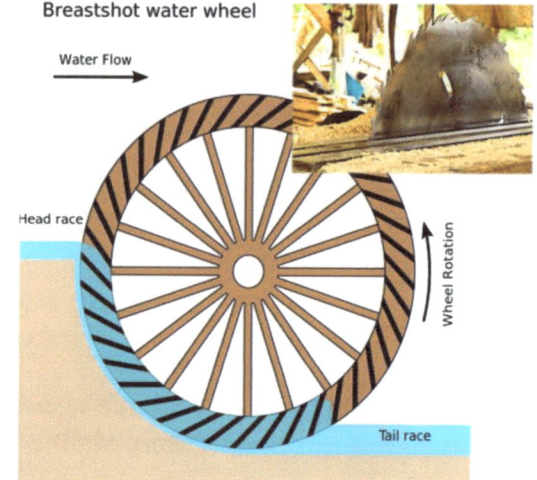

A water wheel is only pushed in one spin direction:

The water fills the buckets and torques the water wheel down always in the same direction due to gravity.

The water wheel can be attached to belts which spin a circular saw.

The idea of synchronized pushes to keep a motor spinning has been applied in many types of motors

No Sideways Force: Balanced Magnets to Stop Net Forces and Vibrations on Rotor

> We want motors to have maximum torque, but also to have no sideways forces so the rotor does not vibrate itself to pieces, or wear down the bearings.

The maximum torque happens because the rotor wants to align its field with the stator field.

A simple stator design suppresses force in two ways, in a double-breakthrough conservative design, described below.

- **Suppress force 1**: There are top and bottom stator magnets, and both attract the rotor in equal and opposite directions. These two forces cancel each other at any angle of the rotor.
- **Suppress force 2**: At the exact angle of the sideways 90° orientation of the rotor, there is no force from either of the two stator magnets. It is not a matter of cancelation, there is simply no force. This is because there is no change in energy if the sideways rotor moved up or down.

When you look at or listen to a motor spinning, you don't hear it rattling or jumping all around. That is because the rotor is nicely balanced and does not experience forces to the side.

"I feel no net force."

(This arm ripping is actually an exaggeration. Horses can rip the arms apart. For magnets, there is no force or strain on any molecule of a magnet in a uniform field. There is only twist.)

No force on rotor, Breakthrough #1: Force cancelation from top and bottom magnets, even when rotor is not exactly sideways

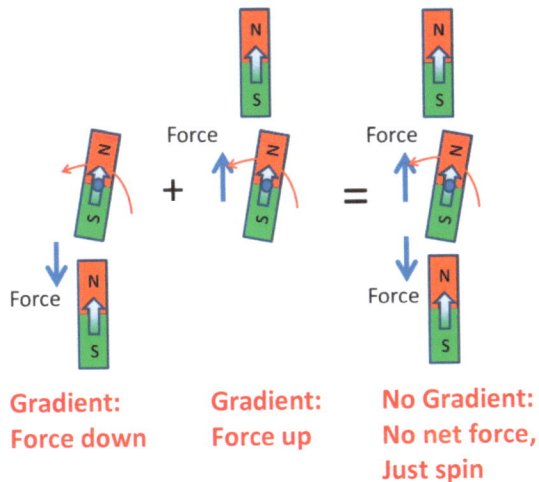

No force on rotor, Breakthrough #2: No force when rotor is exactly sideways

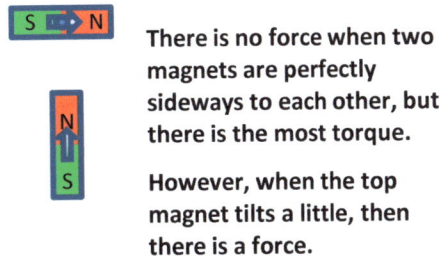

Gradient: Force down   Gradient: Force up   No Gradient: No net force, Just spin

There is no force when two magnets are perfectly sideways to each other, but there is the most torque.

However, when the top magnet tilts a little, then there is a force.

Force on magnet in a magnetic field, with balanced forces and unbalanced force:

If the forces on the rotor were not perfectly cancelled, or zero, then the rotor would have forces to the side that will wear one the bearings. Also, there can be a resonance frequency where the rotor just keeps getting more wild and more wild, oscillating back and forth.

Any motor design should avoid any forces that can cause bad resonances. *See Appendix G, equation 4*

Motor bearing wear and tear:

Levitating shaft (no sideways force), only spin

Permanent magnets on both sides are a good choice, both in the same direction:

- There are opposite forces from both sides that cancel, so there are no sideways force on bearings up or down, no matter the orientation of rotor

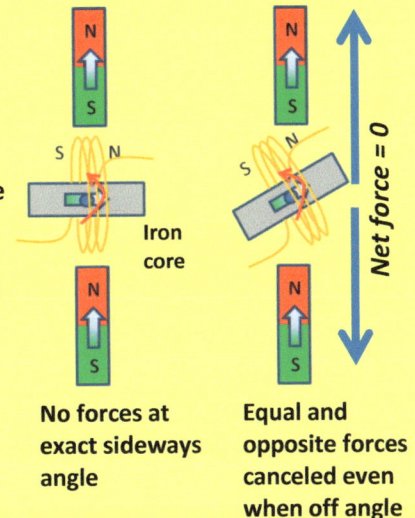

Iron core

Net force = 0

No forces at exact sideways angle

Equal and opposite forces canceled even when off angle

Permanent magnet on just one side is a bad choice:

- Have wear and tear and vibrations on bearings from sideways force when rotor is a little off sideways

With an unbalanced stator field, there is a force pulling down on the shaft, in addition to spin

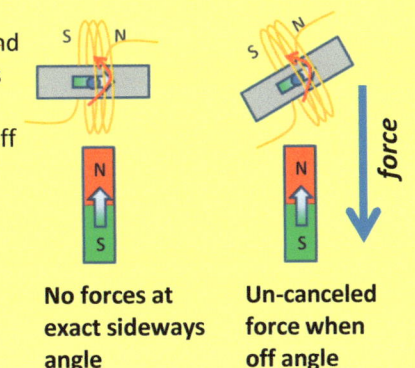

force

No forces at exact sideways angle

Un-canceled force when off angle

> **Vibrations and bearing wear are greatly reduced by avoiding any forces on the rotor.**

No Sideways Force: Stator Field at Rotor, Between Two Magnets

The rotors in the motor are there to provide smooth spin, and not get dragged against the bearings of the shaft through some sideways force. The shaft does not need the unnecessary wear.

To avoid drag and force against the gearing, we want to have magnets on the top and bottom to cancel the attractive and sideways forces. The attraction of the rotor to each permanent magnet is canceled by the other permanent magnet, so the rotor is doing a balancing act.

To avoid force and drag, another way of saying 'no net force' is that we want a uniform magnetic field (the same through out, no gradient) at the rotor. In a uniform magnetic field, the center magnet has the same field and energy no matter where it moves, so there is no force, just torque.

"Nice uniform magnetic field around the rotor, for friction-free spin, like floating on a cloud."

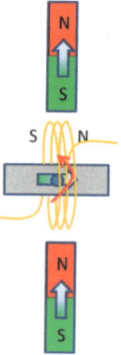

Permanent magnet in uniform field experiences only torque, no force.

The sum of the two magnetic fields of permanent magnets create a uniform field in the middle.

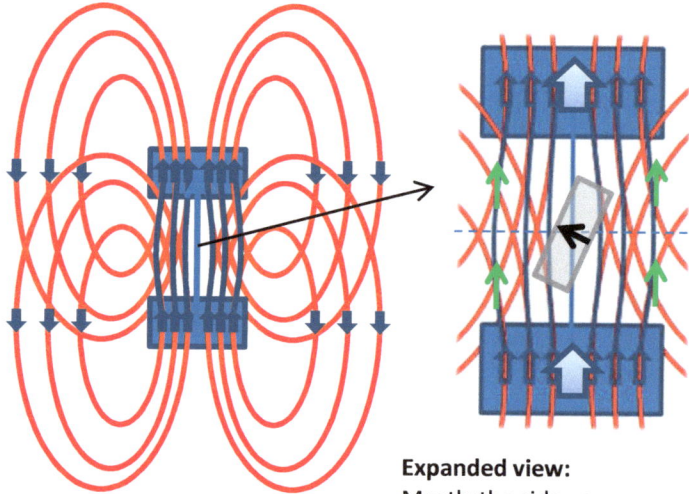

Expanded view:
Mostly the sideways magnetic fields (red) from the two magnets cancel, leaving the uniform field (green).

- Top and bottom stator magnets have equal and opposite force on rotor, so there is no net force.
- Said a different way, there is no gradient in the field in the direction between the magnets, along this middle plane. The field is straight like prison bars.
- Sideways fields cancel from the two permanent magnets, so no sideways force either.

Electro-magnet rotor in uniform field experiences only torque, no force.

Weaker torque,

No force because of cancelation

Strongest torque when rotor field is sideways.

No force because rotor is perfectly sideways, and cancelation

Weaker torque,

No force because of cancelation

B field stator

B field

B field

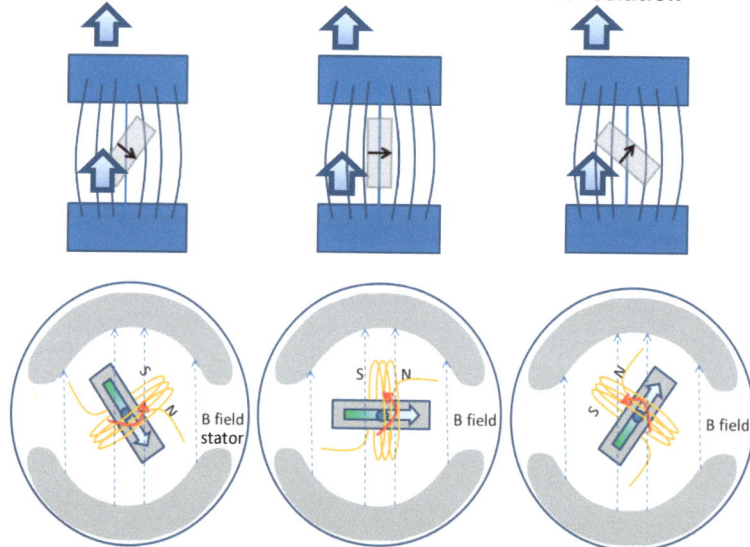

Give electric current to whichever coil is angled sideways.

- Attraction of rotor to each permanent magnet is canceled by the other permanent magnet, so the center rotor is doing an easy balancing act.
- Using circular shaped magnets, the stator field is uniform in center between two permanent magnets. Uniform means the field does not change as you go up or down.
- The rotor only feels twist or torque, no sideways attraction to any magnet, so there is least wear on the bearings.

See Appendix G, equation 4 and 6

Two magnets on top and bottom create a uniform field in the middle, with no net force on middle magnet.

No Sideways Force in a Uniform Magnetic Field

Let's not just talk about a uniform well-designed stator field of a motor. Let's talk about big likes the Earth's magnetic field.

There are other examples of where there is no force on a magnet in a uniform magnetic field. There is no energy change by moving in the uniform magnetic field, so there is no force to causing moving.

A motor has a uniform field by the symmetry of using both top and bottom stator magnets, and the Earth has the uniform field by being extremely large.

"Life would be completely different if everything magnet got forced to the ground, but fortunately the Earth's magnetic field does not do that."

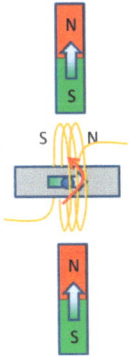

Compass in Earth's uniform magnetic field

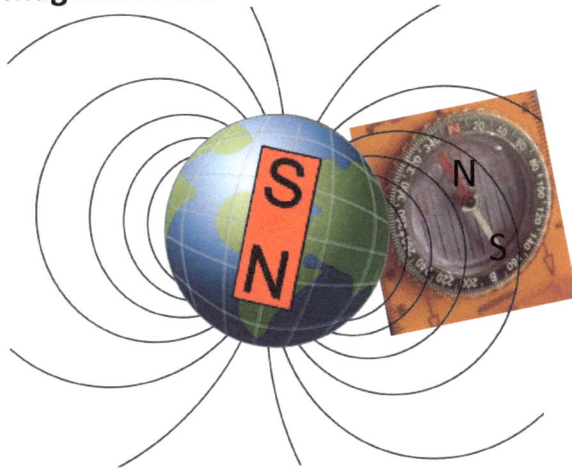

Needle is not yanked out of your hand.
There is no force on the needle in the Earth's uniform magnetic field
– torque yes, force no.

Iron Needle Rotation: Compass needle lines up with Earth's magnetic field, where opposite poles attract

- Iron Needle Forces: There is no tearing of the compass needle out of your hand. Instead, the needle just rotates to align. This is because Earth's magnetic field is mostly constant or uniform compared to the small size of a compass.

Raw Iron in Earth's uniform magnetic field

Magnetite or Lodestone, attracting iron nails

Iron ore is not pulled into the Earth's core.
There is no force on the iron ore in the Earth's uniform magnetic field
– torque yes, force no.

All the iron in the Earth's crust is not pulled into the earth due to magnetic fields. There is only the regular gravity force.

- The large magnetic field of the Earth is uniform, and causes torque but not force.
- Think how we would not have iron in the Earth's crust if all the iron got magnetized and sucked into the Earth's core.

Magnetized ship in Earth's uniform magnetic field

De-magnetizing ships: In the Earth's magnetic field, the iron hull of ships get magnetized and can set off magnetic mines. Here the submarine is getting de-magnetized using a large coil.

Iron ships magnetize and are not pulled under by the Earth's magnetic field.
There is no force on the iron ship in the Earth's uniform magnetic field
– torque yes, force no.

Iron Ships naturally magnetized in the Earth's magnetic field. The ships do not get sucked down below the waves because of magnetic forces.

- The large magnetic field of the Earth is uniform, and causes torque but not force.
- If the Earth's mostly uniform magnetic field actually caused a net force, then all the ships would be built from aluminum or fiberglass, not iron.

Two magnets on top and bottom create a uniform field in the middle, with no net force on middle magnet.

Current Loop Explanation of Torque:
Forces That Rotate Magnets Causing Torque on Magnet

Why is there torque, you ask? Lets look at the forces and torque on the sideways current loop below. This figure shows there is a torque on a coil with current or magnet in a uniform external magnetic field, when the coil or magnet is sideways, all due to magnet force on each part of the current. The electric current in a loop represents any magnet.

"Turn to align the magnetic fields, Padawan."

Maximum torque on vertical loop or magnet, both forces on opposite side of loop add to spin about shaft

No torque on horizontal loop or magnet, neither force on opposite sides of loop causes spin about shaft

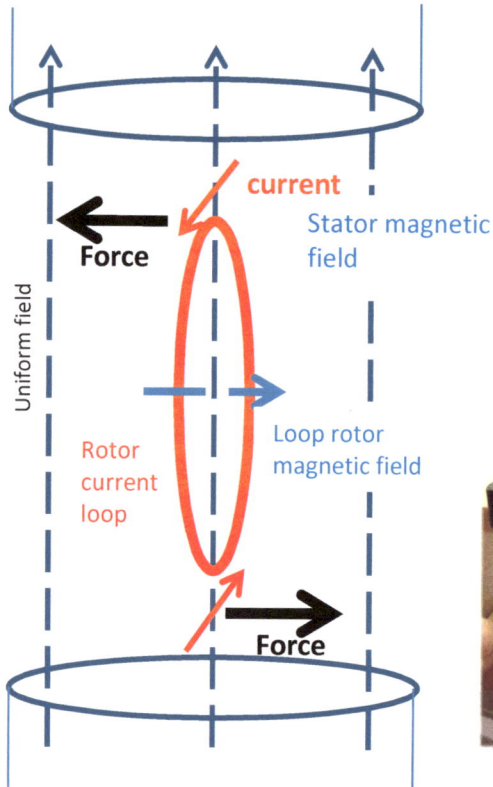

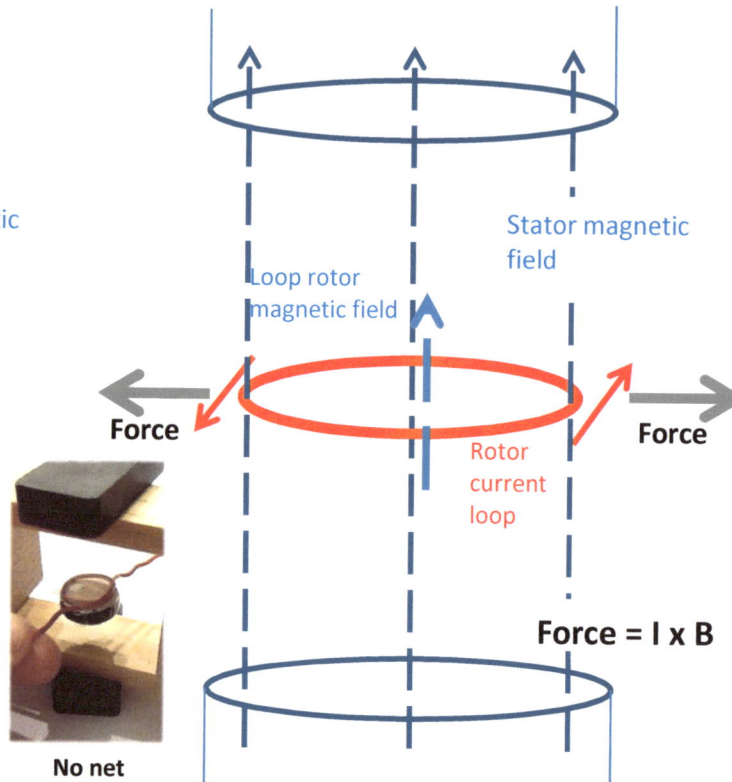

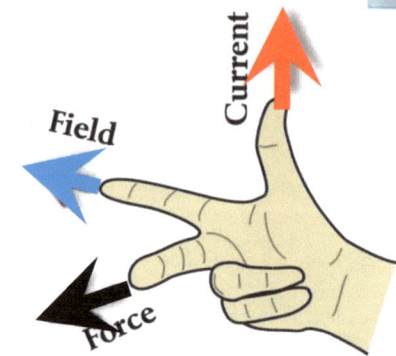

current

Force

Stator magnetic field

Uniform field

Loop rotor magnetic field

Rotor current loop

Force

Stator magnetic field

Loop rotor magnetic field

Force

Rotor current loop

Force

No net force, just twist

Force = I x B

Field

Current

Force

Fleming's Right Hand Rule

- Coil gets net maximum torque when current loop is sideways to external magnetic field
- Forces spin together around pivot point, to create strongest torque.
- There is no net sideways force, due to the top and bottom balanced stator magnets.

- Coil gets no net torque when current loop is aligned with external magnetic field
- Forces cancel around pivot point, to cause zero torque.
- There is no net sideways force, due to the top and bottom balanced stator magnets.

Current loop representation:
All permanent magnets or electro-magnets can be represented by current loops.

Net torque or twist:
An external magnetic field, even without a gradient, causes opposite forces on current going in opposite directions; that is, opposite sides of the current loop. These opposite forces act to twist the current loop.

With both top and bottom magnet, without a gradient, there is no net force, only torque.

Diagram of forces on a rotor current loop (red) at different angles, inside a balanced stator magnetic field

A sideways current loop in a magnetic field gets a net torque.   An aligned current loop in a magnetic field has no torque, all based on the magnetic forces on currents using the 'Right Hand Rule'.

Current Loop Explanation of Unbalanced Sideways Force:
Forces That Cause Magnetic Attraction using Current Loop

Magnetic forces are explained by magnet forces on electric currents. An electro-magnet has obvious currents. A permanent magnet has 'effective' currents around it's surface, like a current loop, to create the external magnetic field.

current loop = magnet

Why is there attraction, you ask?
A diverging magnetic field will create a net force on the current loop, for the forces don't exactly cancel. Part of the force on all segments of the loop will add up either away from or toward the magnet. The force on the current is the product of the current times the magnetic field, where we also need to keep tabs on the direction of the force using the 'Right Hand Rule'.

A uniform magnetic field does not cause a net force, and the forces exactly cancel. Consider a compass needle. A compass needle only gets a twist and not a force moving it sideways in the Earth's magnetic field, which is relatively uniform.

Another interpretation is that there is a higher magnetic field closer to the surface of the magnet and this causes an energy change as a second magnet gets closer. Energy changes mean a force.

> *"Magnetic field falls off with distance, which enables a force, Padawan."*

Attraction or Repulsion of magnets or current loops: Net force together or apart for horizontal loop

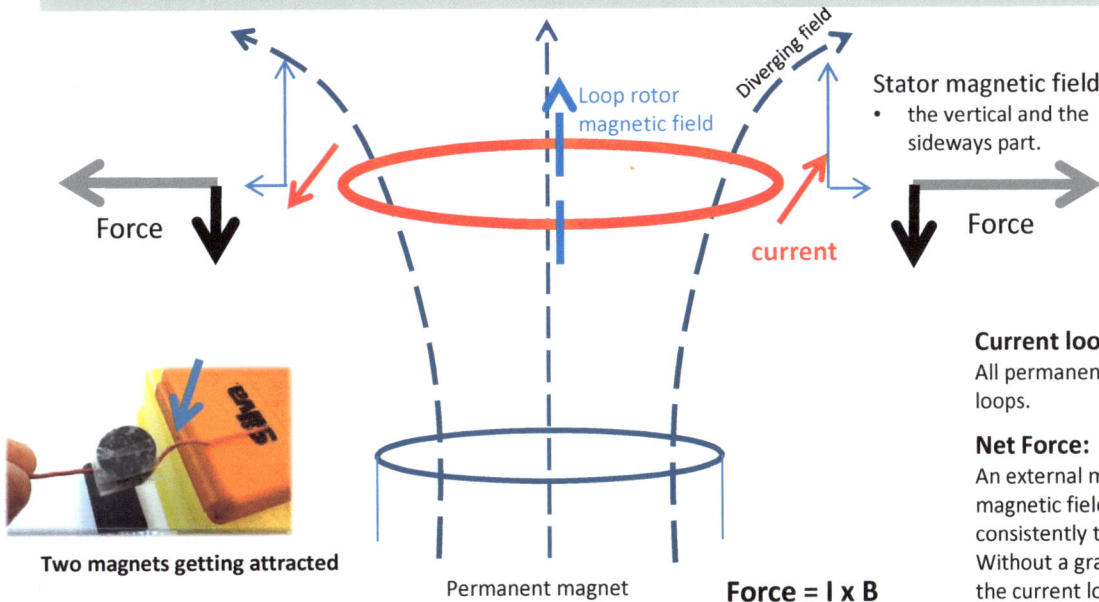

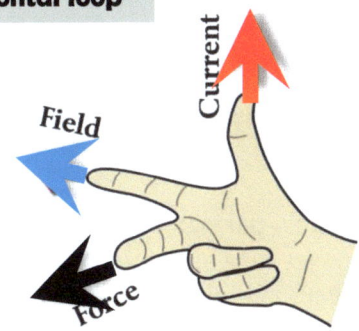

Loop rotor magnetic field

Diverging field

Stator magnetic field
• the vertical and the sideways part.

Force

current

Force

Current
Field
Force

Two magnets getting attracted

Permanent magnet

Force = I x B

Current loop representation:
All permanent magnets or electro-magnets can be represented by current loops.

Net Force:
An external magnetic field with a gradient has sideways components to its magnetic field, not just vertical fields. These sideways components act consistently to attract or repel the current loop.
Without a gradient, there is no sideways field and no net force up or down on the current loop.

Diagram of forces on a rotor current loop (red), above a single unbalanced magnetic field; that is, a magnetic field with a gradient. The forces up or down (attraction or repulsion) add together to produce a net force. The forces to the side from the vertical part of the magnetic field cancel each other so there is only a force up or down.

See Appendix F for fundamental forces on a moving charge.
See Appendix A and B for energy explanations of forces and torques

With a single magnet, the magnetic field is not uniform and is diverging, like most fields are near a magnet. The main vertical part of the magnetic field just wants to compress or expand the current loop, without any net force. The diverging sideways part of the magnetic field on both sides wants to pull the current loop toward or away from the diverging magnetic field, using the 'Right Hand Rule' for force.

Electromagnets enable electric motors which move things around us and run our lives. Electric motors enable car starters or electric cars, water pumps, dishwaters, tools. Electromagnets also enable speakers which help telephones, TVs, and music. But let's focus on electric motors. Let's explore how the basic rotation or torque between two magnets allows mechanical rotation, and how electromagnets enable continuous spin.

Motor fundamentals: field alignment and balanced stator field:

We're going to take ideas – twisting for field alignment and a balanced force on the permanent magnet - from the previous section 3A and make a dc motor.

The rotor magnetic field wants to align with the stator magnetic field. There is the most twisting torque when the rotor magnetic field is sideways to the stator magnetic field. For brushed motors, the rotor can not be a permanent magnet, or the rotor will just rotate at most ½ turn to alignment and stay there, with no more rotation. We need the rotor coils and the switch or brushes to the sideways coil to keep the rotor magnetic field always sideways and keep a perpetual torque or spin in one direction.

Also, the stator field needs to be balanced or uniform, with a permanent magnetic on both sides of the rotor, so there is no sideways force on the rotor. We just want the twisting torque for a motor, but not any sideways force causing a load and drag on the shaft at the bearings.

Motor particulars: brushes, poles, parts, and measurable fields:

Brushed motors use brushes to get current to the sideways coil. There are no active electronics figuring out the switching. The switch is simply these brushes making contact with the sideways coil using the split rings.

The number of coil or poles on the rotor can be many. More poles means a steadier torque, where the excited coils are almost always at the perfect angle of sideways to the stator field.

Take apart a toy dc motor. Disassembly is easy. Motors are made to be easy to assemble, so they are also easy to take apart. There are only three pieces - the stator, the rotor, and the commutator with the brushes.

We can measure the direction of the stator magnetic field using a compass.

We can excite the motor with a battery. For a dc motor, the current then flows only through the coils of the rotor. When the rotor is rotating inside the commutator, the current is changing coils to the sideways coil and the rotor magnetic field is staying in one direction, even though the physical rotor is spinning. Just rotate the rotor inside the commutator and place a compass next to the rotor. The compass direction holds steady.

See, all these ideas about magnetic fields can actually be seen and measured using a compass.

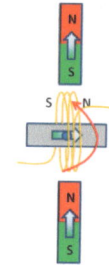

Brushed dc motor with commutator and electromagnet rotor.

Many poles with split ring coils to get sideways rotor field.

Three simple parts of a dc motor: rotor, stator, and commutator.

Compass points along permanent magnetic field on outside of stator

Electric motors generate motion, and are a very practical use of magnets. These motors are all around you.

Motors, the Basic DC Motor Using Electro-Magnets

If we wanted to have a top two competition for best application of electromagnets, we'd probably select electric motors and speakers. But let's focus on motors. Let's explore how the basic rotation of two magnets allows rotation, and how electromagnets enable continuous spin.

1st place: Motors, Most practical award

Electric Motors are one of the greatest successes of using magnetic fields for the cause of humanity, and electric motors enable industry like power plants, industrial machines, car starters and motors, and every day critical stuff like refrigerators, air conditioners, and computer hard drives.
The electric motor is the topic now.

2nd place: Speakers

"I am motor, hear me rip!"

Motors are used everywhere. Motors spin because magnets twist to have their magnetic fields be in the same direction, with opposite poles closer together.
If you want to talk 'why', then we all get back to aligning these magnetic fields.

Flipping: Concept to practical motor using electro-magnet and commutator

Magnets want to flip

People need and want motors

Flipping and force, a basic magnetic phenomena:
In motors, magnets want to align their magnetic fields, so they rotate to align the fields.

Flipping and no force with balanced DC field of the stator:
Electric motors, big and small, rely on magnets getting aligned to each other to rotate.

Practical motors with electro-magnet rotors:
Electric motors are used in computers, cars, shop tools, elevators, winches, generators, elevators, robots, anything that can move.

Speakers: Serious 2nd runner up to great successes of magnets.

To the runner up … Music speakers are a great success too, but they don't power your house, move cars and pump heating oil

…No offense intended to the music audiophiles out there for not getting top billing. Speakers, thankfully, did give us music everywhere, movie sound, and the TV and telephone.

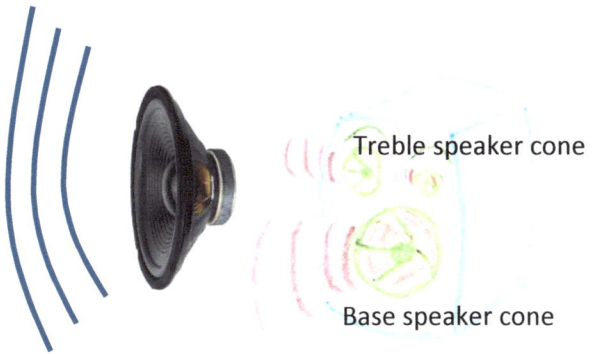

Treble speaker cone

Base speaker cone

Speakers are a great use of electromagnets too.

Electric motors generate motion, and are a very practical use of magnets. These motors are all around you.

Field Alignment: Why Motors Work for Direct Current (DC) Motor

Electric Motors are possible because …drum roll please …two magnets want to align their magnetic fields AND electro-magnet rotors and commutators can keep the rotor's magnetic field always stationary and sideways. The rotor can keep perpetual motion by changing the excited coil to whichever coil is sideways so the field always wants to rotate in one direction.

Instead of allowing the top magnet to fly away, or snap down, just allow it to rotate as a motor demonstration. To see this alignment, attach the top magnet to a shaft, like a bent paper clip.

Flipping:
Permanent Magnets and Alignment

Flipping:
Electro Magnets and Alignment … it's the same concept

Magnets want to flip and be aligned

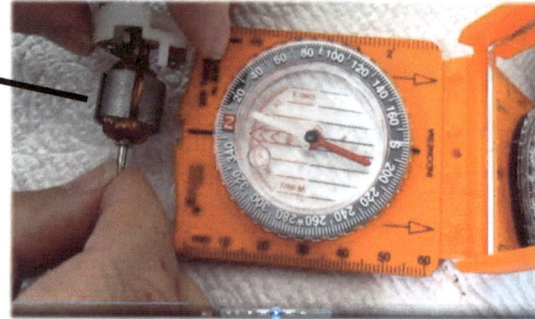

Field of rotor:
Constant direction rotor magnetic field, as spin it inside commutator.

The rotor keeps rotating, unlike a permanent magnet, because the electric current in the rotor is only put through whichever coil is facing sideways to the permanent field of the stator magnets, due to the commutator.

Take apart toy motor to view the three main parts: rotor, stator, and commutator.

A Keep-It-Simple-Silly (KISS) motor demonstration:
This top magnet will rotate so that the field's are aligned, which is the lower energy state. In a motor, the top magnet is the rotor, which has a magnetic field controlled by electric currents. The magnetic field of the rotor is kept sideways to the permanent magnet, to keep the most torque on the rotor.

We can test all these fields – stator and rotor – with the convenient compass. We don't need fancier magnetic field sensors – Hall probes or giant magneto resistance sensors. We just need the simple everyday compass.

Field of permanent magnet
The fixed stator magnet is the field the rotor wants to align with.

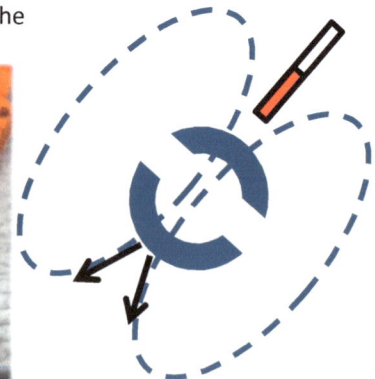

Magnets want to align, or rotate to align, so we can have motors.

Why Motors Work for Direct Current (DC) motor

An electric motor! A voltage applied to coils, with an iron core for strength, will always excite a rotor's magnetic field sideways to the stator's permanent magnetic field, which makes the rotor want to twist, or rotate to align the magnetic fields.

"Dude, he's getting a lot of mileage out of that one toy motor.
Keep paying attention to motors, and give some long overdue homage to things that move like Blu-ray disks, toy trains, and real electric cars."

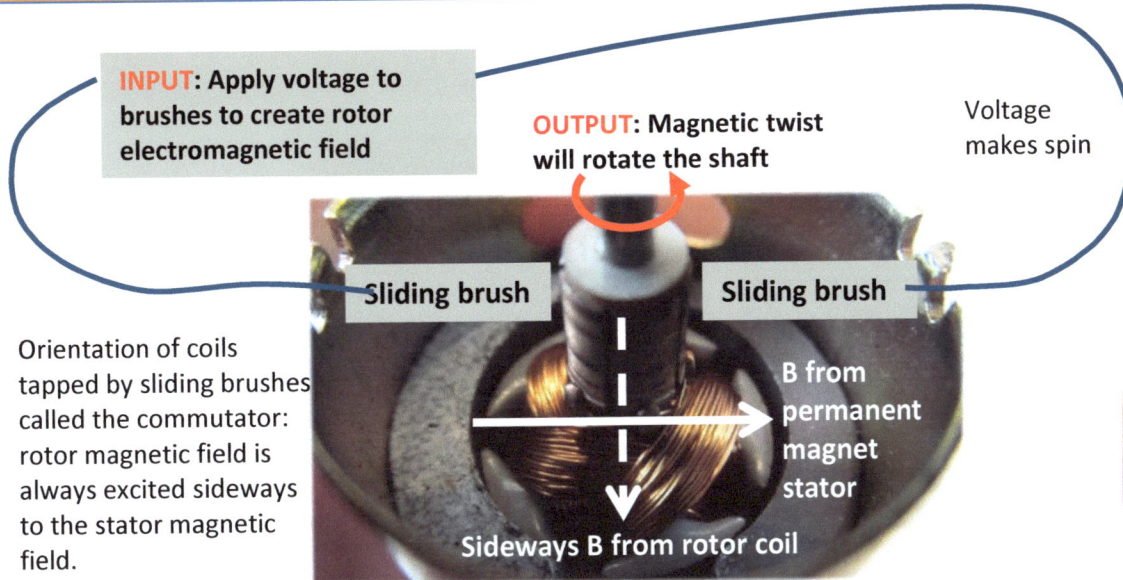

INPUT: Apply voltage to brushes to create rotor electromagnetic field

OUTPUT: Magnetic twist will rotate the shaft

Voltage makes spin

Orientation of coils tapped by sliding brushes called the commutator: rotor magnetic field is always excited sideways to the stator magnetic field.

Sliding brush

Sliding brush

B from permanent magnet stator

Sideways B from rotor coil

Sliding connectors or brushes always excite the correct coils to keep the same rotor sideways magnetic field direction.

- Create a constant sideways direction magnetic field from rotor, by switching current to the coil that is sideways to the field of permanent magnet, to always have the most twist (torque)

These two opposite magnets are ready to flip to align the fields. Magnets want to flip, and the shaft is dragged with it. A rotating shaft is the conversion from electrical to mechanical power.

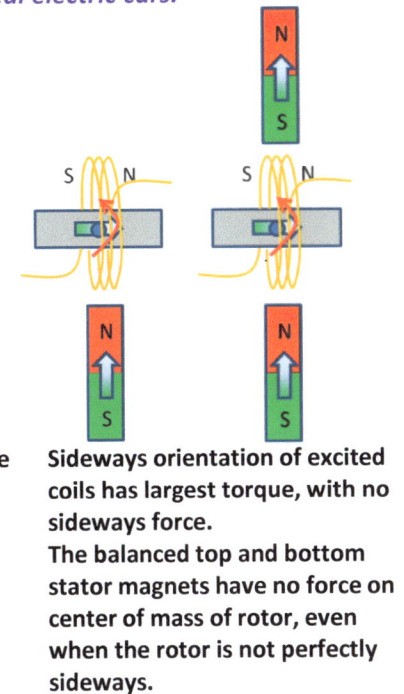

Sideways orientation of excited coils has largest torque, with no sideways force.
The balanced top and bottom stator magnets have no force on center of mass of rotor, even when the rotor is not perfectly sideways.

Alignment and lowest energy:

Here is the main reason the shaft of the electric motor rotates. Magnetic fields want to be aligned.

Twist to align magnetic fields:

Here is the main thrust of the story: the magnetic field of the rotor needs to be sideways to the stator magnetic field, and, bam, you have torque. You want more torque, then use more current, more iron, larger magnetic fields. The magic of the sliding brushes is that the field of the rotor is always pointing the same direction, always offering the same optimum torque. Even though the rotor is spinning, the rotor's magnetic field is not spinning.

Why is there a twist or torque? Because the stator magnetic field and rotor magnetic field want to align, to get to a lower energy state.

For the most constant sideways direction of the rotor field, many poles are desired, which means many coils are desired, with the many pads on the rotor shaft in a split ring design. The commutator keeps switching current to the next sideways coil.

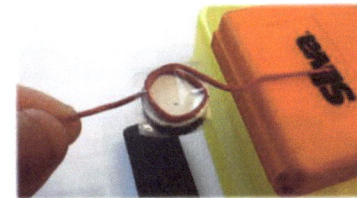

By applying a voltage to whichever coil is sideways to the stator magnetic field, there is a twist force on the rotor, which spins the shaft.

Why Motors Work for Direct Current (DC) motor

Let's look at the rotor design.

The pads on the rotor connect to individual coils wrapped in the grooves of the iron core.

These are really high torque electric motors. The coils have many turns to generate a large rotor magnetic field. A lot of hand wrapping and soldering go into building a rotor.

More pads and coils means that each coil can be excited for a shorter time, when the coil is perfectly at the right angle of 90 degrees to the stator magnetic field and the torque is the largest.

"Look at all those pads for the brushes to slide over."

Examples of commutator pads or split rings on rotor, to excite only the sideways coils. The brushes slide across these pads.

By applying a voltage to whichever coil is sideways to the stator magnetic field, there is a twist force on the rotor, which spins the shaft.

Toy Permanent Magnet DC Motor, Driven and Dissected

Here's an experiment following the time tested rule of 'Show the obvious'.
Show that motors work when batteries are powering them. The electric current flows through the rotor, creates its own magnetic field, which then wants to rotate so that 'opposites attract' with the field from the stator.
Now, if the motor does not start when a battery is applied, either the battery is dead, or the rotor coil or commutator is burnt out. If the motor heats up without turning, then there is an electrical short and you should disconnect the battery.

"Dude, I can power my toy boat with that DC motor."

Experiment 3.3: Spin a DC motor with current from a battery

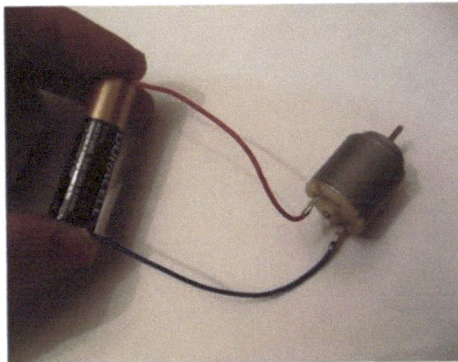

Power a spinning motor

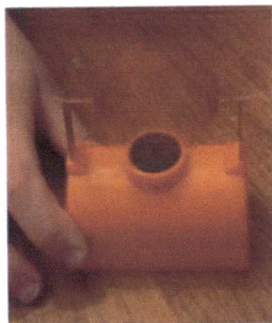

1-pole demonstration motor: Spinning rotor coil

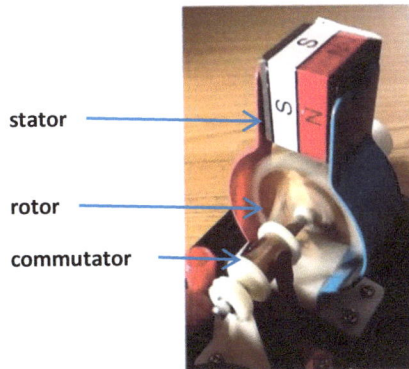

stator

rotor

commutator

2-pole demonstration motor: discussed in Chapter 4.

Fun facts about motor size and strength

Motors can be made in a huge range of sizes, for more or less power, but same concept.

4 foot motors:
- High torque industrial power

1 inch motors:
- Remote Controlled car racing

1/2 inch motors:
- Micro motors for slot cars, or cell phone vibrators, or shaving.

Uses:

Machine tools: milling machine (or drill press, lathe, or band saw)

Here are toys that use common electric motors.

RC cars

RC boats

Slot cars

This might be obvious—dc motors spin when attach a battery—but it is always good to confirm the obvious.

Toy DC Motor Dissected

The toy motor used here is easy to slide apart. The three parts are completely separate—the rotor, stator, and commutator.

Experiment 3.4: Take apart a toy motor, and show the guts of the motor:

<u>Demo 1:</u> Show permanent field in 'stator' using compass
<u>Demo 2:</u> Show field in rotor using a battery and compass
<u>Demo 3:</u> Spin rotor inside the commutator and show the rotor's magnetic field does not change direction (does not rotate with the rotor), using compass

"Easy Peezy, just un-bend a few tabs and the motor slides apart."
"Dude, that motor looks so simple to put together that I bet I can take it apart and put it back together, unlike my old typewriter."
"Good practical ideas have a way of getting used."

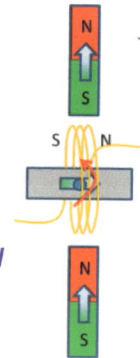

Motor with only three parts, easy to slide apart

Here is the toy motor apart. There are only three pieces:

Piece 1: Rotor
- Yes, it spins, and turns the axle or shaft.
- The rotor needs to spin, so it can only slip around inside the stator and commutator.

Piece 2: Permanent magnet stator
- The stator is something for rotor to push against.
- The stator is there to provide the stator magnetic field. A DC motor has a permanent magnet for the stator.
- The stator is there to provide the bearing for the rotor.

Piece 3: Brushes, called a commutator
- Brushes allow electricity to pass. The sliding brushes provide the electrical contact between the commutator and the rotor.
- The commutator has conductive sliding brushes that just push against the rotor pads. The brushes typically are made of a soft metal, like graphite or copper. Soft metal allows good contact even as it wears.
- The commutator section can also have a ball bearing for heavier loads, but this toy motor just has a plastic sleeve.

These three pieces are completely separable, with no wires soldered between them.

1) Rotor with 5 pads or poles or coils

2) Stator permanent magnets
- **Balanced magnetic B field across rotor, with no gradient.**

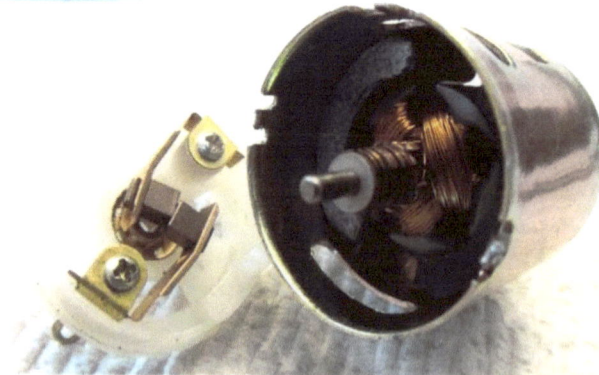

3) Brushes or electric sliding contacts to rotor, called a commutator

What toys and technology use these DC motors?
- DVD or Blu-Ray drives, for spinning your favorite movie
- Quad-copters, for crazy flying
- Remote Controlled cars, for racing around hallways.

Experiments: Look at three parts to a motor: rotor, commutator, stator

Really, this electric motor came apart super easy. The rotor needs to spin easily, so the rotor just slides out.

Toy DC Motor Stator and Rotor Magnetic Field Directions

The commutator keeps the torque going. The commutator keeps the magnetic field of the rotor always sideways to the stator and always wanting to twist. The commutator allows the coil facing sideways to the permanent magnet stator field to get the electric current. The brushes are spring loaded to push against the rotor pads and allow current into whatever pads are touching the two brushes.

"Dude, those motors were invented over 100 years ago, replacing horse drawn taxis or carriages with electric trolleys. Kids in the 1950's were putting those motors on model boats."

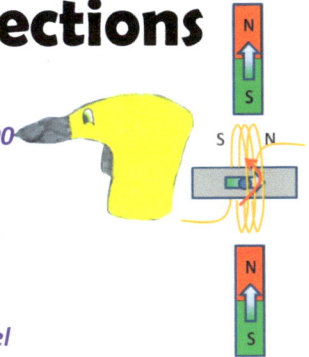

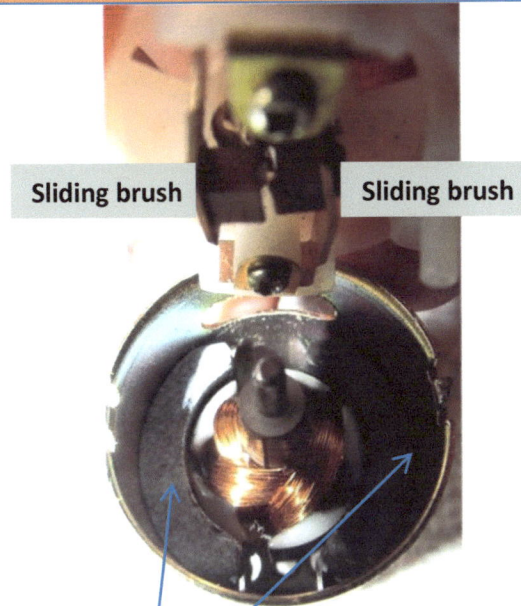

I never thought I'd care so much about 90° angles between two magnetic fields.

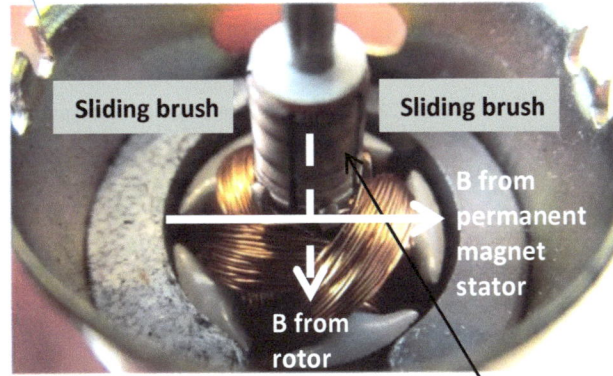

Sliding brush

Sliding brush

Permanent magnets for stator

Sliding brush

Sliding brush

B from permanent magnet stator

B from rotor

Sliding connector excites correct coils, for commutator.

Cartoon of stator and rotor field directions:

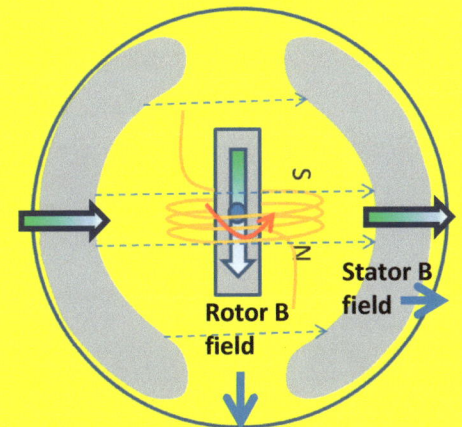

Stator B field

Rotor B field

Magnetic fields inside motor:
Recall the excited rotor coil and stator fields are perpendicular to each other, and so have the largest twist force.

Commutator before insertion around rotor:
The commutator brushes are notched into the fixed position that allows the rotor magnetic field to stay sideways to the stator permanent magnetic field. The more turns of the wire for each coil generates more magnetic field, for equal current.

Conductive metal pads for sliding brush to excite correct coils:
Here is the final assembly of rotors and stators. The fields are optimum to create lots of twist, or torque.

The brushes need to be wide for large contact area and low resistance. Brushes need to be long, so that as they wear down there is still more left over to create contact with the pads on the rotor. The brush's metal also needs to be softer than the pads on the rotor, so the brushes wear down instead of the rotor.

Note that the coil wire is magnet wire, which means a metal wire with an enamel coating (not a rubber cover or insulator like speaker wire). The enamel coating stops the current from shorting out between the wire turns or loops.

As the rotor spins, the magnetic field B from the rotor does not turn and always stays constant in a sideways direction to permanent field.

View the Permanent Magnet Field from DC Motor Stator

The stator magnetic field is no secret. With a compass you can observe the stator permanent magnetic field, which goes outside the motor as well as inside. The magnetic field can not just stop at some material, and needs to wrap around and go back in the other side.

Experiment 3.4: Take apart a toy motor:
Demo 1: Show permanent field in 'stator' using the needle of a compass

The stator magnetic field is uniform on the inside and then curls or wraps around like a dipole or bar magnet on the outside. Magnetic fields can never stop or terminate. They wrap around in a loop.

"Stator has a mostly uniform field inside, so there is no up and down gradient. The rotor feels only torque in a uniform field, and no force on center of mass."

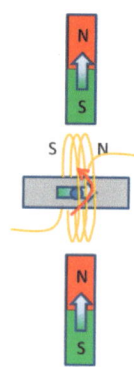

Permanent magnetic field of stator, inside

Outside stator field, sideways to motor

Compass needle points along permanent magnetic field on outside of stator

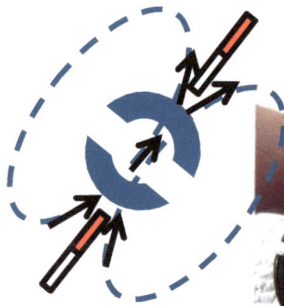

Compass on top of motor, to measure outside stator field: The stator magnetic field crosses sideways to the motor shaft, and so does the rotor magnetic field.

Outside stator field, looking on the axis of motor

This stator magnet field has two top and bottom magnets and is uniform on the inside for two reasons*.

- One, the fact that there are two permanent magnets on either side of the housing means that the magnetic field inside does not fall off across the rotor.
- Two, the uniform magnetic field means that the rotor experiences no sideways force, but a lot of rotation torque. Sideways force would increase friction and wear and tear on the shaft. Sideways forces require a slope or gradient to the permanent magnetic field, while torque does not require a slope.

The compass magnet needles always stay aligned with any curved, or wrapped around, field of motor magnet.

*See chapter 3A as well

The compass needle follows the magnetic field of the stator permanent magnet. The permanent magnet has the typical dipole magnetic field pattern. This verifies that the rotor is getting a good magnetic field looping around it.

Verify the Rotor's Magnetic B Field of Single Coil

You can observe the rotor magnetic field by seeing a compass needle move when apply a battery and current to the rotor.

The rotor's magnetic field should be much larger than the Earth's magnetic field, which of course means a strong torque.

"My center electro-magnet rotor has a strong magnetic field when current applied, sideways of course."

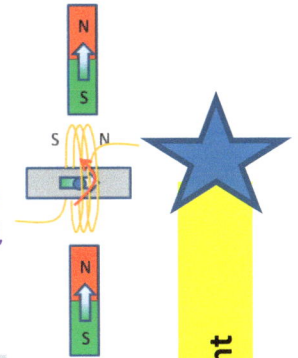

Experiment 3.4: Take apart a toy motor:
Demo 2: Show magnetic field of the rotor using a battery and compass.   Don't move the rotor inside the commutator.

No rotor magnetic field without current:
No current through electro-magnet rotor,
No contact with battery

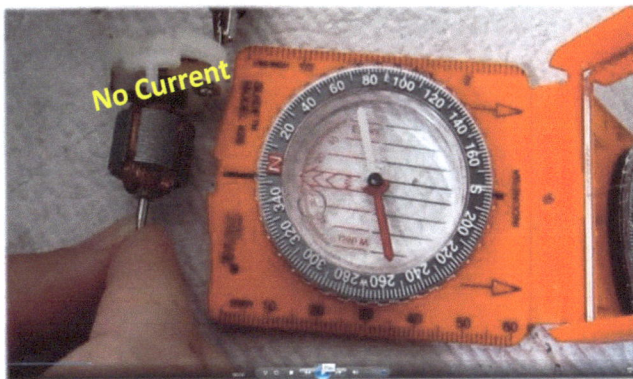

No Current

Earth's field before a current is applied to the rotor, shown with compass needle.

Rotor magnetic field with current:
Apply current through electro-magnet rotor,
Contact with battery

Attach leads of the commutator to the battery

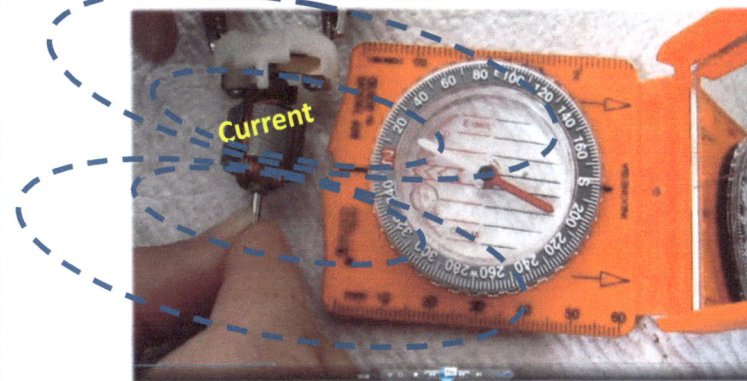

Current

Rotor magnetic field dominates after current is applied to the rotor, shown with compass needle.

Here we demonstrate what an electric current through the rotor does to the magnetic field around the compass. The compass needle is deflected from pointing along the Earth's magnetic field, toward pointing along the rotor's magnetic field.   All those turns of wire in each coil, and all that iron inside the rotor, really make a strong magnetic field.

When the compass turns, we know two things: one, the direction of the magnetic field of whatever (the rotor) is closest to the compass, and, two, that this field is larger than the Earth's magnetic field (about 0.5 Gauss).

An electric current through the rotor creates a magnetic field, which turns a compass needle because fields want to align.

Experiments: Show magnetic field of rotor when apply current

Keeping the Rotor's B Field in Constant Direction, Using All Coils

You can demonstrate that the rotor magnetic field stays in the same direction, even though the rotor is spinning. The magnetic rotor field is designed to stay sideways to the stator magnetic field by switching current to different coils using a commutator, to create constant mechanical torque. When you rotate the rotor but keep the commutator fixed, the rotor's magnetic field stays in one direction. It is a motor!

Experiment 3.4: Take apart a toy motor:
    Demo 3: Spin rotor and show field does not change direction due to commutator, using the needle of a compass

Excite one coil only, current fixed:
No commutation, using just a single coil:
Hold rotor and commutator rigid together, to defeat the role of the commutator.

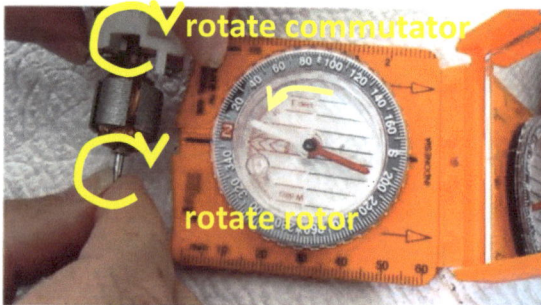

Rotate both the rotor and commutator rigidly together until needle points to rotor.

Rotate the rotor and commutator together.
- You are keeping the current flowing through the same fixed coil, without allowing the commutator to switch coils.
- When the needle points most toward the rotor, that is where the dominate rotor B field is pointing from the excited coil.

When rotate the commutator brushes and the rotor together, we can tell when the compass needle points the hardest toward the rotor. That is the rotor angle where the magnetic field is pointing.

Excite all the coils in sequence, current switches:
Allow commutation, using all coils:
Let rotor slide or rotate inside commutator, to enable the commutator.

Rotate the rotor but not commutator, just like a motor: Compass needle holds steady as rotor turns.

Rotate the rotor inside commutator
- Current is switching coils in rotor, to keep magnetic field in same direction.
- Compass stays stable, pointing at rotating rotor:
    →rotor magnetic field is not rotating.
    →rotor magnetic field is always sideways to the stator permanent magnet field, when inserted into the stator.

Rotate the rotor, but hold the commutator brushes at the same angle. Then the rotor magnetic field still stays in the same direction. That is the point: the commutator keeps the current flowing only in the coils that keep the magnetic field sideways to the stator magnetic field.

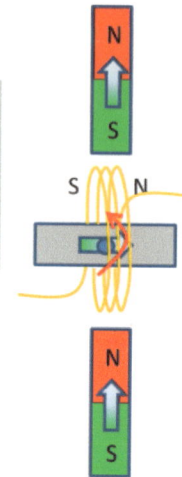

Sideways magnetic orientation has most torque, or most need to rotate, and no force. The electro-magnet rotor keeps this constant sideways magnetic field, thanks to the commutator, even when the rotor is physically rotating.

Experiments: Show constant direction of rotor magnetic field as turn the rotor, using commutator.

This steady rotor magnetic field is the beauty of the commutator: the rotor magnetic field stays pointed sideways for constant torque.

Summary of Basic Demonstrations of Motors

Let's demonstrate some cool electric motor stuff, as described in the following pages. Hands-on work or play time with motors will get you used to any electric tools, and make your knowledge deep.

The inner rotor spins because the current flows through the coils and wants to align it's field with the stator field. The rotor has many coils and is an electromagnet where the commutator or brushes switch the current to only the sideways coil.

Go ahead, make a copy. Then check off the list.

Experiment 3.1: Show magnet wanting to flip over other magnet, the basic motor
- Show that magnets want to rotate so that 'opposites poles attract', by attaching a bent paper clip to the top magnet (and use tape if attraction to paper clip is not enough), and see the top magnet rotate. It can not snap down, or fly away, but it can rotate.

Experiment 3.2: Show that uniform magnetic field will only cause a flip, not a force. This flip is the basic idea behind a spinning magnet.

Experiment 3.3: Spin motor shaft with electric current from battery
- Show the motors work when batteries are powering them. The current flows through the rotor, creates its own magnetic field, which then wants to rotate so that 'opposites attract' with the field from the stator.

Experiment 3.4: Take apart a toy motor:
  Demo 1: Show permanent field in 'stator' using compass
  Demo 2: Show field in rotor using a battery and compass
  Demo 3: Spin rotor inside the commutator and show the field does not change, using compass
- Show the guts of the motor. The toy motor used here was easily taken apart, and the parts are easily separate.
- The rotor has multiple coils for multiple poles.

Experiment 3.5: Show other electric motors around the house:
  Some electric motors are powered by Direct Current (DC, from batteries or 5 volt power supplies in electronics) and some are powered by Alternating Current (AC, from wall socket)
- Show all the electric motors around the house, or school. Some are DC, like this toy motor with a stator magnet, and some are AC, which takes the AC voltage from the power outlets in the walls.

"Dude, I always wanted to do this stuff. Why wait for college, where you spend a bucketful of money to do the same thing?"

Magnet flips to align (twist or torque) but also feels force up or down

Magnet flips to align and feels no force up or down, just a twist

Rotor constantly tries to align using commutator.

Rotor, stator, and commutator inside a simple DC motor.

Stator has a permanent field in DC motor

AC motor inside an electric drill plugged into wall, which is measured outside the stator motor body

Experiments: Magnets flipping and motors are related

Motors work because magnets want to flip over and have opposite poles attract, with aligned fields.

Forever Twist and Shout, More than an Ice Skater, Q and A

Motors are designed to have all twist, and no sideways force on the bearings. For large torque, the electromagnet rotor is given an electric current that switches coils as the rotor spins, to always want to rotate in one direction. To avoid a net force on the bearings, the rotor feels an attractive force to the top stator magnet, and an attractive force to the bottom stator magnet, and these forces cancel.

What did we cover? Mechanical motion ... the other applications

Workhorse of industrial Age

Lifting forces
1. Holding iron objects
2. Refrigerator magnets
3. Speakers and relays

Twisting forces
1. Motors creating mechanical motion, when electric voltage and current applied.
2. Compass

Generators:
1. Motors creating electrical energy, when shaft is spun mechanically (a generator or dynamo).

Magnets rotate to align their magnetic fields

Question 1: Does a magnet in a uniform magnetic field feel twist? Yes, magnets want to align.
- Yes, the magnet's internal field wants to align with the outside (stator) magnetic field, for lowest energy. Recall there is no force in a uniform field, just twist.

Question 2: How is a sideways force (translation force) avoided on the rotor from the stator, so there is no pressure on the bearings from a rotor getting pulled? Balance, uniform stator magnetic field.
- The stator magnetic field needs to be uniform, with no gradients, or no changes in magnetic field at different locations near the rotor.

Question 3: Can you see the outside magnetic field from the stator, outside of the motor housing? Yes, magnetic fields can not just stop. They must loop around to form a loop.
- Yes, just put a compass around the permanent magnets of the stator to see the outside magnetic field. Magnetic fields need to loop around (have no start and stop), so the field inside the motor needs to loop around back to the other side.
- You can't literally see a magnetic field. But you can see a compass direction, and you can see magnets getting attracted to each other, and you can see the pattern of iron filings.
- The forces from magnetic fields are visible. For example, magnets snap together and motors rotate. When looking at the sun, cameras can see explosions where flying charged particles follow a curved path along a magnetic field, and cameras can see changes in the different absorption lines of the gas depending on the magnetic field.

Rotor feels twist in uniform stator field

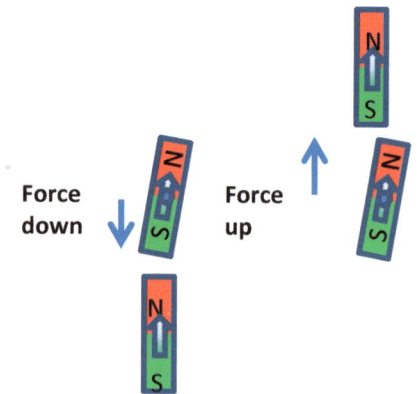

Question 4: Can you feel the pull, or attractive force, on the rotor magnet, using an unbalanced lopsided stator magnet (only a permanent magnet below, without a top magnet)? Yes, because there is a field gradient.
- Yes, the inner rotor wants to pull toward the stator magnet, as well as twist to be aligned. This would be bad for vibrations and wear and tear on the bearings. This net force is canceled when there is a top and bottom stator magnet.

Question 5: Can you feel the pull, or attractive force, on the rotor magnet, using a balanced stator with permanent magnets on both sides of the housing? No, because there is a uniform field and no field gradient.
- No, the inner rotor is easy to move around the inside of the stator, without getting jerked to any side. The inner rotor feels a twist to align with the magnetic fields, but it does not feel a sideways force toward the stator walls.
- The inner rotor does not feel a difference if its center moves around because the stator field is uniform, so there is no energy change when the center moves and no force in any direction.

Force down   Force up

Cancelation of forces when have both top and bottom stator magnets.   There is only twist in a uniform stator field. The twists add, but the forces cancel.

There are no limitations between using a DC motor and an AC motor. Both types of motors are strong, both can get stronger with more current, and both reverse direction by flipping current direction to the rotor.

An AC motor might be less expensive because there is only iron, and no permanent rare earth magnet. A DC motor using a stator electromagnet (only iron) has more starting torque because more current flows when the rotor is barely spinning.

A brushless motor is a DC motor with transistor switches, not brushes, for the commutator. The DC motor has a permanent magnet for the rotor, but the stator fields rotate around with electromagnets and switches, no brushes. The middle rotor does not have any electrical contact with the power supply.

Lets look at the similarities between dc and ac motors.

First of all, they both work due to a sideways coil getting torque from the stator, using a commutator.

What is the main difference between dc and ac motors?

The dc motor has a permanent magnet for the stator. Electric current only flows in the rotor coils. The rotor is an electromagnet with an iron core.

An ac motor has electromagnetic for both the rotor and the stator, so current flows in both the rotor and stator. There is actually more loss in an ac motor because the current is flowing through more wire, and because the iron has some loss when it keep flipping its magnetization.

In an ac motor the current can keep oscillating but the relative direction of the magnetic fields between the rotor and stator doe not change. The torque stays in the same direction no matter the direction of the electric current. Current is flowing through both the rotor and stator in series, so whatever is happening to the rotor is also happening to the stator.

So, can motors reverse direction? Of course.

We simply need to reverse the current in the rotor relative to the stator.

For the dc motor, there is a switch that reverses the dc voltage to the commutator. The permanent magnet field of the dc stator of course does not change. The sideways magnetic field of the rotor will now be pointed in the other direction.

For the ac motor, recall that the stator and rotor are both electromagnets, and that they are excited together in series. To reverse the spin, just have a switch that connects the commutator brushes to the opposite contacts of the stator coil. Now the sideways rotor field is pointing in the other direction relative to the stator field. They both are oscillating rapidly, completely independent of the spin rate of the motor, but they are oscillating together. We just care about the relative fields, and the oscillation together stays in series.

Manufacturing millions of motors:

For manufacturing of a motor, the motor assembly process can be quite automated or quite manual, depending on the quantity. After all, there are a lot of coils wind and the coils are fit in very tight grooves in the iron housing.

Brushless motors:

In addition to motors with brush commutators, there are also brushless motors with transistor switches for commutators. Again, the sideways fields are required to get torque. These brushless motors are discussed in the next section 3D. These brushless motors do use active transistor switches. There are magnetic field sensors around the stator to trigger the switches and keep the fields sideways. The rotor is actually the permanent magnet, and the stator field is spinning around to be sideways to whatever the angle of the rotor is.

Same torque concepts apply to DC and AC motors ('universal' motor), because the stator magnets and the rotor iron magnets both are powered by the same current supply and both reverse or alternate direction together.

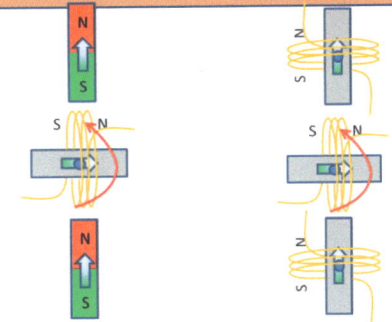

Brushed dc motor with permanent magnet stator and ac / dc motor with electromagnet stator and rotor.

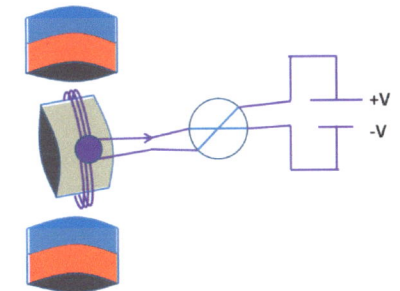

DC current is changed direction to the rotor to change spin.

Electric car brushless motor

DC and AC Motors

Let's compare DC and AC motors. Both are used widely. A DC motor is typically plugged into a battery or inside electronics. Think cordless drills and cd drives. An AC motor is typically plugged into the power grid wall outlet. Think dish washers, clothes washers, refrigerators, and corded drills.

DC motors usually have a permanent magnet, but DC motors could also be just an AC universal motor with no permanent magnet, but that has electromagnets for the stator and an iron core. AC universal motors are called universal because they can run on AC or DC current.

AC motors rely exclusively on iron electro-magnets, where everything gets magnetized by the current. The magnetization direction of the iron follows the current, for both the stator and the rotor.

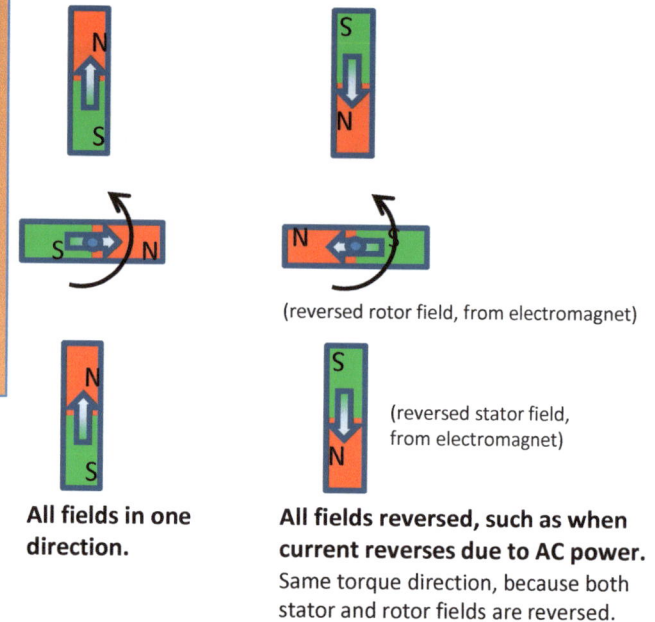

(reversed rotor field, from electromagnet)

(reversed stator field, from electromagnet)

All fields in one direction.

All fields reversed, such as when current reverses due to AC power. Same torque direction, because both stator and rotor fields are reversed.

Rotor constantly tries to align using commutator. The top and bottom magnets cancel the equal up and down forces, so there is only torque.

Point 1: For both DC and AC motors, the rotor's magnetic field is always sideways to the stator's magnetic field, using the same concept of torque.
- Yes, the sideways field has the most desire to rotate to align with the stator field. This means a sideways rotor field provides the most torque for any type of motor.

Point 2: A DC motor type uses a permanent magnetic field for the stator, and an AC motor type uses an electromagnet for the stator.
- The DC motor uses a permanent magnet for the stator.
- A DC motor can also use an electromagnet for the stator, just like an AC motor. In fact, up until the last few decades when permanent magnets were improved, this electromagnet stator design with iron core was dominant

Point 3: For an AC motor, the stator and rotor AC fields are always oscillating together. The fields are always sideways together and the motor keeps a constant torque and keeps turning.
- The stator and rotor AC fields oscillate together, and always keep the sideways rotor field relative to the stator field. The torque is the product of the stator and rotor field, so if both are in reverse direction (negative) or positive direction the torque is still in the same direction.
- Because the stator is also an electro magnet, the stator field can reverse direction along with the rotor field, so the same direction of torque is preserved. The oscillating direction of the rotor field in the sideways direction is canceled out by the oscillating stator field.

Point 4: An AC motor will run on DC current.
- Yes, both AC and DC current will work. The AC motor does not care if the current is oscillating or not. The AC motor is designed to keep the same torque direction regardless of the current direction, because the same current flows through the coils of the stator and the rotor. Put another way, the ac motor does not have a permanent dc magnet stator.

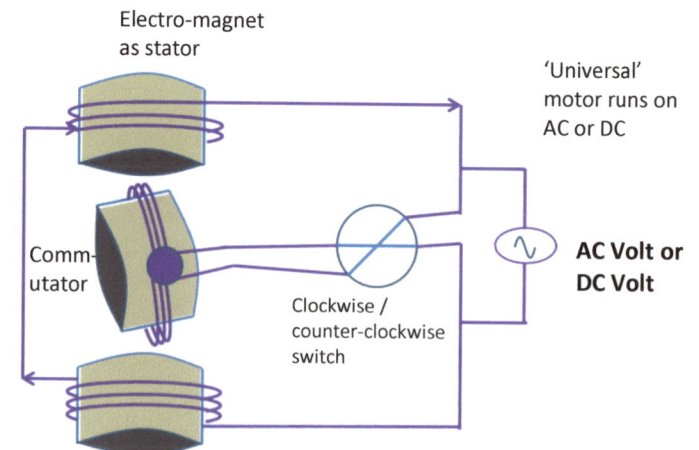

Electro-magnet as stator

'Universal' motor runs on AC or DC

Commutator

Clockwise / counter-clockwise switch

AC Volt or DC Volt

AC Motor: AC Voltage changes the same for both rotor and stator electromagnets, keeping a consistent alignment of fields and torque.

Direct Current (DC) and Alternating Current (AC) Motors, Still a Sideways Rotor and Not That Different

Again, there are no limitations between using a DC motor and an AC motor. Both types of motors are strong, both can get stronger with more current, and both reverse direction by flipping current direction to the rotor. An AC motor might be less expensive because there is only iron, and no permanent rare earth magnet. A DC motor using a stator electromagnet (only iron) has more starting torque because more current flows when the rotor is barely spinning.

DC motor design: permanent magnet for stator

DC power: direct current, not oscillating. Computers convert wall power to 5 Volt DC for electronics.

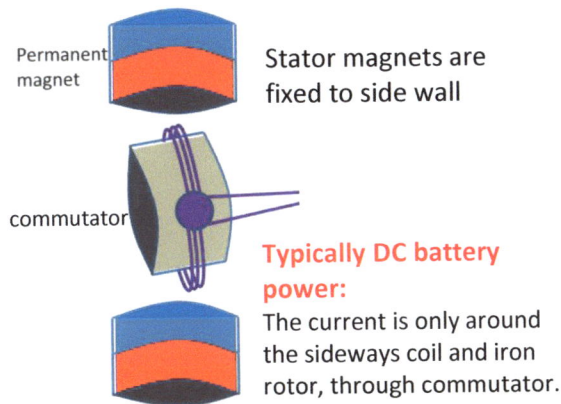

Permanent magnet

Stator magnets are fixed to side wall

commutator

Typically DC battery power:
The current is only around the sideways coil and iron rotor, through commutator.

DC voltage for power:
permanent magnet for stator

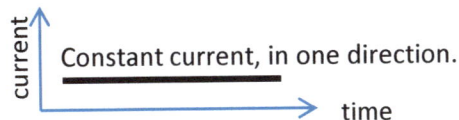

Constant current, in one direction.

current | time

DC or AC motor design: no permanent magnet anywhere, just magnetization of iron cores that follow the DC or AC current

AC power: alternating current, oscillating at 60Hz. All your wall outlets from the town are AC power, at 120 Volts.
DC power: the current through the stator electromagnet stays constant.

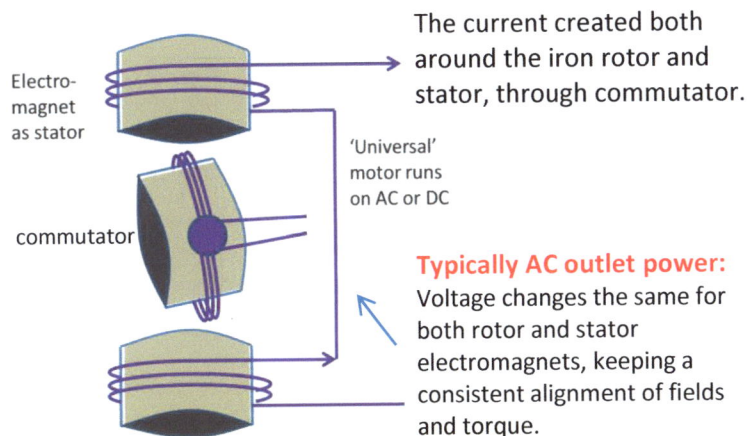

Electro-magnet as stator

commutator

'Universal' motor runs on AC or DC

The current created both around the iron rotor and stator, through commutator.

Typically AC outlet power:
Voltage changes the same for both rotor and stator electromagnets, keeping a consistent alignment of fields and torque.

AC voltage for power:
electromagnet and iron for stator

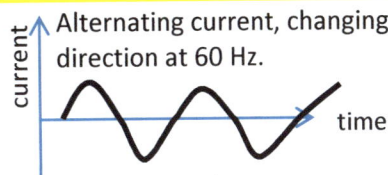

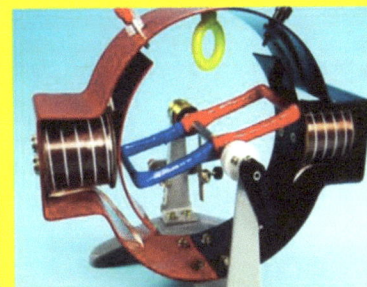

Alternating current, changing direction at 60 Hz.

current | time

Both DC and AC motors use sliding commutator switches for current directions to keep rotor magnetic field sideways. For an AC motor, the stator fields and iron magnetization reverse in synch with the rotor fields.

Same torque concepts apply to DC and AC motors ('universal' motor), because the stator magnets and the rotor iron magnets both are powered by the same current supply and both reverse or alternate direction together.

AC motor demonstration, using 2 poles:
When current reverses, both stator and rotor fields reverse, so torque direction between them (clockwise or counter-clockwise) remains the same.

For AC current, the fact that current is changing at 60 Hz does not matter: All currents go up and down the same. What matters is relative currents between the rotor and stator, keeping the best coil excited at the right times.

Permanent magnets can not be used for AC motors, because the stator dc field won't track the ac current and the fields need to always create same torque.

How to convert a DC to an AC motor?

- Alternating voltage or current, from your wall socket, needs to power electric motors, and it does. Life is good for the Universal AC motor.
- Here is where electro-magnets come to the rescue. The permanent magnets of the DC motor **stat**or (from **stat**-ionary magnets) are replaced with electromagnets. The stator magnetic field reverses due to the AC current, but the rotor magnetic field also reverses. The torque then stays in the same direction.

Reversing Spin Direction Using DC and AC Motors

If you've reversed direction of an RC remote controlled car, you've reversed direction of an electric motor. DC motors are used in RV toy cars, and in electric cars for people: the car can move forward or backward by switching the DC or AC voltage to the rotor, while keeping the stator field the same. The flipped current in the rotor creates an opposite magnetic field, which reverses the torque. For AC motors, the rotor current is also simply flipped compared to before, or compared to the stator field.

Reverse spin: Reverse DC current to rotor

- For DC motor with permanent magnet, switch the leads to the motor to reverse direction.

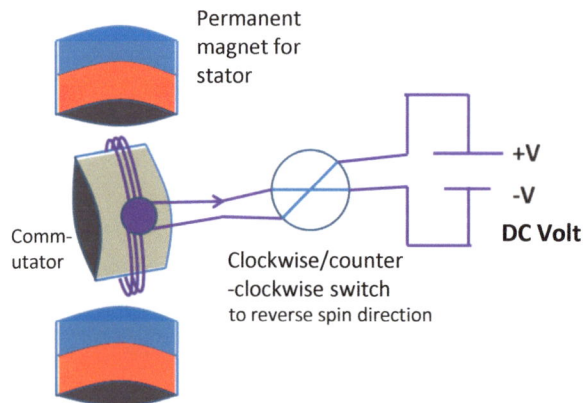

Permanent magnet for stator

Commutator

Clockwise/counter-clockwise switch to reverse spin direction

+V
-V
DC Volt

DC current is changed direction to the rotor to change spin.

Reverse spin: Reverse DC or AC current to rotor, but not DC or AC current to stator

- For AC motor, switch the leads between stator and rotor.

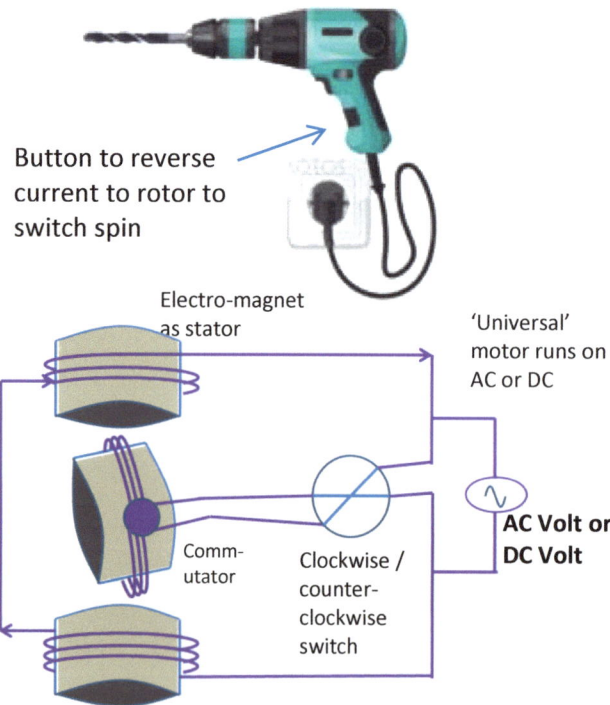

Button to reverse current to rotor to switch spin

Electro-magnet as stator

'Universal' motor runs on AC or DC

Commutator

Clockwise / counter-clockwise switch

AC Volt or DC Volt

AC current is changed direction to the rotor to change spin.

Reverse wires (voltage) to the rotor on the DC or AC motor, and reverse direction of spin, because the electric current in rotor is reversed and its magnetic field is opposite.

DC remote controlled motors (cars / robots): reversing direction

Prius Hybrid Electric

Hybrid cars use large electric DC or AC motors to power wheels, which reverse current in rotor to go backwards.

Lego robotics

FIRST Tech Challenge robot

DC motors let the robot go forward, backward, and turn, by reversing voltages to the motors or rotors. This switching is programmable.

Remote Controlled cars, Lego robots, all are able to move in arbitrary directions, controlled by you, by reversing currents and voltages to the DC or AC motors.

Necessary Applications of Reversing Spin: DC and AC Motors

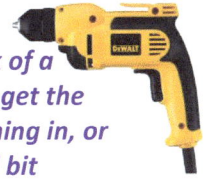

If you've used an electric drill, you've reversed the direction of a motor. Using a drill, people can routinely make a drill bit screw in or screw out. The drill can change spin direction of the motor and the drill chuck rotation with a switch.

"At the flick of a switch, can get the drill bit turning in, or get the drill bit turning out."

When reverse direction of current in the rotor for DC or AC, but not the stator, then the spin changes direction:

Electric cars: Electric cars can use DC or AC motors.

Drills: DC or AC motors are used in drills, depending on wall power or battery power.

Fan: If you've changed the direction of spin of an overhead ceiling fan, from summer to winter, you've reversed the direction of a motor.

Inkjet Printer: If you've printed on an inkjet printer, the printer motor is reversing directions many times, which pulls the printing cartridge around.

3D printers: If you've watched a 3D printer in action, you're watching an electric motor spin in both directions to control the printing nozzle.

What's going on when reverse spin? How do the motors do it?

- A switch can change the spin clock-wise or counter clock-wise (spin the drill bit forward or backwards), just by reversing the current to either the rotor or stator, without changing the current to the other.
- For AC current, the fact that current is changing at 60 Hz does not matter: All currents go up and down the same. What matters is relative currents between the rotor and stator, which keep the sideways coil excited at the right times and torque in a constant direction.

Electric cars, forward and backward

Car motors on axles, with strong torque

Electric Cars:

- Electric cars can use DC or AC motors. Both can reverse direction. The DC motors need some strong permanent magnets inside the motor, and DC motors don't require circuitry to go from DC batteries to AC current.
- This is a significant decision for car designers, whether to use DC or AC motors.

Drills, drill in or drill out

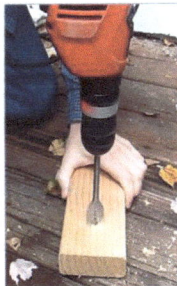

Clockwise or Counter-clockwise

DC or AC motors used in drills:

- Using a drill, people can routinely screw in or screw out a screw or drill bit by changing direction of rotation, by flipping a switch to reverse current to the rotor.

Think how many times you have reversed the direction of a drill without even thinking about it, if you have installed shelves or built anything from wood. Now that is success for electric motors and their usefulness and practicality.

Ceiling fan, to spin blades, clockwise and counter-clockwise

Low cost air conditioning.

Fan direction and the seasons:
Here is the reason to change wind direction in summer and winter, according to popular lore:

- In winter, blow air up slowly, to force hotter air to the walls where it is typically cooler: creates a uniform temperature in the room. Don't spin too fast, because do not want to feel a draft.
- In summer, blow air down to create wind to cool you off by the wind chill effect.

3D Printers, back and forth

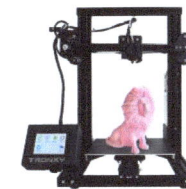

Stepper motor on 3D printer:
turning back and forth, clockwise or counter-clockwise.

Motor variations:

Permanent magnets: inside DC motors.

Many varieties:

- Some motors are good for constant spin (DC or AC), some motors are good for precise rotation angles (servo) with many poles.
- Some motors have the spin equal to the alternating current frequency (synchronous motor) to avoid the brushes.

- **Drills, printers, electric cars need to spin in both directions to be useful.**
- **Combustion engine cars instead need the transmission and gears to change direction.**

Jedlik: Electricity Development ... a Little History

First discovery and early science had to happen, such as electrostatic forces, creation of magnetic fields, the explanations, DC and AC current, and prototype motors and generators. Again, this was in 1820 when magnetic fields were noticed around dc currents. And then the ideas get applied by industrious engineers who see a need, like trolleys in cities for commuting and electric lights for safety and movement at night. Science can take a while before it improves lives. Notice it took 60 to 80 years before mass applications happened.

Anyos Jedlik: (1828, in his 20s)

- Hungarian
- A physicist and a Benedictine priest
- First DC motor, with stator, rotor, commutator. Unfortunately, Jedlik did not publish results of his motor for a few decades, and Werner von Siemens also gets credit for the invention of the motor.   Possibly motors would have become more common sooner had Jedlik published and received credit.
- Motor powered skate board

Coils around steel bar to create electro-magnet for rotor

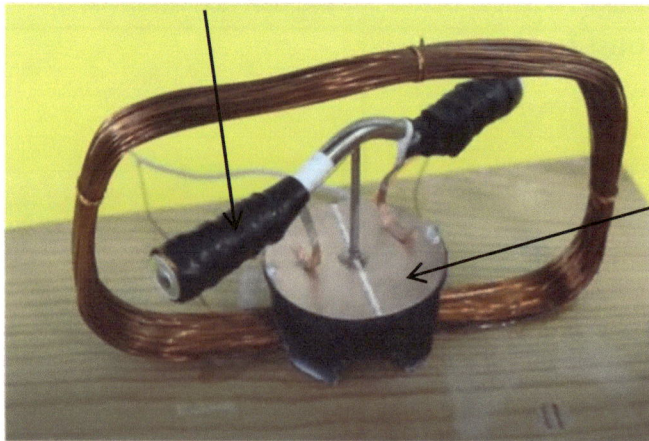

commutator

Example of Jedlik's first motor, in this case a 2-pole DC motor.

A DC current runs through the outer electro-magnet coil. The middle rotor has two coils wrapped around an iron bar which is a iron loaded electro-magnet.

Due to the simple brush commutator, the current in the rotor flips direction halfway through the spin, and keeps the rotor magnetic field pointed to the same side as the outer electro-magnet, to keep the twist or torque in the same direction.

Note that the maximum torque occurs when the iron rotor is in the plane of the stator coil, because the two fields – the stator and the rotor – are perpendicular to each other.

Note also that an AC current could run through this motor, because it is a 'universal' motor, where the stator field is generated by the same current as the rotor current, so these currents are synchronized.

Demonstration of motor powered skateboard

The 2-pole DC motor is connected to gears which power and move the skate board.

A battery needs to be carried by the skate board. The typical lead-acid battery was also available at the time, so the potential to make electric cars was real.

Jedlik investigated many science topics. His 2-pole electric motor was one very practical result, but he did not seek any applications although he might have showed people.

Manufacturing for an Electric Motor

Electric motors need automation and efficiency for production, as millions are made each year and the cost needs to be low. There is a lot of wire to wind into coils for stator and rotor. There are permanent magnets to insert for DC motors.

How are stator and rotor coils wound?

Smaller motors, many produced, and more automation with wire winding machines

Insert unwound rotor into winding machine

Spin the copper wire into the rotor grooves.

Smaller rotors: a machine spins and wraps the wire for the coil around the rotor:

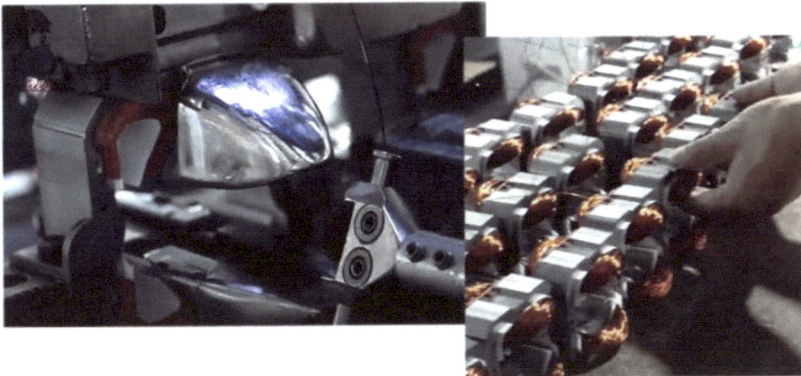

Smaller stators: a machine spins and wraps the wire for the coil inside the stator: the machine goes back and forth like a spindle for making fabric.

Larger motors, fewer produced, and less automation

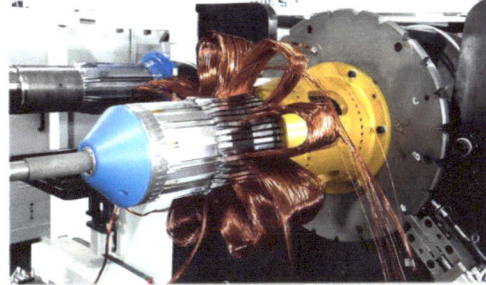

Large rotors: Inserting coils into the rotor by hand, and tighten by machine

Large stator coils by hand: Enamel insulated copper coils in pressed into the insulated slots in the iron stator. This is the labor intensive part of the assembly. After the wires are connected, the whole stator is dunked in resin to help with insulation.

Tightening the stator coils inside the stator: a machine pulls the coils tight.

*How Konesko produces electric motors, for a crane manufacturer.

Electric motors are now built in very automated factories. The metal grooves, the huge length of copper magnet wire, the soldering, and the tight precision and spinning all require a well controlled fabrication.

85

Let's compare DC brushed motors and DC brushless motors. Both are used widely. Both use a permanent magnet and both use DC power. DC brushed motors using sliding mechanical brushes, with friction and wear. DC brushless motors use transistor switches, with no friction and no wear but require sensors for switching.

It may seem advanced to talk about brushless motors in this book. But hey, the 1960s were 60 years ago. Transistor switches have arrived a long time ago, and magnetic field sensors (Hall sensors) are here. Brushless motors are dominating EV cars, electric tools, and EV airplanes. Brushless motors have many advantages. These brushless motors are low maintenance without mechanical brushes that wear, more efficient without drag from friction of the brushes against the pads, and more reliable without current needing to go to the rotor. We can't avoid talking about brushless dc motors.

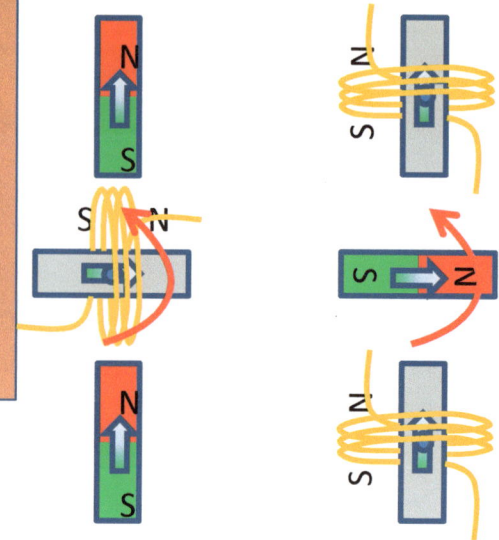

Brushed dc motor, with electromagnet for the rotor.
The permanent magnets are for the stator.

Brushless dc motor, with transistor switches and with electromagnet for the stator.
The permanent magnets are for the rotor.

Compare DC brushed and DC brushless motors:

The brushed DC motor is old school and has the permanent magnet for the outer stator, with the sliding mechanical brushes that can wear away and need to be replaced for high use motors.

The brushless DC motor is 'welcome to transistor switches' new and has the permanent magnet for the rotor. Yes the rotor is still the rotating part. The stator is made from electromagnets. But here is the big difference: the field direction of the stator is rotating using switches to attract or push away the permanent magnet or magnets on the rotor. There are no brushes. There is no electric current crossing over to a rotating shaft.

Transistor switches were developed in the 1960s, so the early inventors of the motor and commutator in the 1880s did not have access to transistor switches to allow a brushless motor.

Brushless motors last much longer than brushed motors, because there are no brushes to wear down. The transistor switches can last longer than carbon brushes, and provide less friction. The transistor switches can be immune to salt or rain on the road for EV cars or for electric bicycles. But brushless motors can still wear out. The bearings can fail, or the motor can overheat.

Brushless particulars:

For brushless motors, the permanent magnet for the rotor is typically a rare earth magnet, due to the higher magnetic fields. The motor can then be made smaller for the same torque. A weaker ferrite magnet could be used due to less cost, but then the torque is less.

The transistors for the controller are triggered by a magnetic field sensor near the rotor. These sensors, called Hall sensors, can tell if the rotor magnet is close or at the right angle for the stator coil to turn on.

Point 1: For DC brushed motor or the brushless motor, the rotor's magnetic field is always sideways to the stator's magnetic field, using the same concept of torque.

- Yes, the sideways field has the most desire to rotate to align with the stator field. This means a sideways rotor field provides the most torque for any type of motor.

Point 2: A DC brushed motor type uses a permanent magnetic field for the stator, and an DC brushless motor type uses an electromagnet for the stator, but a permanent magnet for the rotor.

- The DC brushed motor uses a permanent magnet for the stator.
- A DC brushless motor can also use an electromagnet for the stator, and a permanent magnet for the rotor.

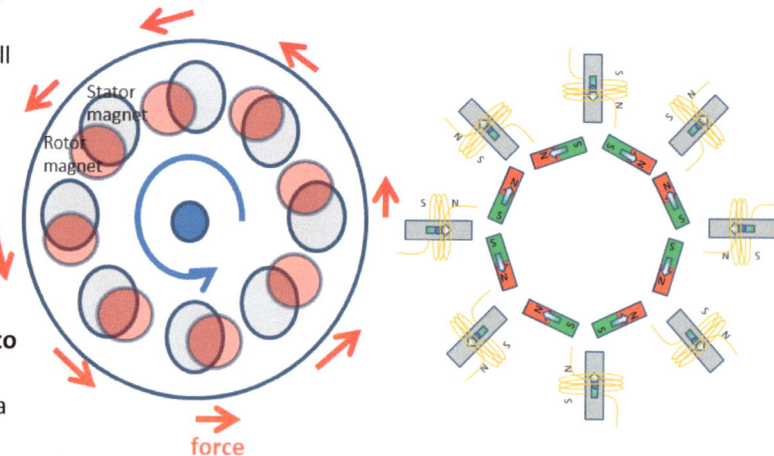

Brushless dc motors, with transistor switches and with electromagnet for the stator.
The permanent magnets are for the rotor and get pulled from stator electro magnet to electro magnet. A linear motor is shown.

Brushless Motor Examples

Not all motors need brushes that wear, and that's a good thing. Brushless motors use transistor switches. However, now the motors need controller units to switch current to the coils at the right angles of the rotor permanent magnet.

The new brushless motors have a reversal of the magnetic field roles for the stator and the rotor. The rotor actually has the permanent magnet, and the stator has the electromagnets.
There is no current flowing to the spinning rotor. Therefore no current needs to flow across a bearing or moving surface.

The concept of energy can provide different answers to motor design, Padawan.

"We just need two magnets to push against each other around a shaft to get a motor."

Electromagnets for the stator

Rotor with fan and permanent magnets

DC brushless ducted fan.
The two coils on the printed circuit board interact with six round permanent magnets in the fan assembly

This motor uses many permanent magnets around the rotor, so a sideways force on the magnets is exploited to get torque around the shaft, not a rotation of the magnets. This sideways force concept is a 'linear motor'.

Recall the force ideas from the previous chapter, and refer to Appendix B for torque and energy.

DC brushless motor for cooling fan.
The outside stator has the rotating magnetic field, and the permanent magnet rotor is pushed around by the stator field.

This motor uses a single permanent magnet on the rotor, so the flipping concept is used, just like brushed motors.

Brushless motors have no current flowing across a mechanical moving surface, so there is less wear and less friction.
The energy idea can lead to different designs for rotating motors, like brushless motors.

Brushless Motors and Less Wear and Tear

Brushless motors are now the cutting edge motor design. The spinning rotor is actually the permanent magnet and the outside is the electromagnet, so the electric function of the rotor and the stator are actually reversed. The rotor still rotates. The stator field also rotates due to the switching transistors and the angle of the rotor, to always keep the fields sideways to each other.

The brushless motor avoids the friction and wear of the commutator and instead switches currents to the stator coils using solid state switches for high reliability. By avoiding a brush that wears away, brushless motors have less maintenance and a longer lifespan.

Brushless motor: Switches control the stator field, instead of a brushed commutator controlling the currents in the rotor.
Applications: Electric bicycles, EV cars

Brushed motor, standard

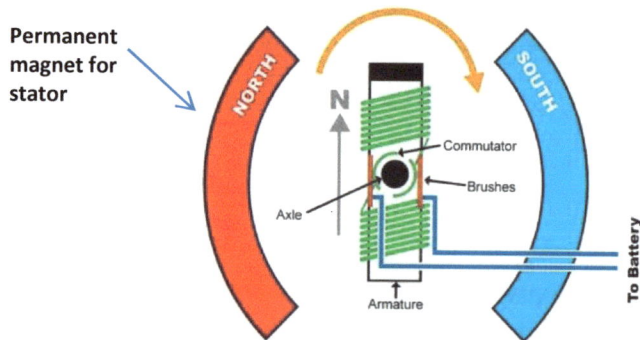

Permanent magnet for stator

NORTH — SOUTH
N
Commutator
Brushes
Axle
Armature
To Battery

Standard brushed motor:
The rotor has coils, and brushes apply current to the right coil to be sideways to the stator magnetic field.

Brush advantages:
- Low cost
- Good low end torque

Brush disadvantages:
- Larger commutator brush friction at higher spin rates
- Wear on the brushes
- Stops working when wet

Brushless motor, new

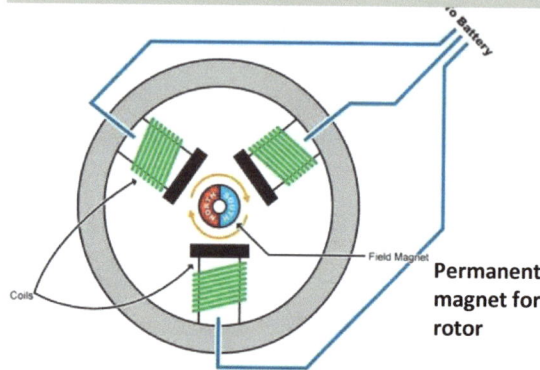

To Battery
Field Magnet
Coils
Permanent magnet for rotor

Main difference of brushless motor:
The rotor is the permanent magnet, not the stator, so no current crosses into the spinning shaft.

Brushless advantages:
- No brushes, so no friction from the commutator brushes, so more power especially at faster spin rates
- No brushes, so no wear of the commutator brushes, and longer lifespan
- No brushes, so less issue with exposure to water
- No sliding brushes, so is a quiet motor

Brushless disadvantages:
- Need electronic controller to switch currents to the right stator coils, where the permanent magnet rotor is sideways to the stator magnetic field. Still, transistor switches are highly reliable.

Brushless motor applications, with permanent magnet in rotor:

Electric motor in wheel hub
battery

Electric bicycle brushless motor in rear wheel

Electric car brushless motor

Sensored Brushless Motor
540 4.5T V3
Designed for 1/10 On-road Drifting Electric RC Racing Car

Electric Remote Controlled car

Brushless motors use electric switches to the stator coils and a permanent magnet in the rotor. All motors still use the concept that the rotor and the stator magnetic field need to be sideways to get the most torque.

Brushless or Brushed Motors

All the many types of electric motors can be used for a car. If permanent magnets are in short supply, then brushed motors can still be used, as long as the brush is protected from the elements. If the availability of permanent magnets is not an issue, then brushless motors can be used. Brushless motors use permanent magnets and dc current because they don't want current to need to flow into a rotating shaft, across some mechanical contact point.

Brushed motors are still used for EV cars as well:

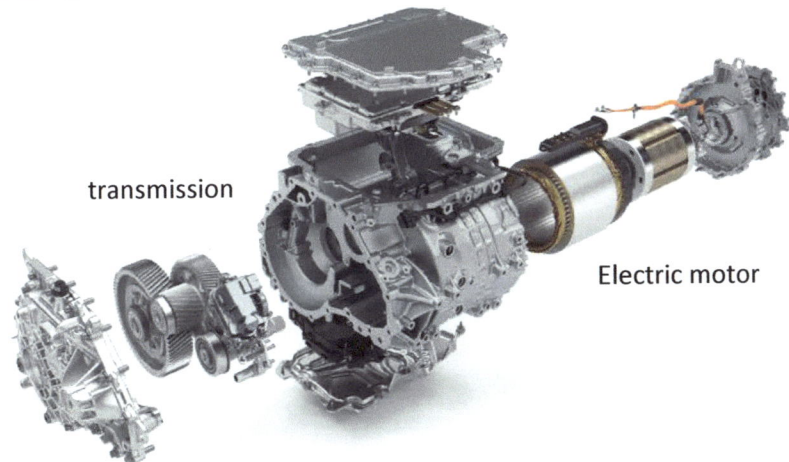

transmission

Electric motor

Brushless motor applications, with permanent magnet in rotor:

11000 RPM +

ESC Brushless Motor

Electric car old school brushed motor, for BMW cars. The sealing needs to be excellent to avoid maintenance.

As a side note, most EV cars do not have a transmission. The electric motor has full torque at high or low spin speeds.

Some higher performance EV cars do have a transmission, for more efficiency at high speeds.

Brushless motors can spin faster because there is no friction from a brush. More power is packed into a smaller volume.

Electric cars and electric bikes can use either DC brushed motors or DC brushless motors.

Chapter 4: 1-Pole and 2-Pole Demonstration DC Motor Experiments

This chapter describes two motor kits that you can build at home, to help understand motors and the commutator that keeps the current switching in the rotor for constant sideways field and for constant torque.

These toy electric motor demonstrations are a great hands-on tool to understand that motors depend on coils and magnet fields.

Understanding this 1-pole and 2-pole DC motor demonstration leads to learning DC motors, which leads to understanding real DC and AC motors used in tools.

After you get hands-on tinkering on 1-pole DC electric motors, you can better understand the far more practical 5-pole toy DC motors, and the even more practical AC motors used in tools. The light bulb moment of 'Yes, I get that' will come easier.

"What's the difference between a magnet and a construction site? A construction site has more poles."

"Figure out the demonstration motors. You can do it, You can figure it out, We have faith in you, Make it happen!"

Toy and demonstrations: Demonstration 1-pole and 2-pole DC motors

DC motor (1-pole) kit

Build one: Cub Scout Webelos engineering requirement

DC motor (2-pole) kit

Scouts could assemble this kit too!

Real: Real 5-pole and more pole motors, with stronger magnets, electro-coils, and iron cores.

DC motor for toys (5-pole)

AC motor

Practical motors have many poles, or coils in the rotor:
Many poles keep the optimal sideways rotor magnetic field, and guarantee starting by avoiding dead angles with no current.

These kit motors do explain a lot about what goes on in a practical electric motor, such as torque and commutation in the rotor coils.

First Step to Understand Electric Motors: Build Kits

Here are the building, magnetic field observation, and spinning steps for 1-pole and 2-pole DC motors.

After building these toy motors, you will learn about keeping the rotor magnetic field sideways to the stator field, about current switching to different rotor coils through commutators to accomplish this steady sideways field, and how the spin direction depends on direction of the permanent magnet and voltage direction.

Here is how to build a simple demonstration 1-pole electric motor.

Homemade or kit? All you need for parts are a battery, a magnet, two metal posts, and a coil, or you can use a 1-pole kit.

1-pole kits: Use the kit 1-pole motor, and build the motor.

The home tinkerer can experiment with reversing the magnetic field of the stator, reversing the coil between the posts, and increasing the number of windings of the coil to get more torque or larger magnetic fields. The home tinkerer can see the strength of the electro-magnet coil by placing a magnetic compass nearby.

	Step 1: Build the coil	**Step 2: Show current in coil makes a magnetic field**	**Step 3: Show spinning**	**Step 4: Show current duty cycle in the rotor**

1-pole motor

 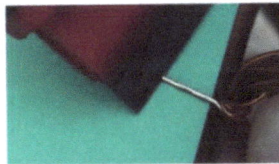   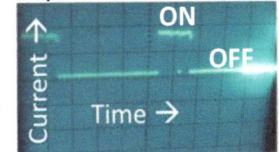

1-pole kit:

1 – pole commutator wire:
- Wind coil.
- Scrape bare contact for commutator.

Demonstrate magnetic field of rotor coil with current flowing

Spinning with commutator

1-pole current versus time

Current flows less than ½ time

2-pole motor

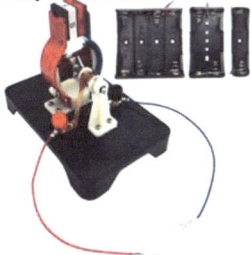    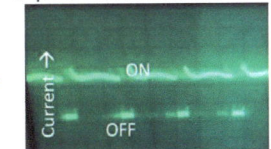

2-pole kit:
EUDAX STEM DIY Simple Electric Motor DC Motors

2 – pole assembled kit:
- Connect stator, rotor, and posts

Demonstrate magnetic field of rotor coil with current flowing

Spinning with commutator

2-pole current versus time

Current flows almost full time

Let's build a simple 1-pole or 2-pole toy motor, and demonstrate basic motor concepts.
Dude, it is so simple. An electric current is flowing through the coil, creating a magnetic field. The coil spins because a permanent magnet puts a force on this current. Basically, using a 1-pole rotor, the current is only on half the time, and generates a magnetic field and torque half the time. A 2-pole rotor will use current almost all the time. Some motors use a flywheel to smooth out the pushes. The current is only applied when the coil (rotor) will give a magnetic field toward one sideways direction.

Here we show how a 1-pole motor is weak with the current on ½ the time, and a 2-pole motor is closer to a practical motor with current on most of the time.

Spinning Rotor on a 1-Pole and 2-Pole Motor from Kits

Watch the kit rotors spin in action:
- When the rotor is spinning, see the blur of the motor (coil). Stop the spin with your finger without much force or pain. But real motors, on electric cars or in machine tools, are much, much stronger.
- The coil spins in a direction that is determined by the direction of the current, and the direction of the stator magnetic field. Reverse one and the coil spins in the other direction. Both the 1-pole kit and the 2-pole kits below make it easy to reverse the magnet and reverse spin.

"For 1-pole motor, the battery supplies current only when scraped metal side touches bare metal post, grasshopper."

Experiments: Build 1-pole and 2-pole motor and have the rotor spin

1-pole: Simplest Commutator, 1 Push per Turn

Experiment 4.1: Get the 1-pole motor to spin by itself, using a 10 turn or more coil, by following instruction on the next few pages.

Not spinning   **Spinning**

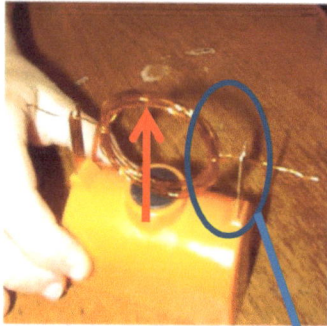

Permanent Magnetic Field

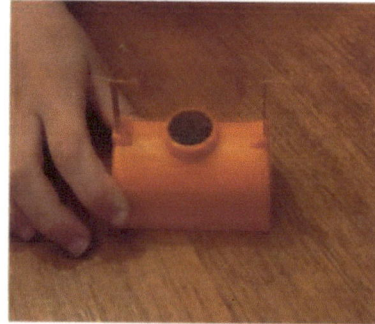

One-pole commutator

- One-pole motor only has current on for less than ½ the time.

1.5 volt battery inside

1-pole current versus time

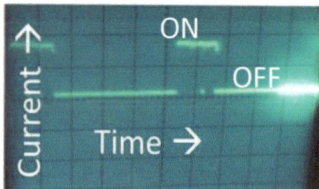

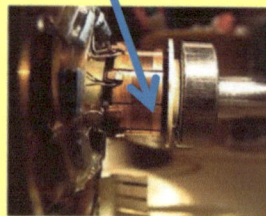

Current → ON / OFF / Time →

One spin

Split ring, practical 10 pole commutator

Our 1 pole demonstration electric motor is equivalent to just one of the coils inside a regular toy motor of 10 poles, or 5 coils. And, as before, this 1 coil needs to get excited when it is sideways to the permanent magnetic field.

See if you can explain the 10 pole, 5 coil, toy motor. For review, the magnets feel the most twist, or torque, when they are sideways to each other. The coil and iron on the spinning shaft is the rotor, and it is an electro-magnet, where the coil that is sideways gets the current.

2-pole: Better Commutator, 2 Pushes per Turn

Experiment 4.1: Get the 2-pole motor to spin by itself, using batteries

Not spinning   **Spinning**

Permanent Magnetic Field

Two-pole commutator

- In practice, use multiple coil commutators, to keep current at all times.

2-pole current versus time

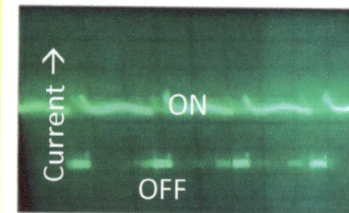

Current → ON / OFF

One spin

Let's make simple 1-pole and 2-pole electric motors. Here are two kits for 1-pole and kit 2-pole demonstration motors, both with magnet, battery holder, posts, and coil.

1-pole Motor Analogy: Clunky 1 Pedal Bicycle

A 1-pole motor is about as welcome and awkward as pushing a bicycle uphill with one leg. More than half the time you are not helping your cause without pushing against something, like pushing against gravity or a shaft.

"Hey, cut me some slack, I'm just a 1-pole demo."

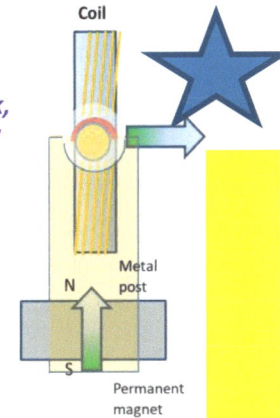

Coil
Metal post
N
S
Permanent magnet

A 1-pole motor is like pedaling a bicycle with only one foot, a bad thing.

pushing with one foot: 1-pole

Coast, No Power, as un-used pedal goes down: Un-used leg, like a one-legged dinosaur. Only one foot touches the pedal

Coast on UP half stroke, because don't have 2nd foot.

Using one leg, the rider better build up momentum from last push down. The rider needs to really push hard and fast for the downward push of that good leg, just to allow the bike to coast enough to get their powered leg back up to the top of the stroke again, and not stall. Nothing gentle about this ride, it takes some grunting effort to succeed in getting up the hill.

Using two legs, this would be your typical gradual hill, with good balance and a slow, steady, relaxed pace.

Do you like an awkward effort to push a bicycle? Do you like abandoning half your muscles, and being forced just to use one leg?

- If 'yes' to awkward, then maybe you want to ride a bicycle **up hill** with only one foot on the pedals.
- If 'no' to awkward, you are sensible, practical, and faster, because you are applying power all the time, instead of half the time or less.

pushing with one foot: 1-pole

Push, Power by pushing with one foot, half the time: Hard, weak, part coasting

Push on DOWN half stroke

Push down and build up enough momentum to coast through the unpowered rise of pedal.

1-pole motor is like using one cylinder engine to turn a crankshaft of a motorcycle or pump.

Oil Field using single-cylinder pumps

Single cylinder motorcycle

'Thumper', 1 push every two turns, so the engine or motorcycle seat can't be a smooth ride

A 2-pole motor is like pedaling a bicycle using both feet, a good thing.

Push with two feet, all the time: Solution for smoother travel up hill:

Two feet with one foot pushing at any one time is like a 2-pole motor, which is easy, strong, and has no coasting.

Imagine having 4 feet and 4 pedals, where one foot is always pushing down when the pedal is sideways, for most torque. This is like a typical car 4-cylinder engine.

Still weak using 2 cylinder car:

Old Russian 2-stroke 2-cylinder car 'Lada'

Weak engine (50HP): This stinky two-stroke car existed for many decades in the former Soviet Union, but it was not a popular car after the Berlin Wall came down.

1-pole motors are not strong. A 1-pole motor is good for a demonstration, but let's acknowledge that the little extra complication to get 2 poles or more is required to get the extra power and torque. Practical motors use 8 or more poles to get a steady sideways field from the rotor, without dead angles.

A 1-pole electric motor is ridiculous and weak with the current on ½ the time or less. Still, some combustion engines have only 1 cylinder too.

Experiments: Ride your bicycle up a hill with only one foot

Increasing the Strength of Motor

Motors might be old and trusted technology, but people need to do the obvious to get more power and torque from motors: namely, more electric current, more powerful magnets, a reasonable number of poles, and tight magnetic coupling between the stator and rotor.

"A 1-pole demo is just a demo. Real motors are super duper strong, with more poles and stronger magnets."

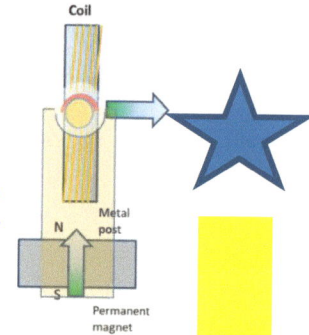
Coil
Metal post
N
S
Permanent magnet

Experiment 4.2: Using this toy DC motor or 1-pole motor, we can increase the torque and spin rate by adding more batteries or getting another magnet.

1. **Add more voltage and the coil should spin faster due to more current and magnetic field or moment of rotor.**
   - For measurements of current, put in a series current meter on alternating current sense.
2. **Hold a 2nd magnet above the 1-pole motor, and see if the motor picks up speed from this larger magnetic field (the stator field).**
   - There is up to twice the torque spinning the rotor around with twice the stator magnetic field, so the motor should show a faster speed of rotor.

Fun fact about spin speed and strength:
Doubling the voltage going to a motor, by adding another battery in series, will increase the speed and torque, unless the motor wire burns up.

Fun fact about new stronger magnets:
Stronger rare earth magnets in DC motors (stator) have allowed higher torque in a smaller package.

There are many other ways to improve this 1-pole motor, which are done to make stronger motors (but not an easy modification of this 1-pole kit):

Pre-optimized practical motor:

Make a multi-pole motor using more poles/coils: Have current all the time, but switch the current to other coils as the coil does the other half of the turn to keep magnetic field pulling.
- Brushes and commutators push on the motor shaft, to switch the current to whichever coil (poles) on the rotor is sideways at the time. Even with a DC voltage applied to the motor, the current keeps going but switches coils, to keep the rotor magnetic field steady sideways to the permanent magnet.
- This sideways field provides a strong torque, instead of just turning off for more than half the time.

1-pole demo motor

Would-be improvements:
1. Use more powerful permanent magnet, and add a top permanent magnet.
2. Add lots of iron core to increase the coil's magnetic field, like in real electric motors.
3. Have very small separation between coil and magnet, where the magnetic fields are the strongest.

1-pole demonstration of motor effect

2-pole demo motor

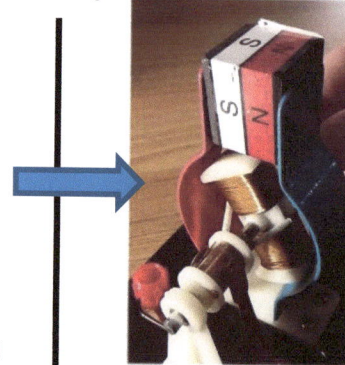

2-pole demonstration of motor effect

5-pole practical motor

How to make more torque in a real motor: More wrapped wire in rotor (more turns), more poles, with iron core.

5-pole practical DC motor

Experiments: Many ways to increase strength (torque) of motor.

This 1-pole demonstration motor is great for showing the motor concept, hands down, but this demo needs to draw or eat more current, to give more torque. This motor needs improvement to do anything useful, like to power a propeller on a toy boat.

Build Demo 1-Pole Commutator for Single Coil

A 1-pole commutator is made just by scraping half the insulator enamel off both ends of the magnet wire, on opposite sides of the coil. This bare metal allows metal to metal contact with the posts, which have the battery voltage, and allows current to flow through the coil whenever the enamel is bare and the coil is facing to one side and only one side.

"Be careful with the razor blade please!"

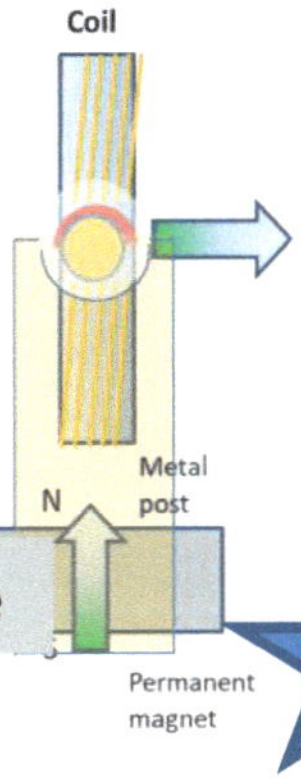

Experiments: build coil for 1-pole motor

Step 1: Scrape enamel off half side of wire

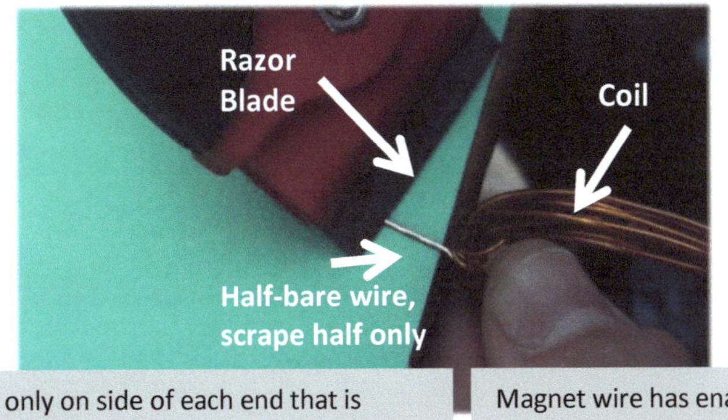

Scrape only on side of each end that is facing down the coil.

- We're building a quick and dirty 1-pole commutator → Current only goes through the coil when the scraped side is touching the metal posts.

Magnet wire has enamel coating to stop shorting out the current between neighboring loops.

- Enamel uses less space and handles heat better than rubber-insulted wire.

Step 2: Rotate both half sides of wire to same angle

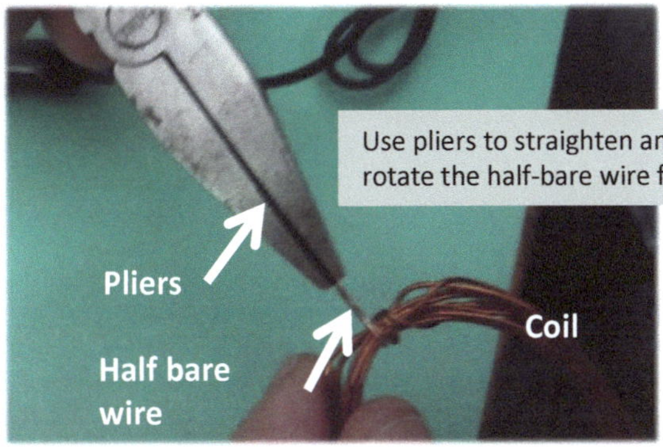

Make sure that the wires that form the axis of rotation of the coil / rotor are well centered, and aligned with the wire axis on the other side of the coil. The current should turn on when the coil is sideways.
If the wire is vibrating and hopping around while spinning (jumping off the post), because it is bent, then the current is not going to flow through it.

Rules for scraping the ends of the wire:
Scrape off the enamel insulator coating over the wire (real magnet wire), but only on the wire side of each end that is facing down the coil. We're building a quick and dirty commutator, where the current only goes through the coil when the scraped side is touching the metal posts.

We do not want all sides scraped off—don't scrape completely around the small circumference of the wire, just one side. We are selective about when the current flows. With all sides, without any enamel and no turn-off, then the electric current would flow all the time and the rotor would align its magnetic field with the stator magnetic field and then stop.

Why care about a 1-pole commutator?
In practice a 1-pole commutator is not used, because it is too inefficient and weak. In the best case, the coil rotor coasts half way around each cycle, and the torque is only strongest for half the other part of the cycle when the rotor field is perpendicular to the stator field.
Fortunately, a 1-pole commutator demonstration does provide a segueway to 2 or more poles. Using many poles, the current changes coils as the rotor spins. The current going through different coils keeps the rotor magnetic field sideways to the stator magnetic field. Then the current flows all the time and we have more average power and more torque.

Magnet wire:
Magnet wire is different than, for example, speaker wire. Enamel uses less space and handles heat better than rubber insulation.

- Magnet wire needs to be wound in coils that have a large number of turns, for more magnetic field.
- Magnet wire can not short out the current from wire to neighboring wire, hence the enamel.
- Magnet wire must conduct the heat away easily, or the coil will melt. Enamel coating allows heat to pass through better than rubber insulation.

With 1-pole, the two half bare wire sides need the same orientation, or same direction against posts.

See Appendix H for best number of turns, with wire resistance more than internal battery resistance.

One Cycle of 1-Pole Motor: Current On when Coil is Sideways

"When done right, the rotor or coil gets a push when sideways."

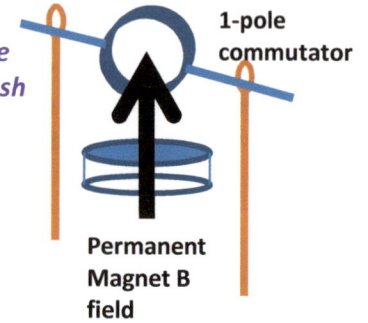

1-pole commutator

Permanent Magnet B field

In a 1-pole motor, current only flows less than half the time, through the exposed wire at the commutator. The rotor magnetic field is only turned on through the rotor that is sideways, in only one direction, so there is consistent twist or torque in only one rotation direction.

1-pole commutator:

Coil or rotor

Below are the equivalent magnets while current is on

Orient scraped wire down when the coil is sideways, to get sideways rotor magnetic field and most torque.

The coil we are building will have 1 pole. When the coil spins and bare metal touches the posts, then current flows.

Bare conductive metal is exposed on the same side, on each end of the coil. Both sides must make electrical contact for current to flow. Both sides must have the same bare metal facing the same way, to get electrical contact at the same time.

A 1-pole commutator uses both posts to get a closed electric current path. A 2-pole or more commutator will wrap the coils around so all the exposed contacts are just on one side.

Retain insulator on one side: enamel

Bare metal or commutator: Orient exposed metal wire so get current when coil facing sideways.

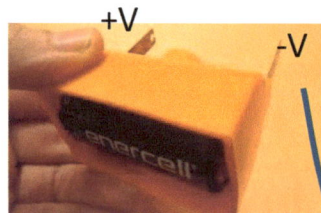

+V   -V

1.5 Volt battery to power rotor coil

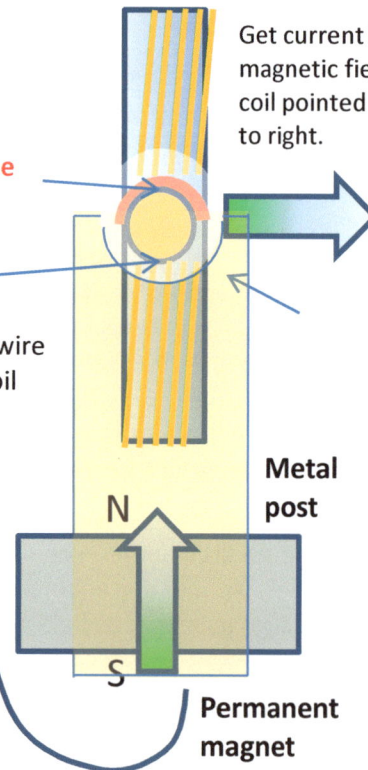

Get current in coil and magnetic field when coil pointed sideways to right.

N

Metal post

S

Permanent magnet

The sideways magnetic orientation has most torque.

S   N

N
S

Rotor design simple coil:
- Coil required for motor so can turn magnetic field off when coil in wrong orientation and keep the spin.
- Permanent magnets don't turn off, so two permanent magnets will not spin, just line up once.

Electric current and commutator:
- When electric current flows, then a magnetic field is generated sideways to the coil. When, ideally, the coil magnetic field is sideways to the permanent magnet stator field, then we get the most torque.
- For more torque, we could have a 2nd stator permanent magnet above the rotor coil to have twice the permanent magnetic field. Then there would be twice the torque, and there would also be no pulling forces up or down (because attraction to each magnet up and down cancels), just twisting forces around the axis of the rotor.

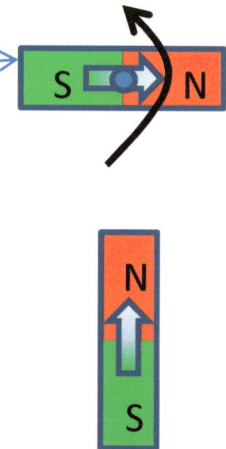

- **With 1 pole, current only flows less than half the time, on exposed wire without enamel.**
- **Electric current flowing near a magnet will cause a sideways force on the wire. This will spin the coil.**

2-Pole Kit Motor, Still with Single Coil

This 2-pole motor has the minimum number of poles to keep a constant current and a nearly non-stop torque a full 360 degrees around the spin.

Yes, there are times or rotor angles when the current is switching directions, and, yes, the rotor is not always exactly 90 degrees to the DC magnetic field because there are only 2 poles, but there is a current flowing most of the time and therefore there is a steady torque.

"Here's another demo motor, from a kit."

Motor parts:

Stator field

Permanent magnet and permanent magnetic field circling around.

This 2-pole kit has a stator DC field. The soft iron channels the DC magnetic field around like a horse shoe magnet.

This 2-pole kit could actually turn some heavier loads. The rotor has an iron core, and the motor has 2 poles, so torque is always being exerted.

The magnetic field of the electro-magnet rotor is always pointing to the same side using the commutator, to keep the same twist on the rotor and shaft.

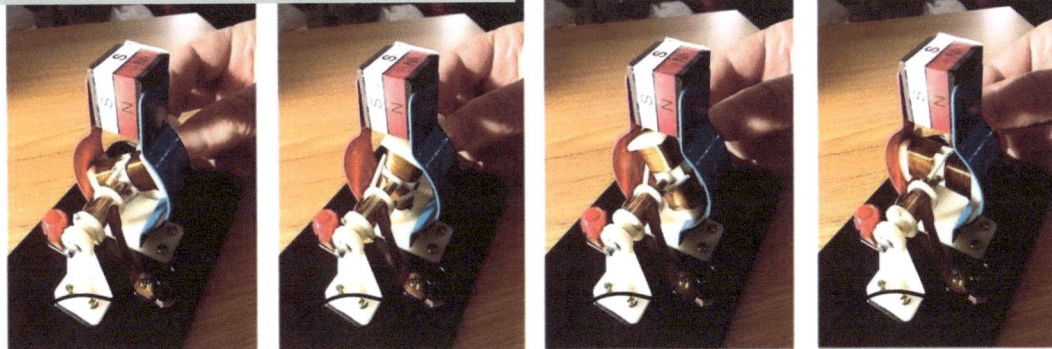

Rotor electromagnet

2-pole commutator   Iron core

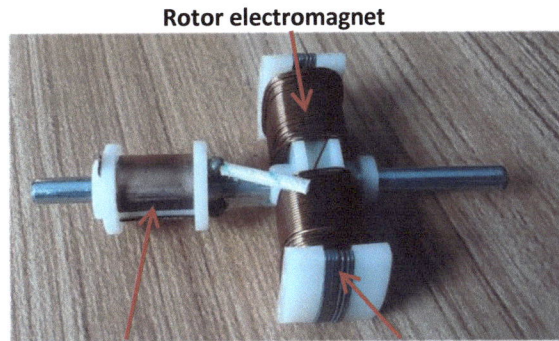

Rotor insert into motor

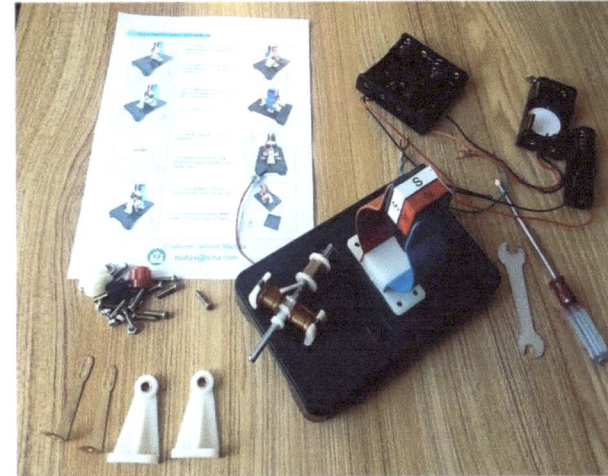

Parts for 2-pole kit, before assembly

Cycle of rotation and electric current:

Strong current   Strong current

No current, switching current direction

Due to the commutator, the rotor coil is at right angle (sideways) to the permanent stator magnetic field when current is excited, to get the most twist or torque.

On the second half of the cycle, the current will flow in the opposite direction and the torque will keep in the same direction.

'IS Icstation 12V DC Electric Motor DIY Kit Model Simple Assemble Kit'

This 2 pole kit motor shows the basics of why a motor spins: commutator to switch current to whichever coil is facing sideways on rotor for a high-torque sideways magnetic field, and permanent magnets on both sides.

Experiments: Buy the 2-pole motor kit and build it

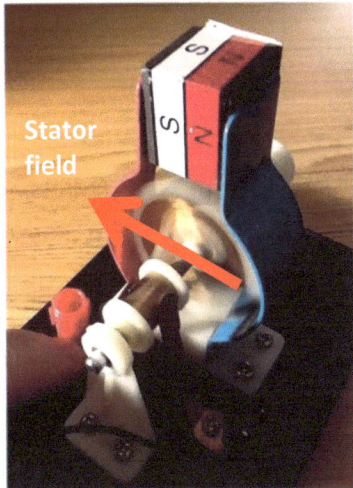

One Cycle of 2-Pole Motor: Current On when Coil is Sideways

"I'm already made as a kit, and I'm closer to a real motor."

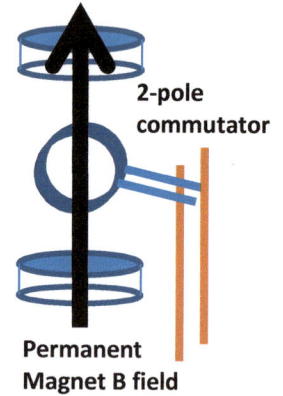

2-pole commutator

Permanent Magnet B field

For 2-pole or more motors, current flows almost all the time, through the exposed wire pads at the split ring on one side of the coil. The rotor magnetic field is only turned on when the rotor coil is sideways, so there is consistent twist or torque in only one spin direction.

2-pole Commutator demonstration:

Commutator switches current direction halfway through turn, with pads all on the same end of the shaft.

Coil or Rotor sideways concept

Excite coil when sideways:
Get current in coil and magnetic field when coil pointed sideways.

Commutator using bare metal against brushes:
Unlike the 1-pole rotor, we have both ends of the coil come out on one side of the coil.

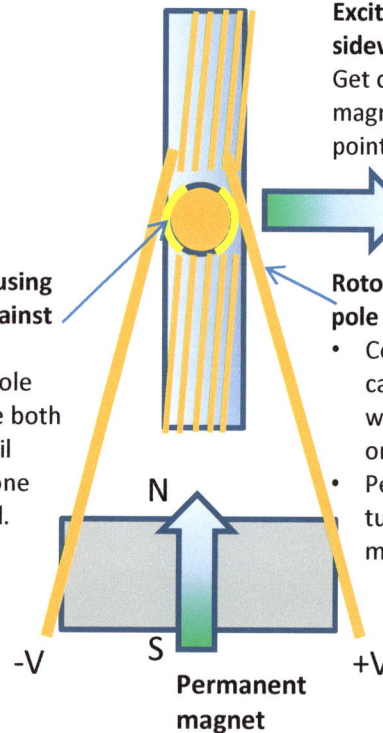

Rotor using single coil and 2-pole commutator:
- Coil required for motor so can turn magnetic field off when coil in wrong orientation.
- Permanent magnets don't turn off, so can not make a motor spin.

N

S

-V   +V

Permanent magnet

Here are the equivalent magnets while the current is on

Sideways magnetic orientation has most torque.

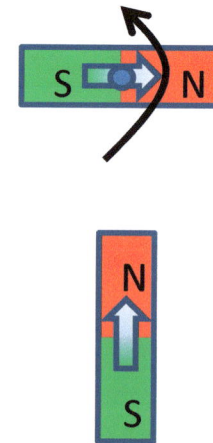

S   N

N

S

Current and commutator:
- When electric current flows, then a magnetic field is generated sideways to the coil. When, ideally, the coil magnetic field is sideways to the permanent magnet stator field, then we get the most torque.

- With 2 or more poles, current flows almost all the time, on pads on shaft, connected to brushes on commutator.
- Electric current flowing near a magnet will cause a sideways force on the wire. This will spin the coil.

Blow by Blow Narration of Current Draw Through One Spin

The current in the rotor needs to reverse at half cycle to keep the same sideways magnetic moment and same twist direction. A one pole rotor needs to turn off current half way through to coast back to the starting point. A two pole rotor can reverse current half way through, for almost steady current draw and torque.

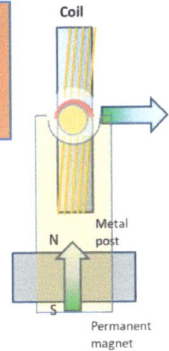

Coil
Metal post
N
S
Permanent magnet

1-pole hands-on learning:

We want the magnetic field of the rotor, for a 1-pole commutator, to be on only when the coil is facing to one side. The sideways coil wants to twist and align the magnetic field with the permanent magnet, which causes the torque.

Without a commutator, if current just stayed on during both the first and second half of the turn, without reversing, then the coil would just align with the magnet at the top and stop, just like the behavior of permanent magnets before. No rotation. Coil would get reverse pull after reached top or alignment of the rotation.

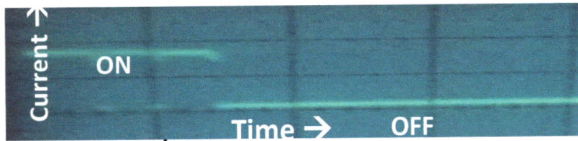

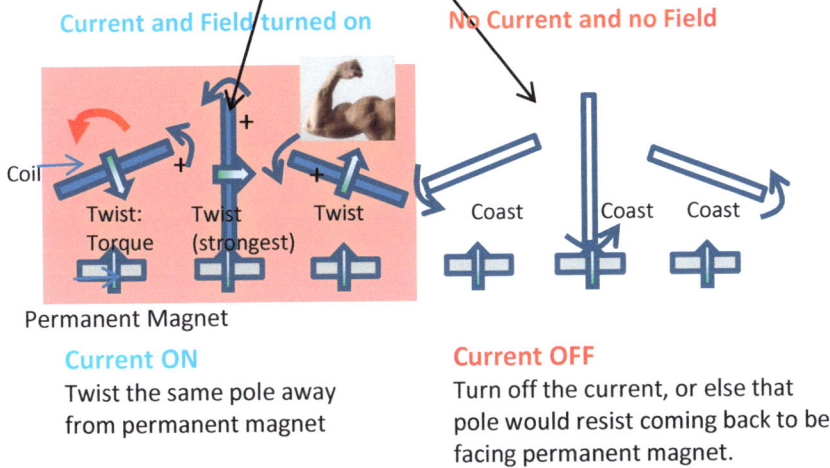

2-pole hands-on learning:

Have current in electro-magnet rotor in one direction to get the roto to flip up, and then reverse the current to get the rotor to flip down. There are two pushes per rotation cycle.

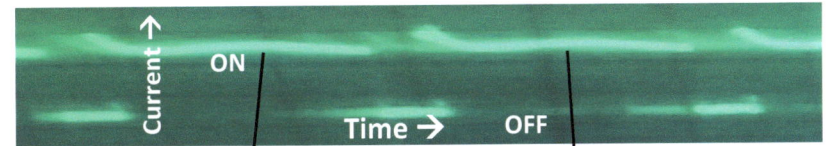

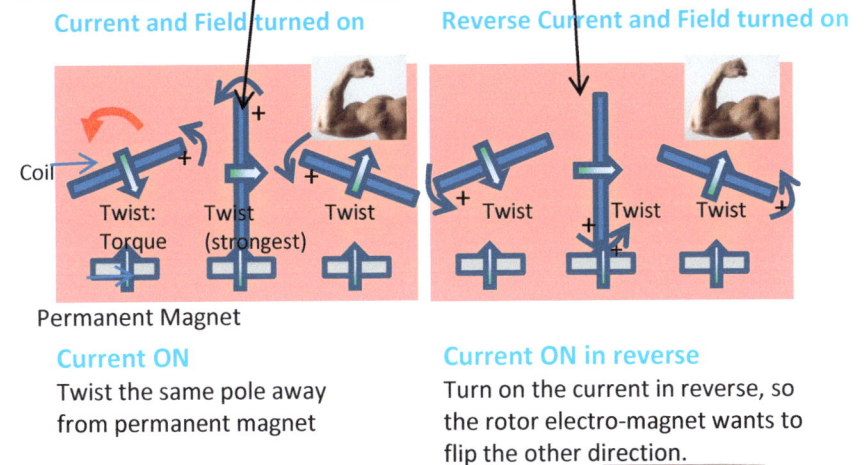

"When it comes to motors, everything can be measured."

Current and Field turned on   No Current and no Field

Coil
Twist: Torque   Twist (strongest)   Twist   Coast   Coast   Coast

Permanent Magnet

Current ON
Twist the same pole away from permanent magnet

Current OFF
Turn off the current, or else that pole would resist coming back to be facing permanent magnet.

Current and Field turned on   Reverse Current and Field turned on

Coil
Twist: Torque   Twist (strongest)   Twist   Twist   Twist   Twist

Permanent Magnet

Current ON
Twist the same pole away from permanent magnet

Current ON in reverse
Turn on the current in reverse, so the rotor electro-magnet wants to flip the other direction.

Like one foot pedaling, pushing less than half the time.

Like two foot pedaling, pushing most of the time.

For a 1-pole motor, the current turns off for 2nd half of the rotation. If the current stayed on for the second half of the turn, then the coil would just slow down and try to go back up and align. A 2-pole motor does not have this issue: current reverses in second half of spin due to the commutator and then the current can stay on all the time with the same direction torque.

See Appendix C for more details on current flow of 1 and 2 pole motors

2 or more poles allows the motor to be pushing all the time, instead of less than half the time with 1 pole.

Summary of 1-Pole and 2-Pole Basic Motor Demonstrations

Here is a summary list of experiments on the 1-pole and 2-pole toy dc electric motor kits, described in this chapter and in more detail in the appendices.

Experiments: Build 1 pole and 2 pole motors

1-pole hands-on learning:

Chapter 4: Build a 1-pole motor and get it spinning.

1. **Build coil**: Wind coil to allow a larger rotor magnetic field, and scrape axis of coil to allow current half the revolution
   - Let's demonstrate how to build the 1-pole commutator.

Observations

2. **Detect coil field**: Show compass needle deflection when have current in coil.
   - Show current only flowing for less than half of spin, using detected magnetic field
   - Let's demonstrate that electric currents in the rotor generate a magnetic field, using a compass to measure the field.

3. **Spin and spin direction**: The current spins in a certain direction, depending on the + side of the battery, the coil winding direction, and the direction of the permanent magnet.

4. **Reverse spin**: Show reverse spin direction when flip permanent magnet:
   - Discuss force on currents.

5. **Stronger torque**: Increase the torque and spin rate by adding more batteries or getting another magnet

6. **See electric current**: Measure current versus time in the spinning coil, using a multi-meter or an oscilloscope.
   - Let's show the electric current as a function of time, by inserting current and voltage meters in the power circuit for the electric motor.
   - Let's show some of the subtleties of reversing currents in the rotor.

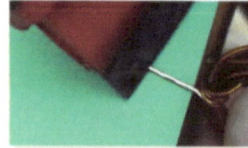

Scrape away half the enamel for 1-pole motor

See rotor magnetic field using compass

White side up

See rotor spin from magnetic field and current

Measure current in 1-pole coil

2-pole hands-on learning:

Chapter 4: Build a 2-pole motor and get it spinning, from a kit.

1. **Assemble kit**: Connect the stator, the rotor, and the batteries.

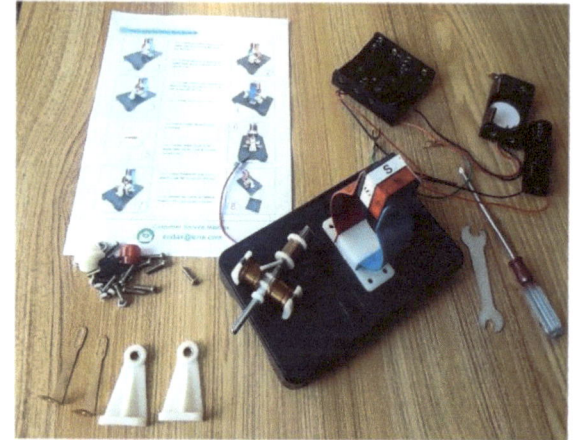

2-pole motor demonstration kit, before assembly

Observations

2. **Coil Field**: Show compass needle deflection as the rotor rotates

3. **See current**: Measure current versus time in the spinning coil, using a multi-meter or an oscilloscope.

4. **Continuous current**: Current flows most of the time, due to switching or commutating between the 2 poles

5. **Stronger torque**: Iron core increases the magnetic field of the rotor

6. **Stronger torque**: 2 permanent magnets in stator increase the torque.

Current is on almost all the time when have 2 or more poles

See Appendix D and H for more ideas for experiments using motor kits.

Wind the magnet wire to make a coil. Deflect a compass with a coil. Reverse the spin when flip magnet. Measure the current in the coil. Dude, this is what you can do. Here's the hands-on stuff.

1-Pole and 2-Pole Motor DC Demonstrations of Current, Q and A

A 2-pole toy electric motor uses the electric power almost all the time, and so has more supplied average power than a 1-pole motor.

Now, even though the 2-pole motor gets current most of the time, the 2-pole rotor is not always in the best sideways position for most torque while current is flowing in it, so we'll need to have 4 or more poles to get the best average torque and most instantaneous sideways magnetic field from coils in the rotor. 4 or more poles means 2 or more coils, because each coil gets current twice around a rotation.

Also, a 2-pole motor has a dead zone where no current can go to the coil, so it may not self start when a voltage is applied. There is only one coil and the two pads are connected to this one coil, so a single brush can not be allowed to bridge the two pads. A 4-pole or more motor with 2 or more coils, with a wide brush that always excites two pads out of the many pads, either 1 or 2 of the many coils, will always self start even when the brush is between two pads.

Question 1: Does the current flow continuously for a 1-pole motor? No, the motor thumps along with jerks once a turn.

- No, the current only flows less than ½ the time, only when the rotor coil is at the sideways angle to get the same spin direction of torque.

Question 2: Does the current flow continuously for a 2-pole motor? Yes, because of commutators or brushes that put the current in the correct coil to get a consistent torque.

- Yes, it can, because the current reverses through the single rotor coil when the coil is facing the other sideways direction.   The torque is in the same spin direction because the rotor magnetic field stays in the same direction as both the coil and the current are reversed.
- However, some of the ½ cycle has the rotor not perfectly sideways but more parallel to the stator magnetic field.   Hence a 4 or more pole commutator is more effective.

Question 3: Which configuration gets more power? A 1-pole motor or a 2-pole motor?   A 2-pole motor is getting constant current, so this is an easy question.

- A 2-pole motor is constantly providing current and a torque, so a 2-pole motor gets more average power, rather than turning off and coasting through half the rotation.

Question 4: Why do practical motors use 4 or more poles (2 or more coils)?   We want the excited coil to be as close to sideways to the stator field as possible.

- For a 2-pole motor, some of the ½ cycle has the rotor not perfectly sideways but more parallel to the stator magnetic field with less torque.   Hence a 4 or more pole commutator is more effective.
- A 4-pole commutator and wide brush will always excite one of the coils, and will self start when a voltage is applied at any angle of the rotor. A 2-pole commutator has a dead zone where it won't self-start when a voltage is applied. For a 2-pole motor, the brush is narrower than the gap between the two pads to avoid touching both ends of the same single coil with the same side of the voltage.

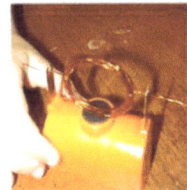

1-pole motor and current ON less than ½ the time

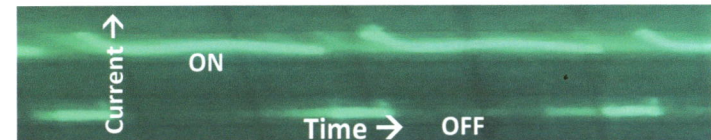

2-pole motor and current ON most of the time, for more average power

8-pole motor:
- **Excite sideways coil when within a narrow sideways direction, for most torque**
- **Avoid dead spots where motor will not start (wide brushes always have a coil or two to connect to)**

See Appendix E for simple circuits to measure current flow 101

Vintage Motor Demos are like Science Prototypes

How did electric motors go from a gleam in someone's eyes to full blown usage?
First discovery and early science had to happen, such as electrostatic forces, creation of magnetic fields, the explanations, DC and AC current, and prototype motors and generators. And then the ideas get applied by industrious engineers who see a need, like trolleys in cities and electric lights. Notice it took 60 to 80 years before mass applications happened, from the 1820s to the 1880s. The first motor inventor, a priest, barely even advertised his invention.

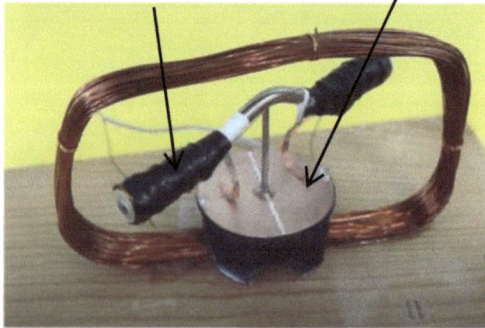

Coils around steel bar to create electro-magnet for rotor   commutator

Example of Jedlik's first motor, in this case a 2-pole DC motor.

A DC current runs through the outer electro-magnet coil. The middle rotor has two coils wrapped around an iron bar which is a iron loaded electro-magnet.
Due to the simple brush commutator, the current in the rotor flips direction halfway through the spin, and keeps the rotor magnetic field pointed to the same side as the outer electro-magnet, to keep the twist or torque in the same direction.

DC motor.
2 crossed coils, so a 4-pole motor.

2-pole DC motor.

2-pole DC motor.

4-pole DC motor.
This is unconventional because the coils are along the perimeter of the circle.

3-pole DC motor.

3-pole DC motor.

Jedlik investigated many science topics, and basically invented a motor demonstration 2-pole kit. Many other 2 or more pole simple motors can be built. The trick is making the motor strong with larger magnets and more poles.

We've shown how motors work, from the twist due to magnetic field alignment and commutators. Now let's show what motors do.

This chapter describes all the practical ways motors are used in our regular lives, like Blu-ray players, pumps in refrigerators, fuel and water pumps for heating and household water, trains, elevators, and starting cars. Section 5A names a lot of items in a typical house that have electric motors. An 'imagination' section 5B asks the question, 'What would we do without electric motors?'. Sections 5C and 5D look into how electric motors move us around. Section 5E looks at the power requirements for electric motors and robots.

Electric motors do make things move, so they are a natural for transportation.

Electric motors have helped produce almost everything that is built – clothes, houses, computers.

Chapter 5: Electric Motors Around the House and Cities:

Let's explore where electric motors are obviously contributing to people's lives. Find all the motors in a house, or in a car, train, or boat. Find future uses of electric motors, like robots and airplanes.

Electric Motors Around the House:

Section A:
How Many Electric Motors in Your House?

A: Can you walk around your house and count how many electric motors there are, before and after reading this chapter?

- Fans
- Air Conditioners
- Refrigerators
- Dish Washers
- Clothes Washers

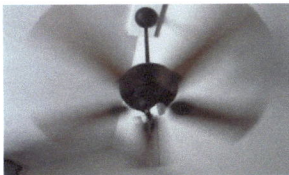

Fan

Section B:
Imagine Life Without Electric Motors and Magnetic Fields

B: Imagine what would your daily life be like without these motors?

- Hand laundry
- Ice cellars for food
- No easy tools in the garage
- Steam trains would still be all the rage

Hand clothes washing

Section C:
People Movers and Cars

C: Do you think cars don't use electric motors?

- Starter Motors
- Generators (Alternators)
- Electric Vehicles are all electric motors

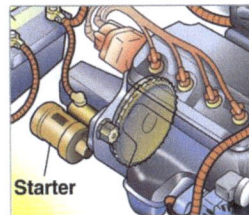

Starter

Starter motor turns flywheel teeth to crank over the engine.

Section D:
People Movers and Boats, Trains, Elevators, Airplanes.

D: What about large ships, subways, elevators, and recent short flight airplanes?

- Electric motors behind propellers on large ships
- Electric lines to subways with no smog
- Short Range Airplanes

The 'Orange' line, Boston, 2013

Section E:
Electric Motors and Robots

E: Robots help with specific tasks for factories.

- Robots help assemble new cars. Each robot is custom designed for a specific task.

Robot welding

Do it! Just walk around and see what electric motors you see.

Magnetic Fields Allow Electric Motors You May Use

Here is a conglomeration of what's to come in this chapter, regarding motors. Look around your house. Electricity is very useful, with refrigerators, computers, anything that moves with motors. There are uses of electric motors everywhere, even if motors are not a thing of beauty. If imitation is the best form of flattery, then electricity and electric motors have some honest respect.

"You depend on me."

DC motors: wherever you have a battery

DC motor applications

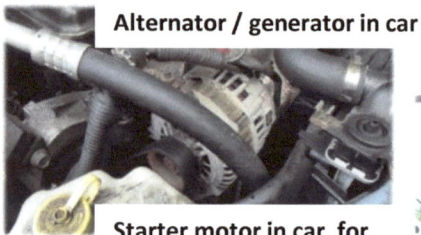

Alternator / generator in car

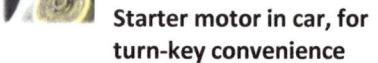

Starter motor in car, for turn-key convenience

Remote Controlled airplane, with a high energy density rechargeable battery (lithium ion)

Starter motor on snow blower, if the engine is heavy enough to make pull cord too exhausting.

The 'Orange' line, Boston, 2013

Tesla Model 3

AC motors: wherever you have a wall outlet

AC motor applications

Compressor inside the air conditioner

Washer / Dryer, spinning

Fan on your ceiling

Compressor in the refrigerator

Treadmill, rotating motor pulls belt

Garage door opener, pulling up door

Oil pump for oil furnace

Water pump for a well

Electric table saw today with electricity

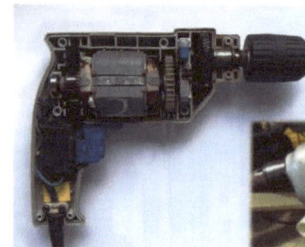

Electric motor inside portable drill

Drill, spinning the drill bit

Tesla Model S, and Chevy Bolt

...And many other applications of motors ...they are everywhere

(see Chapter 5A to identify electric motors in your house)

Now let's talk electric motors, that is, the visible everyday motors around you at your house. Electric motors are all around us and are giving us a lot more time to do other things. Think of all the convenient motors in typical houses that do the laundry, wash dishes, run the refrigerator and air conditioner, pump oil to the furnace, and pump water from a well.

Electric motors have only been around since the 1880s, and that also required the construction of power plants around the city as well. If there was local running rivers, the power plant used water turbines. If there was coal, the power plant used coal. If there was wood, the power plant used wood.

Back then people wanted electric trollies around cities and electric lights. Power plants back then were a bit of a cutting edge investment, starting with power for cities, and people were excited to use new appliances in their house.

Back in the 1890s, power plants can recharge the batteries for the original low horsepower electric vehicles, and power plants can power any light rail going down main street instead of horses.

Probably people in the countryside needed to wait a few more decades for the long distance power lines to get to their house.

The past without electricity, from pre-history to classical times to middle ages to beginning industrial age:

Hey, we existed for millennia without electricity in the house. How? How did we cook, or wash, or stop food from going rancid?

Well, ancient Romans had single rooms for whole families without indoor plumbing or toilets. The outside market in the cramped city was actually the living room, so loneliness was probably not a problem. For lights, they'd use olive oil or fish oil. For cooking, there were outdoor kitchens and fires.

Most people throughout history lived in the countryside in huts, and did farming. They burned wood for cooking and heat, and lived near water. Even in medieval Europe, they'd use wells and buckets for water, or live near a river or pond. The oxen, donkeys, and horses did the manual plowing for the crops.

Women typically were stuck doing very domestic chores, like hand laundry, hand washing, and cooking the meals using a wood burning or coal burning stove.

Food selection was what was only in season. Some fish and meat could be chilled for a few months under some hay and ice in a cellar, but that is for rich people.

Advent of electricity, part way through industrial age in the mid 1800s but mostly in the 1900s:

So what has electricity done? People have refrigerators, with food from around the world. People have cars to drive an average of 13 miles a day, so we're not limited to whatever job is within walking distance. People have air conditioners, so we get a better night's sleep in the summer and think better in the morning. Women can get out of the house and not do time-consuming labor like clothes and dishes. People of course now have electric lights, computers, TVs, and entertainment like movies at home.

Do it! Just walk around and see what motors you see.

How Many Electric Motors are in Your House?

Now let's talk electric motors, the visible and hidden everyday motors around you. Not noisy and polluting combustion engines, but quiet and portable electric motors. Not sweaty manual labor, but electric motors that draw electricity. You can not escape electric motors, because they are so useful. Examples of motors are all around the house.

When forces from magnets and currents were first discovered, most people had no idea how useful they would be. If, according to folklore, the scientist Maxwell said to the Queen in 1820 when asked what use are electromagnets, 'Someday you can tax it', he was so right.

Experiment 5.1: Show other electric motors around the house:
- Some are Direct Current (DC at 5 or 12 volts typically) and some are Alternating Current (AC at 120 volts typically in USA, and 240 volts in Europe)

DC motors: wherever you have a battery or electronics

Stator: Permanent magnets
Rotor: Iron cores, and wire coil for current loops

Toy DC motor

AC motors: wherever you have a wall outlet

Stator: Iron core and wire coil for current loops, no permanent magnet
Rotor: Iron cores and wire coil for current loops, no permanent magnet

Industrial AC motor   **Drill AC motor**

DC motor everyday applications

Car
- Starter motor in car
- Hybrid cars have electric motors powering one or two wheels

Snow
- Starter motor for snow blower

Games / Computer
- DVD drives in a game consul and computer
- Disk hard drives
- RC remote control airplanes and cars
- Auto focusing on cameras.
- Cell phone vibrator
- Electric tooth brush

AC motor everyday applications

Garage
- Garage door opener
- Emergency backup generator

Water
- Water pump for well

Heat
- Oil pump for oil furnace

Cooling
- Compressor in the refrigerator
- Compressor in the air conditioner
- Ceiling fan

Tools
- Drill, table saw, band saw

Cooking
- Deflector blades in the microwave
- Mixer or blender
- Dish washers, using water pump

Games / Computer
- Air pump for bounce house

Cleaning
- Washer / dryer for clothes
- Electric shaver or hair clippers

Exercise
- Treadmill

Your house is chock full of electric motors, most likely. Just go looking for them. Most of the motors are AC motors, because they plug into the wall 120 V AC outlet, which alternate direction of current. AC motors are in your refrigerator and heating system. The DC motors are in the car, the computer, the DVD player, and remote controlled toy cars and airplanes. DC motors typically use batteries, or near the DC power supply for electronics like computers and TVs.

There must be many more ways that motors insert themselves into your lives, where you just take them for granted.

Do it! Just walk around and see what motors you see.

Vintage Home Appliances

Let's look at the start of using electric motors to reduce the monotony of household chores. Yes, electronics were around in the 1920s and 30s. Clothes were washed, water was pumped into homes, and some people had garage shops to build stuff.

1934 Antique Vintage Maytag Ringer Washing Machine:
An electric motor turns the agitator and clothes and turns the ringer.
There was an option for a gas motor to turn the washer as well.

1920 Beatty model A:
Canada's first electric washing machine with an agitator and a ringer.

1930 Water pump, with electric motor:
Made by Westinghouse and McKay equipment.

Large compressor on top.

Westinghouse Refrigerator, 1932

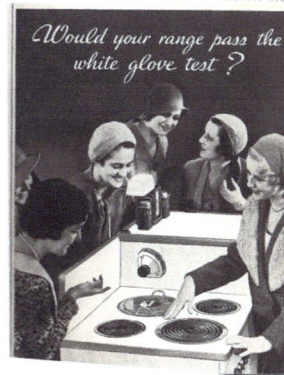

Would your range pass the white glove test?

GE Electric Range circa 1932

Classic Vintage Antique South Bend 9" Lathe:
3 1/2 bed With Stand
The Jeannin Electric Co. 1/4 Horsepower 110v Single Phase Motor

Electronic devices have been around for over 100 years.

Electric Motors You Use in the Garage:
Garage Door Opener, Starter on Snow Blower

Let's look the hard outdoor physical activities people had to do even 50 years ago, and how electric motors are doing those activities for us nowadays.
Uhh, the pain, my back! I'll gladly let a motor take on the task of lifting a heavy door, or heavy snow. Today's people are more protected from the elements than ever before. We use electric motors and combustion engines, so we can drive from garage to garage and almost ignore the weather. Even back in the 1970s, getting out of the car in a snow storm to open the garage door used to be a rite of passage.

"Why lift garage doors by hand? Why hand shovel snow?"

Sheds and Stables

Garage for cars and snow blowers

When I was a young whipper-snapper, I had to walk outside the car in a blizzard to lift the garage door. And we were lucky to even have a garage!

AC electric motor spins a threaded rod or chain to move the garage door up or down.

Why break your back?
Gradual exploitation of electric motors around the house.

AC electric motor to lift garage door

Sheds for horses

Before electricity

1930   *Enabled by electricity*   **1950**   **1960**

Era of electricity

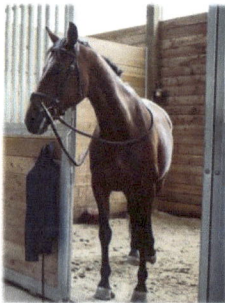

Horses, the pre-1900 mode of transportation (and camels or elephants)

Hand crank

DC motor, the starter motor to turn over and engage the car combustion engine

DC or AC electric car

Electric starter motor to start snow blower
Like a car starter, some snow blowers have an electric motor to get the heavy snow blower combustion engine to turn over.

A few winter storms, back to back, can take the romance out of this back-breaking work.

Physical winter chores, shoveling the driveway

AC motor (starter) for snow blower

Any workhorse electric motor is going to be in the garage. Tools, snow blowers, garage door openers.

Electric Motors You Use for the Basics: Water and Heat

Let's look at the history of water and heat in the house before electric motors and nowadays.

For household water, people once upon a time had no internal plumbing, and baths were taken in streams or after carrying water in buckets to a tub. Nowadays water is pumped from the ground from below the groundwater table, or pumped up into town water towers to have water pressure from gravity at your house.

For household heat, people once upon a time had to carry wood or coal to the furnace to stay warm. Nowadays fuel is pumped to your furnace with no manual effort, by electric motors which spin pumps. Oil is pumped from the oil tank. Natural gas lines throughout the city are kept at a high pressure of the explosive gas using pumps, to allow gas flow to the furnace.

Hand water pump

Bucket and well
Really old well with hand crank, before modern electricity

AC

Water for the faucet and bathroom: Electric motor for water pump for a well

Typical Well Design

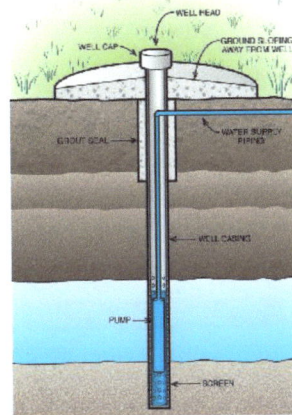

WELL HEAD
WELL CAP
GROUND SLOPING AWAY FROM WELL
GROUT SEAL
WATER SUPPLY PIPES
WELL CASING
PUMP
SCREEN

Electric well with motor pump at bottom

control box
pressure tank
brass T-fitting
pressure switch
1 ¼" steel pipe
5" casing
torque arrestor
check valve
pump
motor

"Electric pumps pump me up!
Houses need electricity to pump oil for heat and pump water from the ground for drinking and for showers and toilets."

An electric water pump is at bottom of well, if use one pipe.
Pump can be above ground if use two pipes going down into well.

- Many country houses have wells, that pump the water from an underground well into the house.
- If the electricity goes out, then the house loses its water, its toilets, its showers ... this is stinky bad.

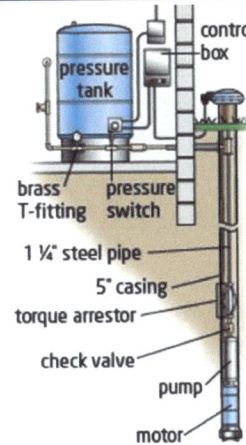

1930   *With modern electricity*   **1950**

Era of electricity

Did you forget about old timers like me? Times were simpler and hard-scrabble back then, just heating our one room house with wood or coal.

Wood burning stove and heater

AC

Heat: Electric motor for oil pump for oil furnace

Oil tank:
- Pump the oil from the tank, through the nozzle, to spray into the combustion chamber.
- If the electricity goes out, the house loses its oil pump and then heat, unfortunately.

Nobody wants to go outside to hand pump the handle of a well to get water for dinner or for a shower.

Exercise: Do you have oil heat? Find the electric motor to pump the oil

109

Electric Motors You Use for Comfort and Food: Ceiling Fan, Air Conditioner, and Refrigerator

Let's look at the art of keeping things cold and preserving food, from ancient times to nowadays with the electric motor.

Instead of using A/C air conditioners, people once upon a time slept outside in the warm summer to avoid the heat build-up indoors, and people still do it. Instead of refrigerators for food, people kept ice in the cellar all year to keep food cold in the summer. A/Cs and refrigerators use electric motors. As an alternative, Native Americans dried their fish and deer meat to make 'jerky' so it did not need refrigeration, by warming the meat in the sun or heating with a small smoky fire.

"Can you hear me spinning those air conditioner compressors to cool off?
A fan uses less energy than an air conditioner."

Spin fan on your ceiling

Spin compressor inside the air conditioner

AC

Cool pipes from evaporated coolant on inside.

AC

INDOORS

OUTDOORS

Fan axle
Condenser coils
Compressor
Fan
Blower
Hot air
Cooled air
Outdoor air
Expansion valve
Cooling coils   Temperature sensing bulb
Indoor air

Fan direction and the seasons for low cost air conditioning:
- In winter, blow air up slowly, to force hotter air from above to the walls where it is typically cooler: creates a uniform temperature in the room. Don't spin too fast, because do not want to feel a draft.
- In summer, blow air down to create wind to cool you off by the wind chill effect.

Air conditioner mounted in window

Squeeze the coolant back into a liquid using a pump.

Radiator for hot pipes from re-compressed coolant on outside.

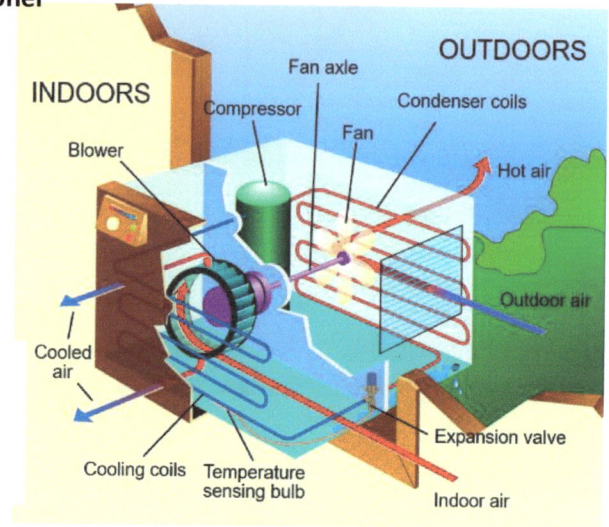

Air conditioner: path of the refrigerant

Fan in ancient Roman times

0   1930   1950   1960

Era of electricity

194

AC

hungry?

Ice cellar to freeze your deer meat, if you live in a climate with winters to get ice.

Motor spins a compressor in the refrigerator

How refrigerators and air conditioners work, with motors:
One electric pump, the compressor, forces a liquid around in a cycle. The electric pump forces liquid coolant out a tiny nozzle, which turns into a gas, evaporates, and cools the coolant. This cold location is inside the place you want to be cold: inside the refrigerator or inside the room for an air conditioner. At another location, the coolant is re-compressed using the compressor pump into a liquid, which heats it up, but a radiator releases the heat to outside the refrigerator, or outside the window. The compressor pump is spun using an electric motor.

Do you take refrigerators in the kitchen for granted?   Do you take a cool air conditioner for granted?

Electric Motors You Use in Basement: Washers and Dryers

Let's look at those chores like washing dishes and washing clothes, before and after electric motors. You want convenience? You came to the right place. Taking care of your household got more convenient when people got electric washers for clothes and dishes. Ask yourself about un-intended consequences: Did electric motors and household conveniences enable the 1960's liberation movements, by freeing up people's time? Luxuries before then were a one car garage and indoor plumbing. Luxuries afterward in the 1960's were more free time.

"Why hand wash clothes or dishes, when you can push a button and walk away?"

Old time dishwasher:
Would you rather wash dishes by hand, like at camp-outs, using the 3 bucket camping technique?

Modern times:
People complain about the 3 minutes to load the dishwasher instead of the 30 minutes to wash by hand.

AC

Electric motor to shower the water around, and to pump water out of a dish washer.

Dishwasher: motor pumps water

1930   **1950**   **1960**

Era of electricity

Electric motor to spin washer and dryer.

AC

Old time clothes washer:
Would you rather wash clothes by hand in buckets or the sink, with washboards?

Clothes washer and dryer: motor moves clothes around
One electric motor pumps the water around, and another turns or agitates the clothes: a belt around tub connects to motor shaft.

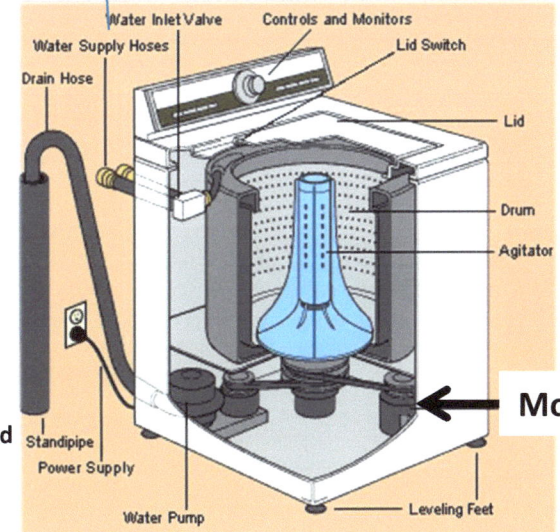

Water Inlet Valve   Controls and Monitors
Water Supply Hoses   Lid Switch
Drain Hose
Lid
Drum
Agitator
Motor
Standpipe
Power Supply
Water Pump
Leveling Feet

Clothes washer: electric motor on bottom spins the agitator

Standard household chores are a lot easier and less time consuming using workhorse electric motors for dishwaters and clothes washers.

111

Electric Motors You Use: Basement Machine Tools

Everyone can build things with electric tools, instead of depending on more skilled and slower hand labor. Your garage shop with electric tools is always waiting for you, without moving near a river and getting a water wheel. The garage shop has electric drills and table saws, while 200 years ago you needed to build a dam to back up water and get a water wheel and belts.

"I am your electricity to command. Open me up and see me in my raw glory."

Good old days and hard labor

How long can you last sawing wood by hand, before collapsing in sweat?

Saws of yesteryear without electricity

- People lived on the banks of a rushing river to turn a water wheel.
- People traded or bartered your cut lumber for horseshoes, wooden barrels, pottery, or wheat.

Handsaw, for nostalgic log cutting competition
Lumberjacks now prefer chain saws.

1800
Breastshot water wheel

Water Flow

Head race

Wheel Rotation

Tail race

1830's sawmill powered by water

1950

Circular saw was developed in America in 1810 by Mrs. Tabitha Babbitt, who was a quiet weaver living in a Shaker community in Massachusetts. That's water power working for you.

Modern days with electricity

Now, with electric tools, sawing wood and metal is more about precision than muscle.

Saws and drills with electricity

- People live anywhere. Just plug into your 120 Volt outlet and start building.

AC   AC

Electric Drill press   **Air power compressor, for air-power impact wrench.**

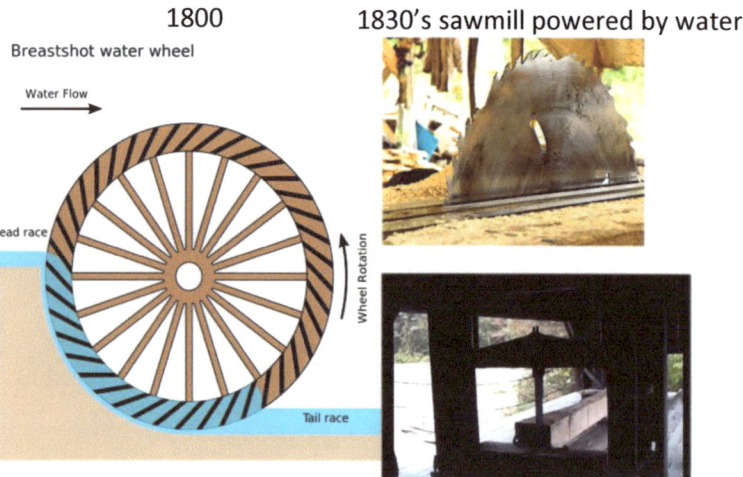

The drill un-dressed!

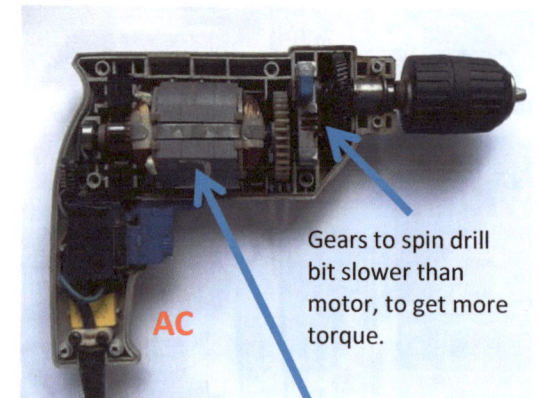

Era of electricity

AC

AC

Gears to spin drill bit slower than motor, to get more torque.

Electric table saw today with electricity:
- One person can work in their own basement, not breaking a sweat.

Electric motor inside portable drill
There's a lot of stamped iron plates to get a thick iron core and more torque.

Electric tools, with workshop convenience, means we can build a lot more things.

Electric Motors You Use for Toys and Entertainment: Miscellaneous Motors Around the House

Let's look at some various motors from all different uses around the house.
In addition to washers and refrigerators and water pumps and oil pumps, electric motors get used on a lot of random products in the house, like DVD players, treadmills, and remote controlled airplanes.

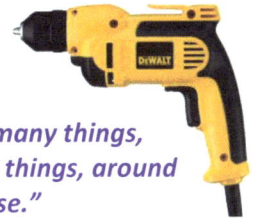

"I'm in many things, random things, around the house."

DC

Motor inside DVD players, to spin DVD.
Show in the multi-pole rotor part of a flat motor. That coil winding process is probably automated to reduce cost.
This rotor has 24 poles.

DVD player with motor inside
Motor needs to be flat and spin at a stable rate.

DC

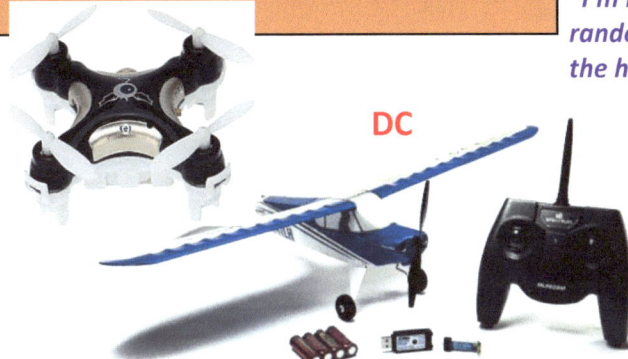

Remote Controlled electric airplane motor
Motors for airplanes are challenging because they need to be light but also high torque.
Thankfully, we have stronger magnets and we have higher energy density batteries, or more energy for the weight, using lithium ion batteries.

1960 — **1990** — **2000**

Electric air pump for bounce house, where motor spins an air compressor.
It only takes about 2 psi of gauge pressure to hold up a person, when their weight is spread around the area of their foot.

AC

Why use electricity to get people to exercise and burn energy?
Normally this treadmill is just an eyesore, collecting dust. Rarely, maybe after wishful new year resolutions in the winter, people burn some human calories on it.

AC

Treadmill for exercise, spun by motor.

Vibrator on a cell phone:
Tiny motor spins an unbalanced weight to cause a vibration

Electric motor spinning water pump inside coffee maker

Which gets more use, the DVD / Blu-ray player or the treadmill?

DC and AC Motors in Your House, Q and A

Electric motors are in your home and business and factory, with many different sizes and types. A motor spinning a computer disk can be 1cm size. An electric motor spinning the water pump in a coffee maker can be 3cm size, to make your morning cup of coffee. Electric motors doing heavier work like household fuel pumps, water pumps, or ceiling fans can be ½ foot in size or more.

Question 1: Did you search high (ceiling fans) and low (water pumps) for electric motors, or inside all your electronics, heat system or kitchen appliances?

- The kitchen has many electric motors. The refrigerator has a motor for pumping the refrigerant gas. The dishwasher has an electric motor to circulate and spray the water around. A Keurig coffee maker has a motor to pump the water through the water heater. The microwave oven has a motor for a fan to bounce microwaves in different directions inside the microwave, which avoids uncooked spots in the food without a microwave field.
- The living room probably has a Blu-ray reader for movies, which uses a tiny electric motor to spin the disks. Computers have motors to spin the magnetic hard drive and optical DVD drive.
- The garage may have a motor to raise and lower the garage door, the 'garage door opener'. If you live in a snowy area, there may be a starter motor on a snow blower. Tools like electric drills and electric saws all have motors.
- The play room might have motors for toy robots, like robot dogs or remote controlled RC cars.

Question 2: Are motors in the house typically DC or AC? Wall outlets are AC and power AC motors, and batteries or rectified voltage are DC and power DC motors.

- If the motors plug into the town grid, like the wall outlets or cables from the electrical breaker box, then those motors are AC motors. In the US, there is 120 Volts alternating current at 60 Hz. Those motors need to be AC, with both an electro-magnet for the stator and a parallel electro-magnet for the rotor, with plenty of iron cores for magnetic field strength.
- If the motors are plugged into a battery, for example in portable cordless power tools or inside electronics like computers, then the motors are most likely DC motors. DC motors have a permanent magnet for the stator, and the current only flows through the rotor. The motor also could be a universal AC motor working on DC current, with current flowing both through the rotor and the stator.

Spinning water pump inside coffee maker

AC electric motor to spin washer and dryer.
Motors spin the drum, pump the water, and spin the clothes.

AC motor to spin compressor inside the air conditioner
Motor pushes the coolant around.

AC motor to spin compressor in the refrigerator

DC motor with permanent magnet stator, powered from a battery
A permanent magnet inside the motor indicates that the motor is a dc motor.

DC motor with permanent magnet stator, inside DVD player

Let's imagine life without electric motors and magnetic fields. That includes power generators because generators use magnetic fields.
People over 100 years ago, and still today, use steam power, gas combustion power, and water wheel power, just to get shafts to spin and do something useful.

"I'm too practical to be replaced."

Substitutes for electric motors and electricity:
Hey, are there other ways we can have modern convenience without depending on electric motors or electricity? We don't want to abandon refrigeration, tools, cars, running water, and music. We just want to stretch our brains and see if there are other solutions to help here.

Well, there are other ways to make a shaft spin to get mechanical motion, like air pressure, combustion engines, and hot steam. This spinning shaft can make clothes washers work, or make a car move, or make a refrigerator work by moving the gas refrigerant around with a pump.

A spinning shaft probably can't make speakers from recorded music. So speakers, electronics and computers won't be able to run without electricity.

Gas driven compressor

Hand crank water pump, crank before your morning coffee

Glamping comfortable living, deep in the woods:
If you are away from the power grid deep in the forest, what would you do? How can people in the woods avoid motors and electricity, without the city power grid?
These people away from the grid probably would not want to avoid motors. Look at isolated wood cabins these days. These cabins have solar panels and power, or geothermal heating. People there still use electric drills and electric refrigerators. The cabin's solar panels can have a large bank of car batteries for energy storage when the sun is not shining.

Water Power

Do you want a combustion engine (stinky loud with exhaust) at your dish washer? Or even one right outside your house?

Imagine Life Without Electric Motors and Magnetic Fields

Do you want quiet convenience, where you just flip a switch and things turn on? Without electric motors, forget about it. Instead, we're talking historic solutions like steam, combustion, and water wheel power, just to get shafts to spin and do something useful.

Or even more historic, we could live in a hut, with a donkey to turn a turnstile, and with a hand crank to lift a water bucket out of a well.

"I'm too practical to be replaced."

Possible replacements to get things spinning, instead of electric motors

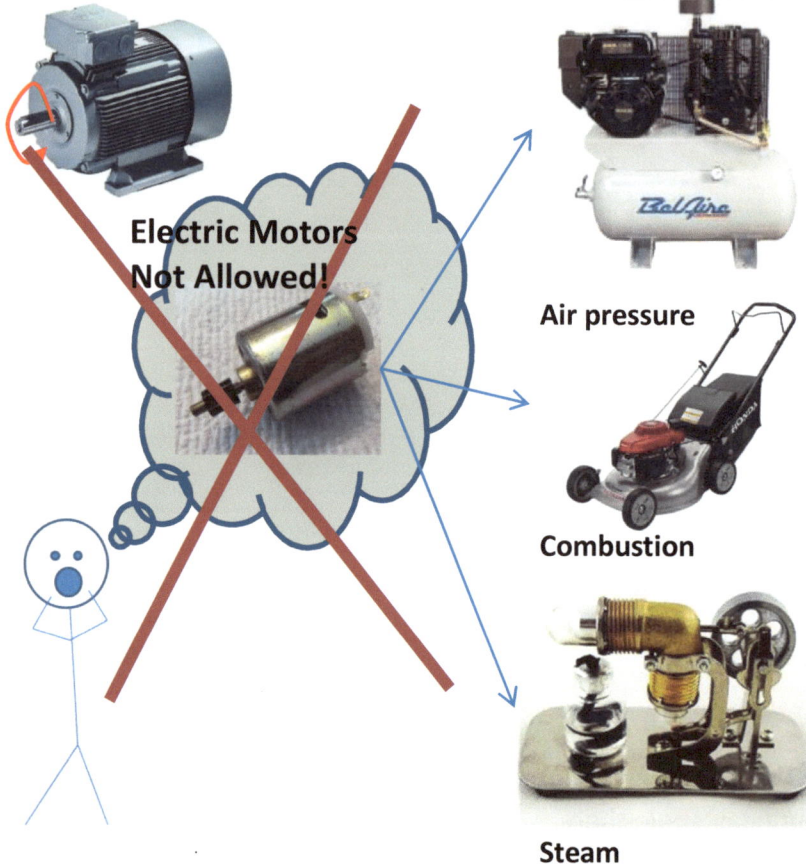

Electric Motors Not Allowed!

Air pressure

Combustion

Steam

All electric motors in house replaced by:

1. **Air motors, where the air pressure is powered by a combustion engine to get a spinning compressor shaft** (high pressure air, pneumatic)
   - Fans / washer/ dryer are converted to air tools, which are themselves powered by a diesel combustion engine to power a compressor
   - This alternative is already like all the high pressure hoses in machine shops

2. **Combustion engines to get spinning shaft (probably diesel because don't need spark plug high voltage, just use pressure)**
   - Use gasoline lines around the house to power combustion engines.
   - Without magnetic fields, gasoline engines could not use a high voltage coil to generate a spark. Instead of gasoline combustion engines, may need to depend on diesel which self ignites with a hot source and pressure, if can still pump fuel into the cylinder.
   - Lots of noise and pollution

3. **Steam engines to get spinning shaft, like 1800's**
   - Need fire to heat the steam, from oil, coal, wood, or natural gas. The steam turns the shaft.
   - Have a long startup time, about ½ hour.

Compressed air (red tank) from engine compressor on ground

This new imaginary life without magnets and without convenience is a made-up alternate reality. Of course, we had this life using steam engines before 1900, before combustion engines were invented, before people had electric power in their homes.

Maybe we don't absolutely require electric motors, but then we put up with noisy combustion engines, pneumatic motors, steam engines, and windmills. These alternatives like combustion and steam engines are loud. Yes, they can replace electric motors with lots of inconvenience. For air motors inside the house, there would need to be a combustion engine outside the house that compresses air, to pipe the high pressure air around the house. Air motors are already exploited for power around work shops, namely, the machine shops and car repair garages around the world.

Do you want a combustion engine (stinky loud with exhaust) at your dish washer? Or even one right outside your house?

Imagine: a Sorry Life Without Electric Motors

Again, do you want quiet convenience, where you just flip a switch and things turn on and do work for you? As a ridiculous exercise in backwards thinking, here are examples of how we could impose <u>less</u> convenience without electric motors, forget about it.

If you are thinking about alternatives, there's a reason there are no combustion engines outside your house, pumping up compressed gas and high pressure feed lines. The combustion engines need gas refills, are loud, require more maintenance, and are less efficient. Electric motors are more convenient.

"You'll rue the day you got rid of me!"

How to get my household labor-saving conveniences back, without electric motors?

Chores with electric motors:
Chores aren't so bad when you have the right tools.

Refrigerator with electric pump for circulator of coolant gas

Clothes washer and dryer with electric pump for water ejection and spinning clothes

Dishwasher with electric pump for water spray and ejection

I can't take it! I'm spoiled by convenience and less physical labor. Yes, I'm softer than 18th century man!

Chores without electric motors:
Oops, now the chores are more effort.

"My washer is not turning. Honey, did you remember to turn on the outside combustion engine to fill the high pressure air?"

All electric motors in the house replaced by historical options:

Fans / washer/ dryer are converted and powered by:
<u>Option 1</u>: Air tools (high pressure, pneumatic),
<u>Option 2</u>: Combustion engines
<u>Option 3</u>: Steam engines, like in the 1800s

Gas driven compressor   **Air hoses**   **Air tools**

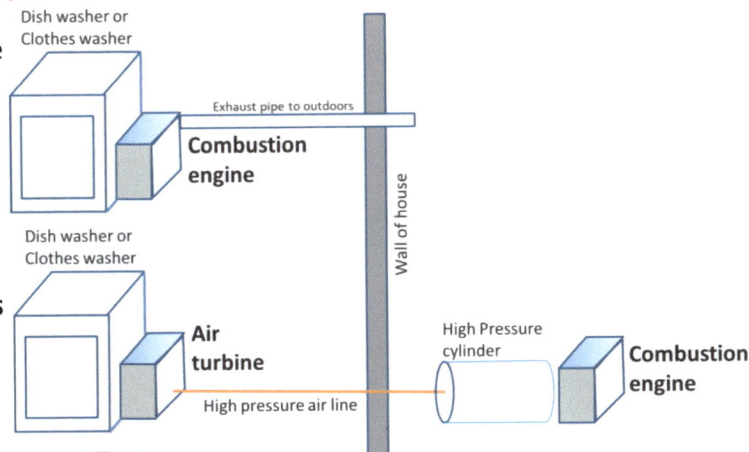

One power system for a typical house could be a combustion engine filling an air tank with compressed air. Then the compressed air is routed through the house, to spin all the fans and oil pumps using air turbines.

The combustion engine could be kept outside because it is so noisy and polluting. Combustion engines are very inefficient compared to having a large power plant.

Alternate reality: Generate compressed air outside the house however you can (combustion engines, steam engines), and have high pressure hoses going through the house. When a motor needs to spin for the dishwasher, then just blow the pressurized air through an air turbine.

List of Sacrifices Without Electric Motors and Magnetic Fields

Let's play a game where magnets or electro-magnets do not exist. People did not have electric motors in their house 100 years ago, so it is possible to survive without electric motors! To keep our standard of life, some items with electric motors just require a different engine: gas combustion, or high pressure steam. Some items even have crystal replacements.

Ice cellar, to trip down ladder before dinner

Hand crank to start combustion engine, out in the cold rain and mud

Household heating, hand deliver wood instead of pump oil

Headlights are real candles, a fire hazard during crashes

Compressor based refrigerators, powered by electricity, replaced by
- ice cellar
- piezoelectric refrigerators (like car portable refrigerators), much less efficient than compressor design refrigerators.
- propane flame creating circulation in refrigerators (like in RVs or campers)
- compressor powered by combustion or steam engine, or pneumatics (like at a machine shop) .

Electric fans to stop heat from collecting, replaced by
- huge open windows to let air in,
- fan with small combustion, steam, pneumatic engine (noisy, polluting)
- moving air using ionized air excited by electric fields.

Starting car engines, instead of a turn-key electric starter motor, started by
- turn a hand crank or rip cord (like lawn mowers), if have small combustion engine,
- rolling down hills, if have manual transmission,
- spring starter, which are available for boats.
- heat a steam engine, to then start the combustion engine.

Electric well pumps replaced by
- hand-powered or bicycle-powered water pumps for well (yes, we'll work for water, gladly)
- combustion, steam, pneumatic engine to power the pump.
- build more water towers above ground, to use pressure of gravity.

Oil furnace, with an electrically powered pump, replaced by
- wood, or coal (you load the wood or coal, like the 1800's)
- using gravity feed of oil, or
- replace oil furnace by natural gas (gas does not need pump, already pressurized in pipes).

Computer disk hard drives, instead of magnetic spinning disks, replaced by
- solid state hard drives.

Treadmill, replaced by
- go walking outdoors
- you'd be working so hard without electric motors, who needs or wants a treadmill?
- just go back to resistive wheel for drag on treadmill, or
- powered by small combustion, steam, pneumatic engine or gravity powered engine.

Lights in cars, replaced by
- burning gasoline in the lamps (mixing fire and gasoline if have a car crash, sounds like a bad idea).

Speakers for sound
- back to more live music and a loud horn section and banjo?
- piezo-electric speaker (vibrating crystal), with much less audio quality (less frequency range). Audiophiles would cringe. Coil speakers provide a full spectrum of sound, but vibrating crystals are quiet, and have narrow frequency and tinny sound.

Hand held fans, no A/C unit

"I'll call you 18th century man. I guess you'd rather spend your day hauling wood, hand pumping water, and sweltering in the heat!"

Hand crank water pump, crank before your morning coffee

You sweat more with no help from motors to turn your exercise machine

Tiny crystal speakers instead of electromagnet cones, forget good sound

No amplifiers for record player, so ask for quiet when music is playing

Humanity lived most of history without electricity and motors, but probably you don't want to.

Some Poor Substitutes, Usually Historical, for Electric Motors and Speakers

Without electric motors, do you want to use a hand pump every time you turn on the water faucet, or wash dishes, or take a bath?

For speakers and listening to music, you'll need live music from pianos, acoustic guitars and singers, instead of recordings. Or can you accept tinny sounds from vibrating crystals, like tiny ear-buds instead of electromagnetic cones?

Old Standbys to spin a shaft: Here are substitutes from history

New Innovations to spin a shaft and make sound: Here are substitutes from modern ideas

Yeah, we built a lot of character in the old days, and now you will too now that electric motors have been canceled!

Hand water pump

Manual sweaty, hard work pump: To fill a water tank on 2nd floor, some richer folks would have a combustion engine, or a steam engine, but most people would manually carry the water up.

Or people could live below a waterfall, or only buy a house near an artesian well, or get water flow like the Romans who built aqua ducts. People after all did survive for millions of years without electricity and electric motors.

Roman aqua ducts

Steam engine transportation

Without motors, use alternative power from heat for refrigerators:

Interestingly, a refrigerator can run on heat, to keep the refrigerant circulation going: just look at RV campers which use a propane flame to circulate the coolant. Also, a combustion engine could push refrigerant with a spinning pump (like hobbyist aircraft combustion engines).

Pre-industrial

Transportation

1800

1950

Laundry: Hand washing, Drying on line

Hand drill

Water Power

Horse and buggy transportation

Caveman: No change of clothes

Without electro magnets, use crystal piezo-electric speaker

Thermo-electric material: temperature difference to weakly generate electricity in cars (1% efficiency)

Going back to pre-industrial nature and a physical lifestyle is good, maybe for a month. But most people like their refrigerators, cars, speakers, and oil or electric heat controlled by a thermostat.

If No Electric Motor and No Magnetic Fields

In the world of make believe, if magnetic fields and electric motors did not exist…
- Put that electric guitar in the closet, and start practicing those trumpets and tubas, and dancing to brass bands.
- You'd fall asleep early, because you'd be 'plum tuckered out' from the day's chores.

"I can't live without music. I'd guess I need to play grand pianos and trumpets now, instead of listening to boom-boxes with batteries and speakers."

Sound with No Magnetic Forces (no conversion to electrical signal)

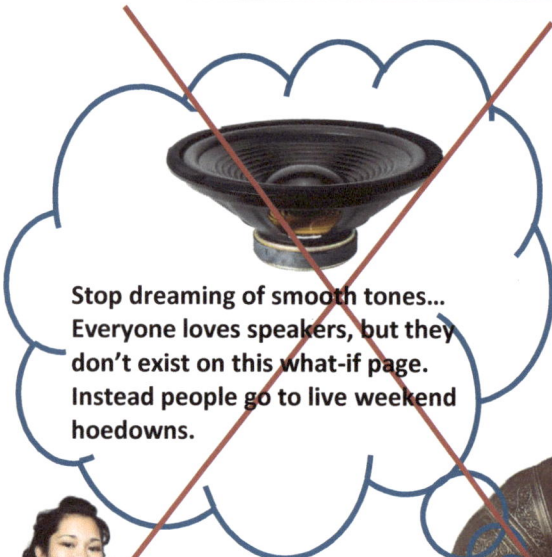

Household chores with No Electric Motor

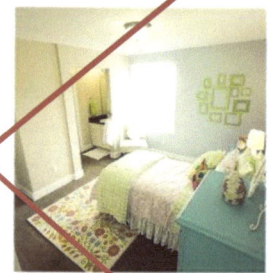

No more thrash grunge parties with electric guitars and distortion set on high.

If its too loud, you are NOT old! Rather, you must be modern and have electric motors and electric speakers.

Stop dreaming of smooth tones… Everyone loves speakers, but they don't exist on this what-if page. Instead people go to live weekend hoedowns.

Ahh, the good old days, when men were men who worked their hands to the bone, and music was a marching band.

Weak signal without amplification

No vacuuming carpets. **Probably people accepted more dirt in their home and hotel rooms.** On the plus side, exposure to dirt can reduce asthma.

No vacuum cleaners so no carpets, wall to wall.

Play your own loud music without electricity

Needle and grooves vibrate a drum, and the sound gets channeled up through a horn, without amplifiers.

Need a hand crank to wind up a spring to keep the turn-table spinning.

Manual Victrola turntable

Instead, like the olden days, people need to haul out rugs, and beat them on a clothes line. Please avoid wall to wall carpeting—it would never get cleaned.

Beating dirt out of rugs

No speakers, no vacuum cleaners! We might as well be back at the American Colonies.

120

Challenging Life Without Electric Motors, Q and A

Even without electric motors, we don't want to go back to the 1800s, with no electricity. But we could go to de-centralized electricity, 'off the grid', using local power generation.

Here is the fortunate real world where magnetic fields and electric motors DO exist...

Question 1: Do cars and trucks use electric motors, even though they burn gasoline?
- Yes, the starter motor turns over the combustion engine to start the engine.
- Cars also use electricity to provide sparks to the spark plugs.

Question 2: Do refrigerators use electric motors?
- Yes, to push the refrigerant around in the tubes. The gas is pushed through a closed cycle of expansion, compression, and radiation.
- If you listen to a refrigerator, you can hear the pump turn on to keep the cold temperature.

Question 3: Do tools like drills and saws use electric motors?
- Yes, to spin drill bits and saw blades, but you could also use combustion engines like chain saws, ignoring that something needs to generate a spark for the spark plugs.
- You could go back to the pre-industrial age and use water wheels, or hand saws and axes.

Question 4: Do speakers use magnetic fields or motors?
- Speakers use electro-magnets (the light wire wrapped around the cone), just like motors. Without electromagnets, you'd be back to playing the old standards of pianos, fiddles, loud banjos and upright basses, and trumpets in marching bands to get music.

Back to horse and buggy if magnetic fields and electric motors DO NOT exist...

People can live 'off the grid', like 200 years ago, with lots of physical labor:

...People truly lived without electric motors, without electric tools.

- Hand wash clothes
- Own a horse
- Tend to their own farm

'Off the grid' was everything   Hand washing

Hand drying

Horse and buggy

Hand drill

Farming with oxen, pulling a tilling blade

Modern 'Off the grid' with magnetic fields and motors, local and currently available

Here is how people would rather live 'off the grid' now, with the help of on-site local generators and still using electric power and magnetic fields:

- Solar power for battery powered tools, and for the refrigerator.
  ...still use electric motors
- Generators using fuel
  ...uses electric generator
- Hand wash clothes
- Own a truck
  ...still use electric starters
- Tend to their own farm
  ...probably use electric starters on tractors.

Solar cells keep batteries charged

Solar power for one house

Refrigerator runs off DC power from solar panels

Battery electric drills

In cars, electric motors once were, and still are, the starter motors, mostly just to turn over and start a combustion engine, as gas vapor is suctioned into the cylinders. Now electric motors are also becoming the engine, powered by a battery or powered by a combustion engine which turns a generator.

There has been quite a history of the competition between ways to power a car, at least up to 1920. Is it cheaper and more convenient to use electric batteries, or combustion engines, or steam engines?

The original competition between steam, electric, and gasoline power cars around 1900:

Is a steam engine practical for cars? After all, steam engines were powering trains across the country back then. The main choice very early on before 1900 was electric batteries or steam engines. Steam engines take a half hour to warm up, and they don't turn off for a quick half our stop, for example at the hardware store.

Back around 1900 there were no powerful combustion engines, and crude oil was not discovered and drilled with abandon. So cars used batteries and weak electric motors, charged by local power plants using coal or hydroelectric power. Electric motors were just getting mass produced to power the car axles. Power plants to charge led batteries were appearing in cities, powered by the coal or hydroelectric power. So electric cars were the next big thing. Local power plants were mostly powering lights and trolleys.

But then crude oil was discovered, using drilled wells, in western Pennsylvania and in Texas in 1880s. Mechanics worked out more reliable combustion engines. Then gasoline ruled fro the next 100 years up until the current day.

Electric motors are still used on all combustion engine cars, mostly as starter motors. No one wants to hand crank and turn over an engine to get it started. That might work for a small engine with 20 HP or less, or a small motorcycle engine with a kick start, but not for typical combustion engine sizes these days.

Now fully electric cars are competing with combustion engines. Electric motors are stronger because magnets are stronger. Batteries have new technology like Li-ion batteries, not just iron-oxide, and people care more about pollution.

Electric motors need to withstand the weather, and the motor needs to be shielded from the dirt and salt on the road. Some metals are better than others for the brushes, like carbon, for wear and exposure to dirt and salts during winters.

DC or AC form of electric power?

Electric motors for EV cars can be DC or AC. Which is better? DC motors have higher low end torque due to two reasons: one, because the dc permanent magnet fields can be higher, and, two, because in dc motors there is no iron magnetization loss which happens in AC motors as the stator iron field would keep changing directions with the applied current.

Some EV cars use DC motors for the low end torque, but then the wires need to be very thick because at low voltage a lot of current is going to the motor. DC motors can also be brushless, which is more efficient and longer lasting. Some EV cars use AC motors to use higher voltages and thinner wires.

Light rail used low voltage dc motors because the power lines or electrified rails are exposed, and we don't want to electrify anyone with a high voltage.

Steam engine transportation

Historic electric tram, or trolley, Germany, started in 1881

One of the first electric cars:

Regular combustion cars: Electric starter motor to get combustion engine started

Fully Electric cars: Chevy Bolt Electric

Electric starter motors are in all combustion engine cars. Electric motors in hybrid and all electric cars are the actual engine, which power the shaft and the wheels.

People Movers and Cars

In cars, electric motors once were, and still are, the starter motors, mostly just to turn over the shaft and start a combustion engine, as gas vapor is suctioned into the cylinders. Now electric motors are also becoming the engine, powered by a battery or powered by a combustion engine which turns a generator.

"Gotta start that car's heavy combustion engine turning somehow, using a starter motor.
For electric cars, we need an even more powerful electric motor to be the engine."

Starting cars:
Electric starter motors to turn over combustion engines in cars

Cars became more practical and less a maintenance burden when a key does all the work: turning the key starts the car, by engaging the starter motor with the battery. Larger engines, especially in trucks, were now possible.

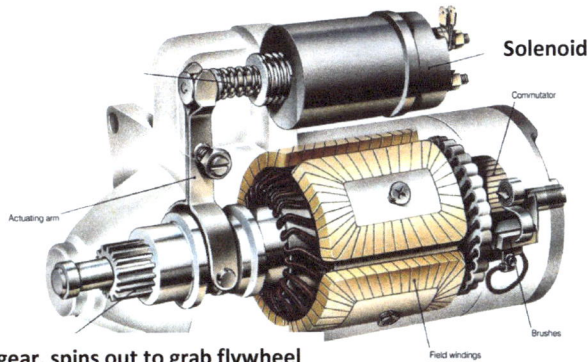

Solenoid
Commutator
Actuating arm
Brushes
Field windings

Pinion gear, spins out to grab flywheel

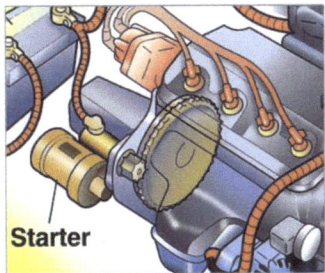

Starter

Starter motor, turned on by solenoid

Starter motor turns flywheel teeth

The 'starter motor' grabs the flywheel teeth and turns the combustion engine and pistons when turn the car key, and allows gasoline vapor to enter and to explode and take over from there. Use the energy stored in the battery to turn the electric motor.

Power car wheels with electric drive motors:

Hybrid electric / gasoline cars

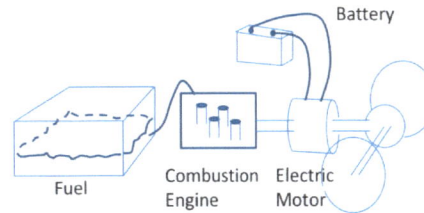

All electric cars

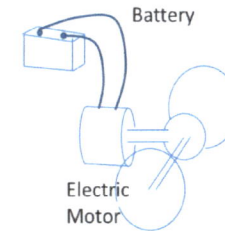

Battery
Fuel
Combustion Engine
Electric Motor

Battery
Electric Motor

For hybrid cars, both combustion engine and electric motor can power the shaft, where the electric motor has two functions, a motor to move or a generator to slow down.

For pure electric cars, the electric motor powers the shaft to move and regenerates energy from the shaft when slowing down. Some EV cars have a paddle on the steering wheel to really engage regenerative breaking instead of heating the brake pads.
All electric cars don't need a transmission because the electric motor has full torque and any spin speed.

Electric starter motors are in all combustion engine cars, just to initially turn the shaft. Electric motors in hybrid and all electric cars are the actual engine, which power the shaft and the wheels.

Combustion Engines Need Starter Motors

Starting cars: This is what your car key does. The small electric starter motor in cars turns over the shaft in the large combustion engine to cause the fuel vapor to get sucked in and exploded. Cars became more practical and less of a maintenance burden when, instead of a hand crank that turns the motor over, a key does all the work: turning the key starts the car by running a huge current through the starter motor by engaging a solenoid switch. Larger heavier engines, especially trucks with even larger motors, were now possible. The electric starter motor has a strong torque.

The small gear spins out and connects with flywheel teeth when motor turns.

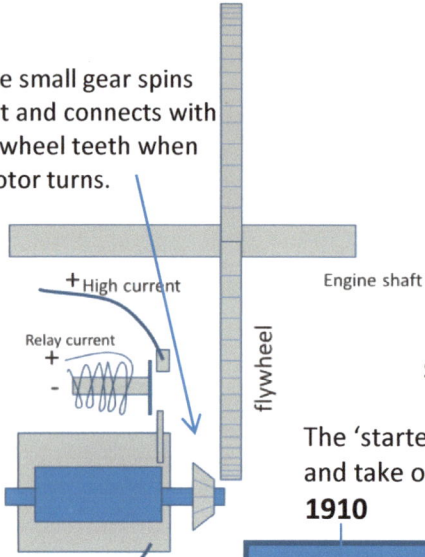

+ High current

Relay current

Engine shaft

flywheel

Starter

Starter motor turns flywheel teeth

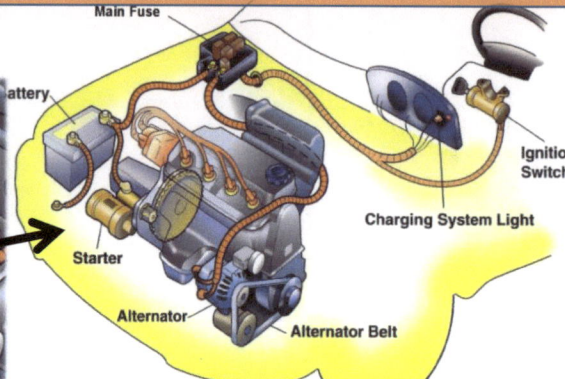

Main Fuse

Battery

Ignition Switch

Charging System Light

Starter

Alternator

Alternator Belt

Car use electric motors to turn-over engine, and use generators to create electricity

"Lighter engines like small motorcycles or lawnmowers don't need a starter motor, and the engine can be turned over with muscle pulling a pull cord or with legs kicking a kick start."

The 'starter motor' turns the combustion engine and pistons when turn the car key, and allows gasoline to explode and take over from there. The starter motor uses the energy stored in the battery to turn the electric motor.

1910   **1930**

Stator motor gear engaging with the flywheel:

A combustion engine can not start itself. It needs the pistons to be moving, rotated by the electric starter motor, to suck in the air and fuel vapor. Due to the large radius of the flywheel, the starter motor spins many times to get the flywheel and engine to turn over once. This allows a weaker fast motor the strength to turn the engine.

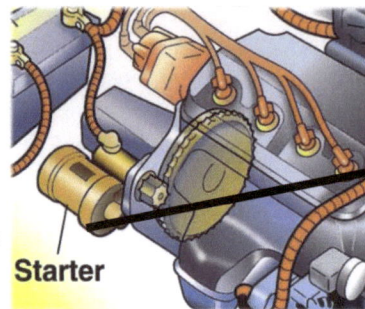

1910, Hand crank

Hand crank to start a car:
The reason trucks were successful in WW1 is that a starter motor allowed the heavy engines to turn over to start.

Improved starter motor to turn teeth of flywheel:
An ignition key triggers the solenoid to start starter motor

From 1930's, Starter motor in car

Cramped location of starter motor:
Near the bottom of the engine, near flywheel.

For light engines, people can still just use a pull cord.

Starter motor gets the pistons moving, so gas and air can get sucked inside to explode. The starter motor is then dis-engaged with no more current so the small gear (pinion) against the flywheel pulls back.

Exercise: Listen to the starter motor engage, and find the starter near the engine

Trucks for WW1 ... a Little History

The battles during WW1 depended on supplies getting to the front line. Larger trucks and prototype tanks want electric starter motors to turn the engine, not hand cranks.

The US government sent out a requisition for trucks to support the war effort in WW1. The US had a large supply of gasoline and did not intend to get stuck in the trench warfare, where neither side has the resources to break through the other sides trench. The US brought lots of heavy armor, including trucks and prototype tanks.

Prior to 1915, all cars had crank starts to get the combustion engine turning so the gas explosions could take over. People just don't have the strength to crank larger heavy engines, so larger trucks were not possible with hand cranks.
The 'Liberty' truck had a top speed of around 15mph.

The other technology to get supplies and soldiers to the battlefield was of course horses. The truck did not instantly replace horses, because horses do not require a supply system of gasoline. In fact, horses were used to some extent by armies all the way up to WW2.

The unintended effect of all these 'Liberty' trucks was that, after the war, returning vets and civilians could get these trucks and deliver milk and other supplies around cities.

Liberty Truck: before and during the war, the engine was turned over and started initially with a hand crank

Liberty Truck: during the war with starter motor, the 3–5 ton truck began in mid-1917

Liberty Truck: after the war, surplus trucks were used for milk delivery.
Milk could also be refrigerated using air conditions, which themselves are enabled using electric motors.

The improvements to trucks for the war effort, such as starter motors, pay off for civilians after the war by enabling larger cars and engines without manual cranks, and making transportation more convenient.

Combustion Engine Versus Electric Motor

Powering car wheels: What is a smoother power source for a car: a combustion engine or an electric motor? A combustion engine has violent explosions per turn equal to the half the number of cylinders (for a 4 stroke engine), offset along the crank shaft. That is, a 4 cylinder engine has only 2 explosions per turn. An electric motor, in contrast, can have 10 or 20 coils, each coil providing their two instants of push per turn, all centered at the same location along the shaft. Yes, an electric motor is smoother.

Combustion Engine, with 2 pushes per turn

Combustion engines need air and a mist of fuel to be sucked into a piston. A spark plug creates a little spark, and the gas explodes, pushing the piston down. The piston pushes on the crank shaft and rotates the shaft. Heavy flywheels are necessary to smooth out the spin between explosions.

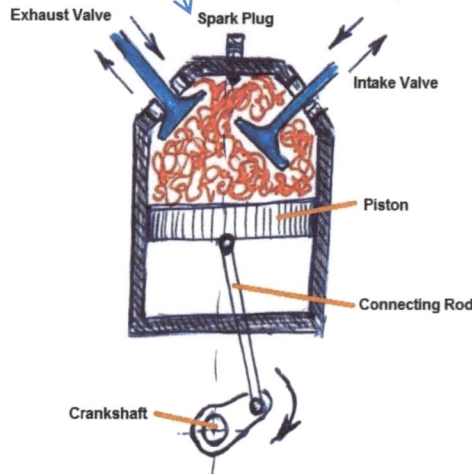

Exhaust Valve, Spark Plug, Intake Valve, Piston, Connecting Rod, Crankshaft

Intake valve opens up as cylinder going down, letting in fuel and air.

V8 engine block
-4 explosions per turn, for a V8 four stroke engine

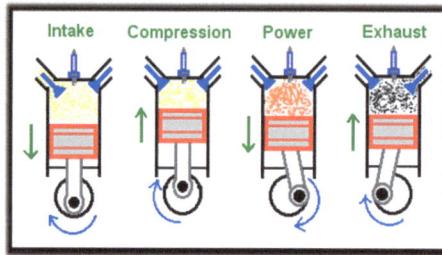

Intake   Compression   Power   Exhaust
4 stages of a 4-stroke engine

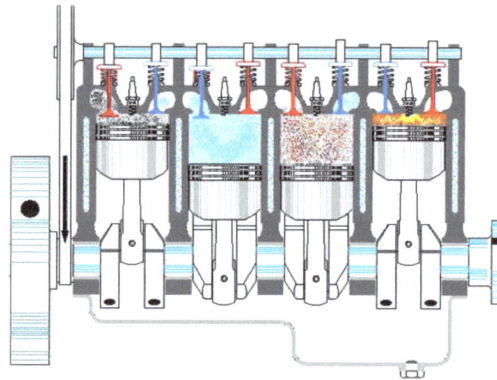

4 cylinders of a 4-stroke engine, each cylinder at a different stroke or stage

Let's look at a 4 cylinder engine. Each of the cylinders has explosions of fuel at synchronized times, pushing the pistons down smoothly and keeping the heavy flywheel spinning. There are only 2 explosions per turn because each of the 4 cylinders only has an explosion every other cycle, for a 4-stroke engine.

Electric Motor, with many pushes per turn

Electric motors do not need air, oxygen or a fuel. Instead, they need an electric current flowing through the coils of the rotor. Of course, this electric current needs to come from some place, like a power plant and the power grid on your street, or even a generator in the car.

A pure EV car needs a source of current in the car, typically by burning coal at a distant power plant at night and charging up the batteries.

Huge car batteries are charged up at night to power the car during day trips.

Anything can create heat and electricity in the power plant, like burning coal, burning gas, or nuclear power.

End use of the current, a motor.

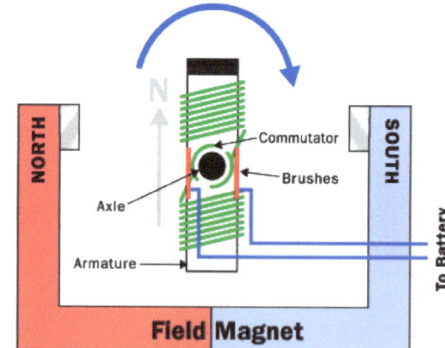

NORTH   SOUTH   Commutator   Brushes   Axle   Armature   To Battery   **Field Magnet**
©2001 HowStuffWorks

Example DC motor driving the wheels
-number of pushes per turn equals the number of pads in the electric rotor.

There is a competitive battle between combustion engines and electric motors for transportation. Electric motors actually have more torque and can beat high end sport cars at acceleration, although electric motors in cars rely on batteries which have less energy per pound than gasoline. Electric motor dirt bikes can also beat combustion engine dirt bikes in races.

Prototype to Modern Electric Cars: 1830

History of electric cars with battery:
Believe it or not, over 100 years ago around 1900, before gasoline was drilled out of the ground with abandon, battery power cars had the lead over combustion engine cars. Battery powered cars, after a few decades, then lost the battle with combustion engines, due to more drilling and newfound cheap gas and weak lead-acid batteries. Maybe now, after the 2020s, battery powered cars will have the last laugh, using stronger motors with stronger magnets and using higher energy density batteries like lithium ion, and batteries with a solid electrolyte instead of a liquid electrolyte.
There is a lot of current research and production of higher energy batteries and chemistry, because batteries have much less energy per weight than gasoline. The limited energy of batteries is the big limitation of electric cars.

"Electric cars started way, way back, more than 100 years ago. Before 1900, electric cars didn't have much competition for a decade from gasoline cars until more crude oil got discovered in Texas around 1902.
Now better batteries and electric cars are making a comeback. "

One of the first electric cars: The first successful electric car in the United States was built in Des Moines, Iowa by William Morrison in 1891. The four-horsepower vehicle had a top speed of 20 mph and could carry up to 12 passengers. It was powered by 24 battery cells that were stored under the seats and needed recharging every 50 miles.
Modern cars, both combustion and electric, have more than 100 horsepower.

Ayrton Perry Electric Trike

Nissan Leaf battery under carriage

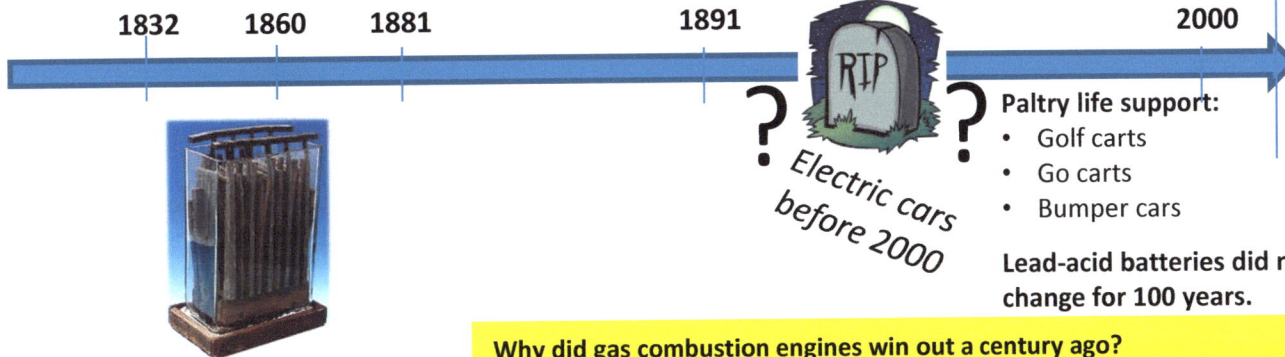

2010
Electric cars come back to life due to new types of batteries, mostly rechargeable lithium ion, with faster charge times and higher energy per weight.

2040?
Electric car powered by hydrogen fuel cell:
Hydrogen fuel cells have more energy than batteries but hydrogen is readily available now.
Need hydrogen re-fueling stations and practical hydrogen production.

1832   1860   1881   1891   2000

? Electric cars before 2000 **?**

Paltry life support:
• Golf carts
• Go carts
• Bumper cars

Lead-acid batteries did not change for 100 years.

Plante's lead-acid battery using alternating lead and lead dioxide plates and an acid (circa 1860). It was initially used to power the lights in train carriages while stopped at a station

http://www.electricvehiclesnews.com/ History/historyearlyIII.htm

Why did gas combustion engines win out a century ago?
The previous decline of the electric vehicle was brought about by several major developments:
•By the 1920s, America had a better system of roads that now connected cities, bringing with it the need for longer-range vehicles.
•The discovery of Texas crude oil, from 1902, reduced the price of gasoline so that it was affordable to the average consumer.
•The invention of the electric starter by Charles Kettering in 1912 eliminated the need for the hand crank, and allows heavier engines during WW1 which only an electric motor can turn over.
•The initiation of mass production of internal combustion engine vehicles by Henry Ford made these vehicles widely available and affordable in the $500 to $1,000 price range ($20k equivalent today). By contrast, the price of the less efficiently produced electric vehicles continued to rise. In 1912, an electric roadster sold for $1,750, while a gasoline car sold for $650.

Electric cars have been around as long as combustion engines, but batteries have been too weak up until now.

Combustion and Electric Cars, the New Rivalry

Hybrid electric cars are more efficient than pure gasoline cars, because the motion of the car can recharge the batteries with regenerative braking. Also, there is no warm-up time, during which a cold combustion engine is very inefficient. When the car is cold, the cold fuel doesn't vaporize and burn as well, transmission oil is not warmed up, and the catalytic converter needs to get warmed up to remove pollutants. Full electric cars are more efficient than gasoline cars because large power plants are more efficient than small combustion engines, because large power plants can use alternative fuels, and because of regenerative braking in the car itself.

Electric cars mean we don't need to depend so much on importing gasoline, and we can use our own power plants with local and/or green energy sources to help charge the car, usually over night.

Although electric batteries take longer to charge than a gasoline pump, there is a fun advantage to electric motors. There is more torque and the car accelerates much faster so the car is more sporty.

However, electric cars and gasoline cars take about the same money to drive the same distance. Even though electric motors are just more efficient than combustion engines, electricity is just more expensive from the power grid than gasoline per gallon. After all, electricity typically came from some fossil fuel too. After the electric energy gets put in the battery, most of that energy goes into forward motion. But it takes burning some fuel to get the electricity in the first place. Home solar panels or cheaper 'green energy' like hydroelectric or wind, will break this comparison, fortunately.

Electric cars can be especially expensive to charge at away-from-home charging stations.

Point 1: Electric cars recharge batteries while braking, which is great for city driving.
- Yes, the spinning wheels can be used to charge the battery, instead of all the energy of motion going into heating the brake pads. Just let the spinning wheel and shaft spin the motor, and a voltage is generated. The motor acts as a generator.
- However, most of the braking at a stop light is done with brake pads, because that is immediate and safe.

Point 2: Fully electric and hybrid cars flourish in the city with more stop and go driving.
- Yes, the batteries can accelerate the car from stops with high torque, and the energy of motion can be used to re-charge the battery while slowing down for a stop light. Gasoline cars are very inefficient and polluting while accelerating from a stop, and even less efficient when idling at the stop light which has zero efficiency.
- The old adage that gasoline cars are less efficient in the city (for example, 15mph in the city, and 25mph on the highway) is flipped on its head for hybrid cars. Hybrid cars are most justified in city driving. On the highway, both types of cars use gasoline engines for the long steady driving after the battery is drained, so there is no difference in efficiency unless the hybrid has a smaller engine.

Point 3: Hybrid cars do not use their batteries much for long trips on the highway.
- For hybrid cars, the few batteries are typically only good to power the car for 10 minutes at most, because there are only a few batteries. Most of the horsepower on the highway comes from the remaining long-range gasoline. If you want more electric power and more batteries, then get a full electric car.
- Plug-in hybrid cars can have more batteries for maybe 30 minutes of driving before the combustion engine is turned on.

Point 4: Electric cars can have the same horsepower as combustion cars.
- It all depends on how many batteries are used, and the current limit of each battery. High end electric sports cars, loaded with batteries, have more horsepower than their 8 cylinder counterparts, because a huge current can be drawn from the batteries.
- High end EVs or any EVs also do not need to be tuned to different air temperatures or pressures at different altitudes because no air is used. In the odd situation where you drive over a mountain, you'll keep the same high performance in an electric car, but not with a combustion engine car.

Point 5: Recharging stations are getting built to recharge the battery of a fully electric car.
- The car needs to be plugged in. Most trips are under 30 miles, and the car can be charged up at night using a regular 120V power supply in the house. Some destinations (work facilities, museums) offer charging stations while the car is parked.
- Long trips are an issue, with 'range anxiety'. There are expected to be many electric car charging stations getting built, but charging a battery still takes a lot longer than filling gasoline into a gasoline car. Longer trips, more than 150 miles, need to be planned out with a break for re-charging the batteries, at a pre-planned charging location. Apps exist on the car and cell phones to map out the route with recharging stations.
- Charging costs at away-from-home charging stations are much more costly than at-home charging.
- Some charging units are slow charge, meaning 120 volts, and some are faster charge, with higher voltage.

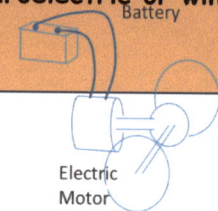

Electric motor can be both a motor which powers the wheels to move or a generator which slows car down and re-charges.

Many more batteries for full EV car, 100s of miles, compared to a hybrid car, for less than 10 miles.

EV Tesla Roadster   **EV Porsche Taycan**

Trucks can take the weight of many batteries without raising the total weight much when the trailer is fully loaded.

Electric Motors You Use:
Maybe You've got an Electric Car

Powering car wheels with electricity: High power electric motors are powering our hybrid and fully electric cars. Smaller, stronger motors are using stronger permanent magnets.

Regardless of the power train, cars are still cars and the shape is still the shape. The engine, or battery, that powers the car forward does not determine the outside shape. The outside shape is mostly determined by the need for low air drag, stability (4 wheels), safety tests using a crumple zone and roll cage, and a typical number of seats.

"I am a hybrid car, and I am a city taxi dream. My electric motor adds to the power of the combustion engine, and I recharge and save some energy in my battery when slowing down for all the many red lights."

Hybrid vehicle examples, with both combustion and electric engines

EV all electric vehicle examples, with electric engines only

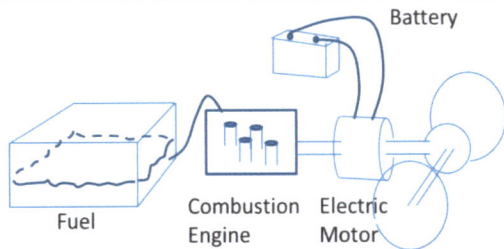

Battery

Fuel Combustion Electric
        Engine   Motor

Hybrid power train, where combustion engine and electric motor both can turn the drive shaft

Battery

Electric
Motor

All electric EV power train, where only an electric motor turns the drive shaft

Motors and generator behind the wheels, to re-capture energy

Toyota Prius and Camry Hybrid Electrics

Hybrid cars look just like combustion cars from the outside ...no big deal.

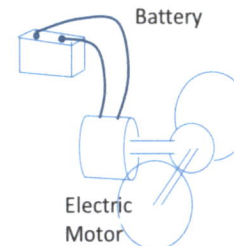

Ford, Nissan, and most other car brands make hybrids.

Honda Accord Hybrid Electric

All electric cars also look just like combustion cars from the outside ...no big deal.

Ford Focus Electric

Tesla Electric

Chevy Bolt Electric

Charging station (need credit card)

Hybrid cars use large electric motors to power wheels. These cars can use either DC or AC electric motors.

These type of hybrid cars, with smaller gasoline engines, batteries, and electric motors powering the wheels, are becoming more common. The electric motors run off the batteries. The electric motors switch roles and act as electric generators during coasting (especially going down hill) or just before braking: the energy of the moving car is used to recharge the batteries. The spinning wheels turn the motor rotor, which generates electricity. This re-charging, or conversion of motion into electrical energy, drags down the speed of the car somewhat, not fully. Brake pads are still needed.

Strong, compact electric motors are a necessity for hybrid cars and pure all electric cars.

Hybrid and EV Chassis

In EV cars, electric motors are the engine, powered by a battery. For hybrid electric cars, the wheels are powered by a combustion engine with gas and an electric motor with a battery. For pure EV cars, the wheels are powered by only electric motors with a large battery pack.

Electric EV cars are very stable against rolling because the huge battery weight in on the bottom.

"For electric cars, we need a powerful electric motor to be the engine.
We also need a large battery pack to get range between re-charging"

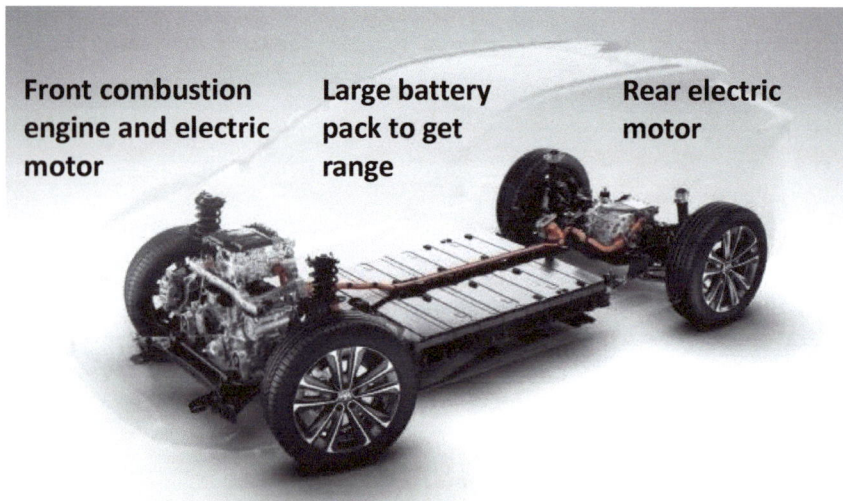

Front combustion engine and electric motor

Large battery pack to get range

Rear electric motor

A Toyota hybrid electric chassis (the bZ4X)

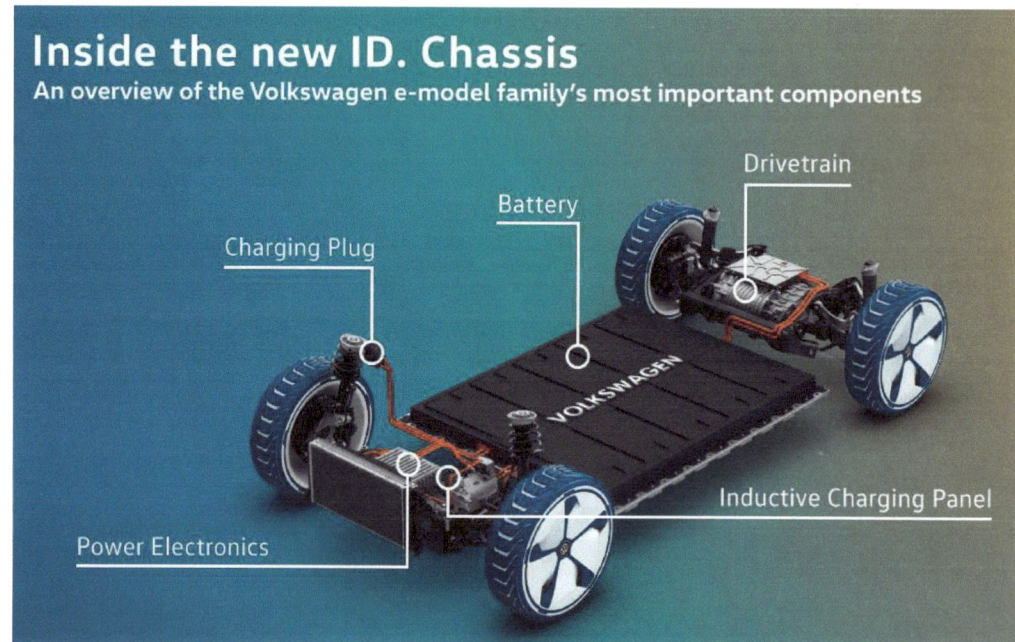

Inside the new ID. Chassis
An overview of the Volkswagen e-model family's most important components

Drivetrain

Battery

Charging Plug

Inductive Charging Panel

Power Electronics

A EV electric chassis

Electric motors in hybrid and all electric cars are the actual engine, which power the shaft and the wheels. The chassis can be common between different types of cars within a car brand.

Ground-swell of Modern Electric Vehicles and Batteries

Powering car wheels with electricity: Why go electric for cars? As higher energy batteries get cheaper and charge up more quickly, the electric vehicle can become the dominant car. Compared to combustion engines, electric motors have more torque, cost less to drive, have less parts, pollute less, and we don't depend on importing gasoline from imported oil.

Hybrid electric/gasoline vehicle, with small battery

Hybrid cars:
Hybrid cars use both high voltage batteries and large electric motors to power wheels.
Typical hybrids have a small bank of car batteries, with about 2 kw-hour energy storage, or a few minutes or miles of operation.

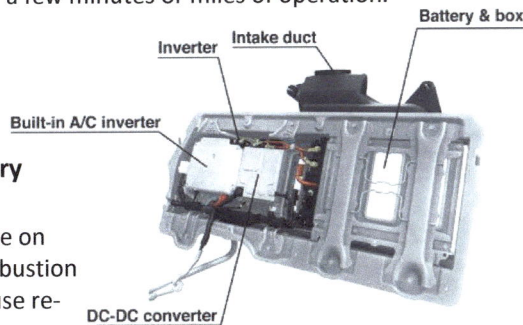

Smaller shared battery pack:
A full hybrid can operate on batteries alone, or combustion engine alone, and will use re-generative braking to recover some energy from braking (in combination with brake pads).

Intake duct
Inverter
Battery & box
Built-in A/C inverter
DC-DC converter

Mass produced hybrid unit: converts 12V to higher voltages, controls when batteries are re-charged and drawing power.

EV all electric vehicle, with large battery

Batteries under the trunk

Mass production of batteries

EV cars:
Full electric cars have no combustion engine, and use full electric to power the wheels. Instead of gas stations, you'll need charging stations and a ½ hour or more during a 'quick' charge.
Without home charging, pure EV cars are more of a nuisance.

Large battery capability:
Large bank of car batteries: costly, heavy, powerful, with about 100 kw-hour energy storage.
1 kw-hour is about 1 HP-hour, so a 50 HP car can drive about 2 hours with the large bank of car batteries.

Q: What kind of car does an electrician drive?
A: A Volts-wagon

"Electric cars have gone a long way, baby!
Now electric motors and batteries are both more powerful."

Ford Focus, All-Electric
Other examples electric cars:
- Tesla Model S,
- Chevy Bolt.

Hybrid Tech whiz-bang! Electric vehicles charge batteries when coasting and drain the battery when accelerating:
The car monitors where the power is coming from and going: Coasting, and re-charging the battery.
The dashboard images tell the driver when the hybrid engine or battery is powering the car, or when the battery is getting re-charged.

Toyota Hybrid Rav4

Energy Monitor 1:45
Electric Motor
Trip Information   Past Record

Charging while coasting: Hybrid, with spinning wheels re-charging the battery while coasting.

Energy Monitor 1:45
Electric Motor

Drawing electric power while accelerating: using battery and gasoline

---

- **Electric vehicles have natural advantages: re-charging battery or re-generation while slowing down or braking, and using the power grid with efficient power plants or home solar panels for energy instead of inefficient gasoline engines.**
- **Gasoline vehicles also have two huge natural advantages over batteries, which is why gasoline cars have been on the roads for a century: Gasoline has much more energy density of fuel so cars can travel farther before needing to refuel. Gasoline can be refueled in minutes at a gas station.**

131

https://en.wikipedia.org/wiki/Hybrid_vehicle_drivetrain

Competitive Fuels for People Movers and Cars

To get a car to move, we need some energy source. We current have liquid fuels, electric batteries, and hydrogen gas.

"There are more car fuel options than ever. Let's see which type gets the lowest cost, best efficiency, and best reliability."

Motor Types and Fuel Efficiency:

Power Source 1: Gasoline
- Gasoline fill-ups take about $1500 per year in fuel gasoline cost per car with regular miles.
- Gasoline is dominant now because gasoline has high energy and can be refueled in minutes.

Power Source 2: Diesel
- Diesel fuel explosions require a heavier engine to get higher compression, but diesel is great for trucks and has the advantage of better efficiency because the fuel explodes at a higher pressure and temperature.

Power Source 3: Hybrid electric
- A combined combustion engine and element motor allow the engine to always spin at its most efficient speed.
- This hybrid electric fills up just as a gasoline car. It refuels with gasoline.
- This hybrid electric is a good kick-around-town driving, with regeneration in stop-and-go-traffic and no warm up.
- There is good acceleration due to the immediate torque of the electric motor.

Power Source 4: Full electric
- Full electric battery power also cost about $1500 per year in charging cost using city grid due to expensive electric power.
- The battery loses 2or 3% of driving range per year, and could be replaced after 10 years
- Power for driving comes from using power plants and better efficiency, not gasoline.

Power Source 5: Fuel cell with hydrogen
- Hydrogen gas may be in the far future hydrogen fuel cells will power the electric motors, with no batteries. Hydrogen can have higher energy per weight, compared to batteries. But right now hydrogen is kept as a gas.

Technology shake-down of gasoline, diesel, electric, and fuel cell:

Which technology will rule in 10 years?
Diesel, Gasoline, Hybrid, All-Electric, Fuel cell (hydrogen) electric? It is not so obvious, when comparing the combined city and highway miles per gallon, or cost per mile.

- Ford Fusion **hybrid** sedan: **42 mpg**
- Ford Fusion **gasoline** sedan: **25 mpg**
- VW Jetta **diesel** 2014:   35 mpg
- VW Jetta **gasoline** 2018:   32 mpg
- Peugeot **gasoline** 3008 2018:   42 mpg
- Peugeot **diesel** 3008 2018:   56 mpg
- **All electric** Bolt: **250 miles** on a charge (30 mpg cost)
- Toyota Mirai fuel cell   400 miles on a fill

Companies believe that hybrid electric is the short term solution, due to the huge gasoline infrastructure like drilling and gas stations and due to the high energy density of liquid fuels. Companies also believe that batteries will improve in cost and energy density, so EV cars are a good bet. Companies say that hydrogen might be the solution over the horizon.

Combustion versus Electric Cars:
Money, Money, Nothing Comes for Free

Electric cars and gasoline cars take about the same money to drive the same distance, but it really depends on the cost of the local electricity. Electricity is just more expensive from the power grid than gasoline per gallon, for the same amount of energy. After all, most electricity originally came from some fossil fuel too. Electric car costs depend on power plant costs and efficiency and power line charges.

The cost benefit of a pure EV car is debatable. The decision to get an EV car is not purely a cost one. Maybe energy independence from oil and geo-politics motivates you. Maybe pollution from fossil fuels motivates you. Maybe the thrill of acceleration motivates you, without paying sport car prices.

Energy Cost to drive some distance:

For gasoline, after the gasoline is put in the tank, only about 30% of the heat generated from the combustion actually goes into the motion of the car. Combustion engines have a theoretical limit to efficiency and there are also many moving parts which have friction which further reduces the efficiency. That poor efficiency does open the door to alternative energy sources, like batteries.

For EV cars, after the electric energy gets put in the battery, most of that energy goes into forward motion. But it takes burning some fuel to get the electricity in the first place. Home solar panels will break this comparison, fortunately, but that requires a large upfront cost to install the panels and connect to the circuit breakers.

If someone makes a car purchase decision purely based on fuel cost, then the best candidate is a hybrid car that gets 50 mpg, not an EV car. If you already have solar panels or really cheap electricity, then the best decision could be an EV car.

Car Maintenance Cost over the lifetime of the car:

Another factor is car maintenance. A gasoline car has more annual repair because there are many more moving parts. A combustion engine has many more parts and more annual maintenance. An EV car has a replacement battery after 10 years or more. An EV car needs new batteries at 10 to 20 year after losing 2 to 3% of range every year.

What does MPG equivalent, or MPGe, mean, for an EV car?

- MPGe does not mean that the car travels 2 or 3 times farther for the same money, when comparing a gasoline car that gets 30 mpg to an EV car that gets 100 mpg-equivalent.
- MPGe is an energy comparison, not a cost comparison. An MPGe of 2 or 3 times better for electricity means an electric car travels farther for the same energy in the car. However, the energy cost 2 or 3 times more, so the costs break even.
  - Gasoline has a large energy of 33 kW-hours per gallon.

The decision to get an EV car is not purely a cost one. Maybe energy independence motivates you. Maybe pollution motivates you.

Combustion versus Electric Cars:
Money, Money, Nothing Comes for Free

Electric cars and gasoline cars both take about the $1 per 10 miles, no matter if gasoline or electric.

Combustion engine cost per mile:
$1 per 10 miles, or about $1000 per typical year

EV car cost per mile, using electricity from the grid:
$1 per 10 miles, or about $1000 per typical year

EV car cost per mile, using local solar panels: Free, but only after pay for solar panels. The solar panels will cost about $500 per year. When average the cost of the solar panels over 20 year lifespan, that is about $1 per 10 miles.

Combustion engines are not efficient.
Exploding gasoline vapor in the internal combustion engine: 30% efficient.
Another 20 to 30% or the energy is lost in friction of moving parts.

Burning fuel to get electricity

Maintenance costs for nuclear power plant

Where is the cost?
Installation and replacement of solar panels.
Also, possible tree removal and possible breaker panel upgrade.

Where is the inefficiency?
The combustion engine

Where is the inefficiency?
The power plant and transmission lines.

Batteries and electric motors are efficient, but getting the energy to the battery from power plants is not efficient.

Install solar panels to get electricity … this costs money.

Batteries and electric motors are efficient, after the energy gets to it.

There are 33kW-hours of heat from a gallon of gasoline, but only 30% of this heat energy of the gasoline goes into forward motion of the car. A typical car gets 30 mpg.

Let's do the math for cost per 10 miles:
- The price per gallon of gasoline is typically $3.
- The cost per mile is about ($3/gallon)/(30mpg) = $0.1 per mile, or $1 per 10 miles.

If a combustion engine car travels 10k miles per year, that is $1000 in gasoline per year.

Electricity costs $0.3 per kW-hour for electricity and transmission fees from the city power grid.
A battery energy charge of 33kW-hours in the battery will bring the car about 100 miles, where more than 90% of the stored energy in the battery goes into forward motion of the car.

Let's do the math for cost per 10 miles:
- The cost per mile is about ($0.3/kW-hour)*(33kW-hours) = $10 per 100 miles, or $1 per 10 miles.

If an EV car travels 10k miles per year, that is $1000 in electricity per year.

Solar panels typically cost about $2000/kW. If the home has 6kW of solar panels, that is $12000 for the original installation.

If the solar panels last 25 years, that is a cost of $500/year.

Of course, if the home already has solar panels, or the home is also using solar panels to avoid using electricity and billing from the city grid, then the cost of the solar panels to add a car does not matter so much.

There is no free lunch. There are some energy losses somewhere in the power production process, either for gas engines or electric batteries and motors.

Combustion and Electric Cars, the New Rivalry, Q and A

Hybrid electric cars are more efficient than pure gasoline cars, because the motion of the car can recharge the batteries when slowing down, instead of just heating brake pads. Also, there is no warm-up time, during which a cold combustion engine is very inefficient. When the car is cold, the cold fuel doesn't burn as well, transmission oil is not warmed up and allows more friction, and the catalytic converter needs to get warmed up and hot to remove pollutants.

Full electric cars are more efficient than gasoline cars because large power plants are more efficient than small combustion engines, and because large power plants can use alternative fuels, and because of regenerative braking.

Electric cars mean we don't need to depend so much on importing gasoline, and we can use our own power plants with local and/or green energy sources (hydro, wind, nuclear, solar) to help charge the car, usually over night.

Question 1: Will all cars be powered by electric motors someday? Electricity off the grid is just as expensive as gasoline, so this is a reasonable question.
- Probably not, because gasoline is like a miracle fuel for cars, with high energy density. The energy density of batteries compared to gasoline is just much lower. So gasoline cars can get longer range, and can be re-fueled in minutes. Gasoline cars are less sensitive to winter cold weather when the energy density of batteries drops. Gasoline cars are also less sensitive to super hot summer weather when the energy density of batteries also drops.
- Electric gas might be more reliable because there are less parts.
- But gasoline cars pollute, and gasoline from the ground is non-renewable, and gasoline money has propped up dictatorships around the world.

Question 2: Are batteries long lasting, with 1000s of recharges? That all depends on if drivers are willing to accept 30% less driving range. The car still works.
- Modern lithium ion batteries are lasting a long time, over 10 years. Rumor is they lose a few percent of their capacity each year due to micro cracks, especially if always charged quickly, so after 10 years they've lost 20 to 30% of their range.

Question 3: Are trucks a good use of electric vehicles? Any transportation that does not care about the weight of the batteries is a good candidate for electric power.
- Yes, because the weight of the batteries is not that important when they are towing a heavy trailer. So lots of batteries and longer driving range can be added for trucks.
- Submarines and boats are two cases where weight of the batteries does not matter so much. Airplanes, in contrast, really care about weight.

Question 4: Are the chemicals used to make the batteries worse than the gasoline?
- Batteries should be recycled, to avoid contaminating the environment and to allow the elements to be re-used.

Question 5: What does MPG equivalent, or MPGe, mean, for an EV car?
- MPGe does not mean that the car travels 2 or 3 times farther for the same money, when comparing a gasoline car that gets 30 mpg to an EV car that gets 100 mpg-equivalent.
- MPGe is an energy comparison, not a cost comparison.
  - Gasoline has an energy of 33 kW-hours per gallon, but only 30% of that energy goes into forward motion.
  - The same energy of electricity will mostly go into forward motion, more than 90%, but that electric energy cost three times as much.
- Also, electricity is not necessarily cheaper than gasoline. The high electrical rates at homes means that gasoline and electricity cost about the same, at about $1 for 10 miles. That is a comparison of a car that gets 30 mpg, to an EV that gets 100 mpg equivalent.
- Different parts of the country have different electricity rates, so some parts of the country are cheaper for EV cars.
- To make the EV car less expensive or free to charge, you would need to get solar panels to avoid the electricity grid, and then the recharge is free after the expense of the solar panels.

Electric Motor history:

Transportation and electric motors share a unique bond. Electric motors are on plenty of transportation besides cars. Electric motors are used to power trolleys and subways, and to start all our cars. Electric motors were originally on light rail, which means subways or metros. We also are gradually replacing diesel buses, powered by stinky combustion engines, with hybrid buses powered by electric motors. Electric motors also go vertical and power elevators.

In the past we had horse pulled trolleys, and horse manure. We replaced stinky combustion engines and horses.

In the future we will have zero emissions busses, powered by lines overhead or batteries or hydrogen. Maybe we'll have electric airplanes, powered by batteries or hydrogen.

Moving around the city:

Light rail in cities were an improvement from horse and carriages. The original light rail used the same fuel as trains, like steam from burning coal or wood. So light rail before 1900 used to belched exhaust plumes. Going through a tunnel was stinky business. Also, the number of cars getting pulled was limited to the number of engines in front, which limits capacity during rush hours.

Well, electric motors change the stench and power limits. Each axle gets an electric motor, so many cars during rush hour have the same performance as a few cars at low travel times.

Thank people like Sprague, an engineer, inventor and entrepreneur who replaced steam engine light rail with electric motor light rail. He decided to learn about new electric motors while in the Navy stationed in Europe in the 1880s. He needed to develop a robust electric motor that can tolerate dirt, rain, and ice and keep running along the tracks in the middle of the roads. That took a few years of trial and error along a demonstration track in Richmond VA. He also needed to build the power plants to power the light rail. The power plant needed to power about 20 fully loaded trolley cars all going up a hill in the city. He needed to develop the control circuitry to power all the electric motors at once in parallel. One control board at the conductor would send voltage commands to all the electric motors on all the cars in parallel.

Moving up and down in tall buildings:

The electric motors did not just help subways. They also help elevators and escalators. That means that buildings can be made taller. The advent of inexpensive steel I-beams also allowed buildings to get taller, so together we have sky-scrapers.

Moving on the ocean:

Electric motors help large ships because electric motors can individually power propellers, and these motors can pivot with the propellers to help turn the ship near port.

Larger boats can use electric motors to drive the shaft because the power source – DC batteries or steam and generator from nuclear power – can be anywhere on the ship.

Moving in the sky:

Electric motors can power many propellers on airplanes. The next practical airplane using electric motors will probably have onboard electric generators. The fuel powers the generators, which then power the electric motors. The generators are turbojets to generate electricity, or combustion engines to generate the electricity. This electricity then goes to many electric motors and propellers on the wings. Believe it or not, many propellers are more efficient than just a few, because the wings have more air flow and more lift, even with a smaller surface area. Less surface area means less drag.

Power subways underground, without choking exhaust.

Present day modern elevator, thanks to steel and motors.

Cruise ship with electric motor at propeller

Electric single propeller airplanes for short distance and training.

Because electric motors spin shafts, that means electric motors can move transportation vehicles.

People Movers and Boats, Trains, Elevators, Airplanes

Electric Motor history:
Transportation and electric motors share a unique bond. Electric motors are used to power trolleys and subways, and to start all our cars. We also are gradually replacing diesel buses, powered by stinky combustion engines, with hybrid buses powered by electric motors. Electric motors also go vertical and power elevators.
Sprague in the 1880s successfully optimized the DC electric motor to handle large starting currents and protected the brushes from rain and dirt. He used graphite brushes and current control for a DC motor with an electromagnet stator.
In the past we had horse pulled trolleys, and horse manure. In the future we will have zero emissions busses, powered by lines overhead or batteries or hydrogen. Maybe we'll have electric airplanes, powered by batteries or hydrogen.

"The electric motor has had a long, practical, successful history.

In the 1880s an electric motor improves from a weak, impractical prototype concept to 'World Look at Me Now'. Then, oops, I'm so successful I'm taken for granted.'

Jedlik's "lightning-magnetic self-rotor", 1827, the world's first DC electric motor.

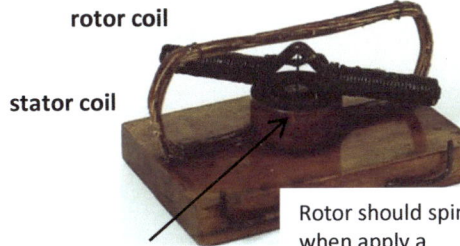

rotor coil

stator coil

Before motors: Toy electro-static devices, not motors (charge builds up, like a shock you get rubbing feet on carpet in dry winter air)

commutator

Rotor should spin when apply a battery

Electric Boat in Paris

Electric Trolley in Germany

Electric cars

American Revolution

American Civil War

World War 1

Now

| 1740 | 1821 | 1827 | 1880 | 1886 | 1887 | 1912 |

First demonstration of magnetic force from currents, by Faraday: Magnetic force on current causes metal rod to stir the liquid.

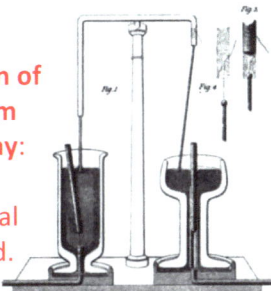

Electric tricycle

Practical DC motor in Richmond, VA
Sprague motor in the trolley, using 600 volts.
First practical motor that draws more power when it needs more power.

Starter

Quiet trolling motors on boats, to not scare away fish

Power subways underground, without choking exhaust.

Elevator with 1 ton counter-weight

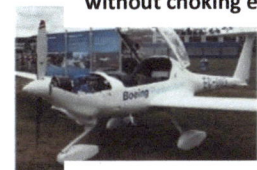

Short range airplanes without pollution

1740: Gordon: simple electrostatic devices
1820: Ampere: a wire carrying a current in a magnetic field has a force
1821: Faraday: conversion of electrical energy into mechanical energy
1832: Sturgeon: first commutator DC electric motor
1886: Sprague: first practical DC motor, as important as Ford's model T car and Wright Brother's first airplane.
1887: First electric trolley system in 1887–88 in Richmond VA
1912: Electric starter for cars invented, eliminating the hand crank.
Today, electric motors stand for more than half of the electric energy consumption in the US, mostly in factories.

Because electric motors spin shafts, that means electric motors can move transportation vehicles.

137

Electric Tram, or Subway, Magnets Underground

: The first trams or electric trolleys in large cities were the great motivation for building electric power plants, and replacing horses. Cities around 1900 were getting buried in horse manure and over-packed transport.

Sprague made electric heavy trains better than steam powered front engines, by installing electric motors on all the axles of each car. The train has the same performance or spunk no matter how many cars are attached, because all cars are powered. This same performance is great for heavy commuter times with longer trains and high passenger load.

"The first practical electric motor was for trolleys, that drew more power automatically when the trolley needed it to go up hills."

First few cities with electric trolleys:

- Electric trollies first demonstrated on a city scale in USA in Richmond VA in 1890s, by Sprague Electric, to replace horse drawn trollies. A new coal power station (~400 horsepower) had to be built to guarantee that all the trollies could climb up hill carrying all passengers at the same time, along with overhead wires.
- Cities (Scranton, PA) that had lots of energy (coal, waterways) for power plants could consider electric trams.
- The success in Richmond VA convinced politicians and unions in Boston that electric trollies had the power to work and also get rid of the horse manure and congestion, so the massive rail/tunnel/power station investment was warranted.

Experiment: Ride a subway

The 'Orange' line, Boston, 2013

Electric power comes from the voltage in the rails below the train. Combustion engines are not used because exhaust from combustion engines for the subway would be quite stinky in the tunnels.

Cities with huge electric subways:

- New York,
- Moscow
- Tokyo
- Beijing
- London

Historic Electric Tram, or trolley, Germany, started in 1881

1881

Richmond VA demonstrated that a whole city can have trolleys going up hills at once.

1890

New York, 1920, last vestige of horses

Modern electric subways

Beginning electric trolleys

1885

Electric power from voltage in overhead lines, like bumper cars at amusement parks.

Horse and Buggy

Double Decker Tram, England, started in 1885

600 DC Volt, Sprague motor for trolley, Trolley museum, Scranton PA
Both stator and rotor are electromagnets

To control speed and power to the wheel's motors, old trolleys used a resistive power divider.

Experiment: Play with toy electric car and train

Toy electric cars are trains.

Different countries use different trolley names:

The names for electric trains are Tram, Streetcar, Subway, Underground, the 'T', the Metro.

Electric mass transportation is alive and well in the big cities.

The advantage of trams is that power is supplied by the city power lines, instead of heavy batteries on board or the pollution from gasoline or diesel engines.

Experiments: Ride a subway underground, or play with electric toy trains

Electric Elevator, Magnets Near Roof Tops

"Electric motors allowed clean, quiet elevators, better than a combustion engine in a building."

Elevators and Building Height: Elevators have saved lots of legwork going up stairs, and had the unintended effect of changing our cities skylines, with taller buildings and more people per square mile. The rich want the top floor now with the great view, not the bottom floor.

Old human power to lift supplies and sails

People used capstans to lift baskets of food and animals

Year 0 AD

People could only walk up stairs.

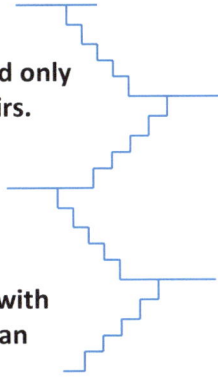

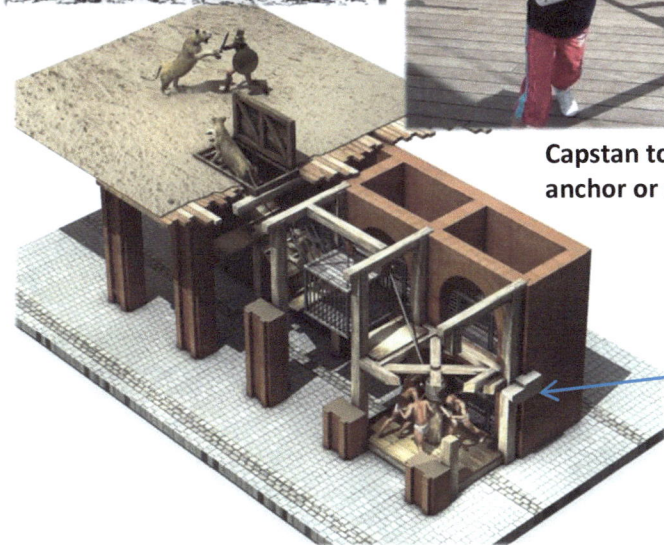

People raised an anchor with raw muscle using a capstan on old sailing ships.

Capstan to lift anchor or sails

In these barbaric times, slaves did the work of electric motors to lift the lions up to the arena for fighting the condemned.

Roman Colosseum elevator with capstan crank to bring lions up onto the arena. The Romans might also have use capstans to lift flood gates to flood the colosseum and have mock naval battles.

Modern electric elevator

1926

Electric elevators made the top penthouse apartment the most posh, prized location in the building, for the great view.

Before elevators, only poor people owned the penthouse! Poor people were forced to walk up 10 stories carrying groceries. But, hey, this is actually a great workout and the poor people probably were stronger.

Way-back   Now

Elevators have electric motors on top floor, where one end of cables goes to the elevator car, and the other end of cables goes to a balancing counter-weight.

1 ton counter-weight to avoid huge stress on elevator motor.

The birth of the modern elevator
The Woolworth Building, 1926
The World's Tallest Building

Without the elevator, no one would walk up 50 floors.

Present day modern elevator, thanks to steel and motors.

Experiments: Ride an elevator

Electric motors have changed the height of buildings and how we get around them using elevators.

Pure and Hybrid Electric Boats, Magnets on the Water

Electric boats: Electric motors on bigger boats are commonly used to turn the shaft, on cruise ships, aircraft carriers, and submarines. The electricity can come from batteries or generators powered by combustion or steam engines (or nuclear power in the Navy), located anywhere within the hull.

- **Pure electric, without a gasoline generator:** Water has a huge drag on the boat hull, so battery power is very short term, and only rowboat size fishing boats use just a battery.
- **Hybrid electric:** For larger ships, the ship uses a combustion or steam engine (or nuclear power aircraft carriers) to create the electricity. Giant cruise ships use electric motors near the propellers.
  - The hybrid electric design allows the diesel generators to be placed in any non-conventional location, and free up the ship design, using wires to bring the power to the electric motors without shafts.
  - Huge ships need super powerful electric motors. More torque is generated in the same size motor when the motors use superconducting wire in the motors, to get huge current and magnetic fields in the smaller motor.

"Electric motors right at the propeller can free the design of big ships. A diesel generator can be put anywhere in the hull, to power the electric motor."

Cruising the Seine in Paris with an electric motor

Modern pure electric boats, using battery and propeller: used for quiet trolling motor for fishing

Small inflatable skiff with electric power

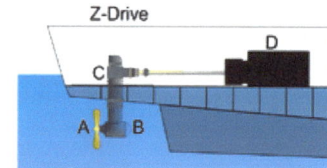

Z-drive, which can avoid electric motors and can still rotate the propeller in any direction.

Z-Drive

1880   1940   **Modern hybrid electric boats**   **Cruise ships have gone hybrid electric**

First electric boats:
Trouve and three friends cruised the Seine in Paris in silence aboard a 17-foot launch with an electric motor powered by DC batteries (two bichromate of potassium batteries). By 1890, Trouve motors powered 100 boats in Europe

Today's batteries have many varieties of chemical combinations, but the basic car battery is alternating lead and lead dioxide plates and an acid.

The other choice of power at that time period was a steam engine or a sailboat.

Diesel-electric submarines (1910)

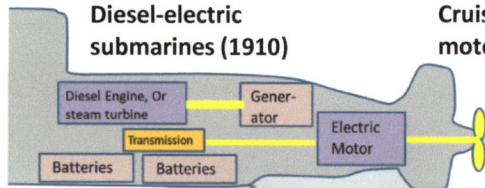

Diesel Engine, Or steam turbine — Generator — Electric Motor
Transmission
Batteries   Batteries

Nuclear-electric aircraft carriers (1960)

Cruise ship with electric motor at propeller

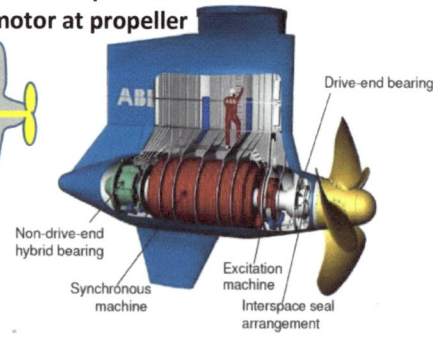

Drive-end bearing
Non-drive-end hybrid bearing
Synchronous machine
Excitation machine
Interspace seal arrangement

Electric motor directly behind propeller, allows full rotation for steering

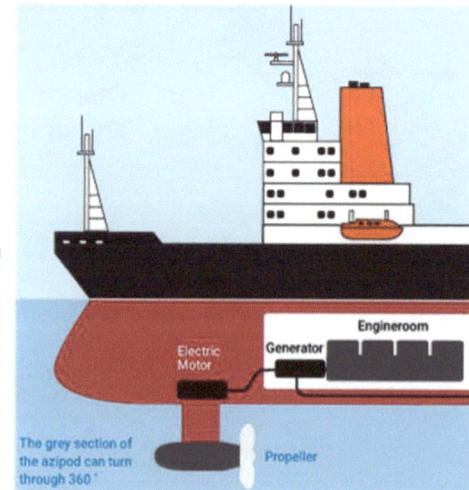

Engineroom
Electric Motor   Generator
The grey section of the azipod can turn through 360°
Propeller

Diesel-electric power, similar to hybrid cars.
The ship's propellers, or 'screws,' are turned by inverter-fed electric motors that get their energy from diesel-powered generators

Boats are one of the most inefficient ways to move, due to huge water drag. But the hybrid approach works - combustion engines powering generators that power propellers - and allows more flexible design locations for the combustion engine and the propellers.

Batteries and Weight for Airplanes

Electric Airplanes: Powering airplanes with electric motors is a challenge due to heavy batteries and low energy density. Yes, there is an obvious reason today's airplanes use jet fuel. Jet fuel for common turbo jet engines have much more energy per weight than batteries. Motivations for electric motors in airplanes are many, even with low energy density batteries. We need the energy density to improve before long flights are possible.

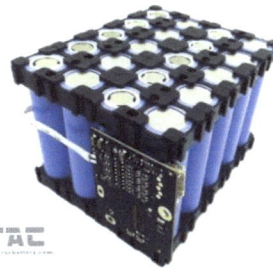

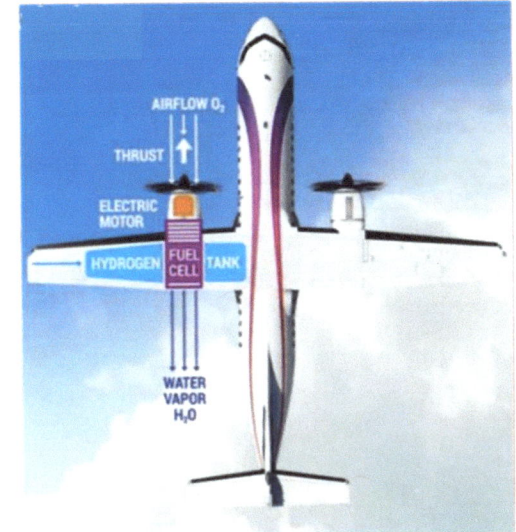

0.1 kWh/kg

0.3 kWh/kg

13 kWh/kg

Energy density of an iron battery

We don't even attempt to make electric cars, let alone airplanes, out of iron batteries. They are too heavy.

Energy density of a Li-ion battery

Electric cars weight about 2 tons (4000 pounds). Half that weight are the batteries.
The driving range is typically about 300 miles.

Energy density of gasoline or a hybrid system using liquid fuel

Cars take about 12 galloons, at about 70 pounds.
The driving range is also about 300 miles.

- **Batteries**: Electric airplanes have been impractical because of heavy batteries with low energy, for very short flight time. Now, electric airplanes are looking up or more practical with the invention of new kinds of higher energy batteries, and rechargeable solar panels.

- **Hybrid approach**: Use the electric motors with fuel cells, which have the same energy as jet fuel, or with a combustion engine and generator. Skip the batteries. Fuel cells use hydrogen and just produce water when reacting with oxygen, so fuel cells do not pollute. Jet fuel would still pollute, even though it is a hybrid air plane.

- **Electric motors**: Also, newer rare earth magnets allow lighter motors with high torque.

- **One driving challenge for flight is the batteries. High energy batteries need to store enough energy while staying light and safe. Only recently have modern more powerful batteries allowed electric motors for flight, using lithium ion batteries instead of iron acid batteries.**

Now and Future Electric Motor Airplanes in the Atmosphere

Electric Airplanes: Experiments and prototypes and real electric airplanes are getting tested. Motivations for electric motors in airplanes are the following, even with heavy and low energy density batteries. Electric motors are quiet, for city taxi flights. Electric motors are reliable and have low maintenance. Even if one motor goes down, multiple rotors for the helicopter or quadcopter increases the safety. Electric motors do not contribute greenhouse gasses with no carbon dioxide CO2 exhaust or methane exhaust, if these electric motors are powered by batteries, solar panels, or fuel cells.

"Batteries are ozone friendly...no jets releasing greenhouse gas emissions (carbon dioxide). New motors and lightweight batteries have made electric flight more practical."

Airplanes now use jet fuel and high energy density, with around the world flight:
- **Combustion engines**
- **Turbofan engines**

Solar powered plane, prototype: Long, slow flights might be possible by recharging with solar panels in flight during the day and gaining altitude. At night just use the batteries and let the plane glide and lose altitude.

Gasoline generators or hydrogen fuel cells have more energy density than a battery, for a hybrid solution.

History of electric flight: none...batteries were too heavy and didn't store enough energy to power propellers.

Now: higher energy batteries (Lithium Ion)

NASA Solar Power airplane, Helios

Hybrid airplane engines, with either batteries, generator, or fuel cell.

Hydrogen fuel-cell powered electric motors

Past → Now → Future?

Experiments: Fly a toy helicopter with lithium batteries and electric motors.

Will future airplanes find some batteries with high enough energy density?

Drone toy quadcopter

Stability of 4 propellers, not 3: A 4 motor design can survive one motor failing without flipping. That is, the 4 motor design will not flip over and crash.
- In contrast, a 3 motor flying disk would tip over if one motor failed.

Electric single propeller airplanes for short distance and training.

NASA airplane concept, battery powered: vertical launch with rotated wing.

Uber electric taxi concepts

- **One driving challenge for flight is the batteries. High energy batteries need to store enough energy while staying light and safe. Only recently have modern more powerful batteries allowed electric motors for flight, using lithium ion batteries instead of iron acid batteries.**
- **Another driving challenge for flight is software. Software constantly controls the multiple propellers, with control software for a stable flight, for most efficient use of power!**

Electric Boats, Trains, Elevators, Airplanes, Q and A

A lot of transportation depends on electricity, like city metros (light rail), boats and submarines, and newer electric cars. Boats and submarines don't care about weight, so batteries are fine. Even newer electric cars don't care so much about weight, just mostly care about the energy and cost and volume of batteries to go 300 miles.

Electric airplanes are a new push. Electric airplanes have no pollution contaminating the upper atmosphere. Pollution at that altitude near the ozone layer reduces the ozone and lets most of the harmful UV radiation from the sun pass through to the ground. Also, the many propellers with precision control of each propeller spin rate helps to avoid some flaps.

However, the batteries are heavy for the little energy they provide per weight compared to jet fuel. Hybrid electric airplanes can fly long distances using some liquid fuel and avoid batteries.

Historic electric tram, or trolley, Germany, started in 1881

Question 1: How long have subways been electric? Over 100 year. Electric subways cleaned up cities, and gave people more transportation.

- Short range subways in large cities have been common for 100 years. These trains run on electricity which is carried through the metal rails.
- All the light rail subway cars have electric motors at each axle, so long or short trains have the same acceleration.
- Nobody wants the exhaust from gas or diesel engines inside tunnels where people breath.

Question 2: Are there long distance electric trains? Some trains have diesel engines that create electricity for electric motors at the wheels.

- Yes, hybrid high speed trains have electric motors at the wheels. Diesel engines spin the generators which then power the electric motors at the wheels, so these trains are hybrid electric.
- Most long distance trains are not the hybrid electric case. The most common train cars are still powered by a diesel engine up front, and the cars behind are just pulled.

NASA Solar Power airplane, Helios

NASA airplane concept, battery powered: vertical launch with rotated wing.

Question 3: Are there electric airplanes, even for short range flights? Electric planes may not have long flight times, but they are clean and have less parts than a combustion engine airplane.

- Yes, people have built prototype battery powered airplanes that have limited range, that can fly about 20 miles over the English Channel. However, modern chemical batteries only have about $1/50^{th}$ the energy density as gasoline. Gasoline is still the obvious winner for airplanes.
- An electric airplane is very quiet, and the design of many electric propellers along the wing can avoid control flaps, have precision control, and allow smaller wings with less drag.
- There are also hybrid airplanes, where the electric current is produced by a turbofan engine using gasoline, or produced by hydrogen fuel cell. That allows the design improvements of using many electric motors and propellers, and allows the same high energy density as gasoline airplanes. The flight range is much farther for a hybrid airplane with gasoline and generator rather than a pure electric airplane with batteries.
- A long distance solar airplane can power itself using sunlight, rising to high altitude during the sunny day and gliding and using the battery at night, losing altitude. Currently the solar planes are almost gliders and do not travel fast or have much power.

Question 4: Why don't elevators use combustion engines? People want peace and quiet in a building, and they don't want to smell exhaust.

- Electric motors avoid stinky and unhealthy exhaust inside a building.
- Electric motors avoid noise.
- Electric motors rarely break down, and this reliability means buildings can avoid maintenance of a private generator or combustion engine. For maintenance, let the power plant experts instead stay focused on a single site and maintain a large power plant for everyone.

The birth of the modern elevator
The Woolworth Building, 1926
The World's Tallest Building

Tall buildings can not happen without elevators.

143

Sprague and Kettering: Electricity Development ... a Little History

Before the common application of motors and generators began, first discovery and early science had to happen, such as electrostatic forces, creation of magnetic fields, the explanations, DC and AC current, and prototype motors and generators. And then the ideas get applied by industrious engineers who see a need, like better transportation around cities - trolleys - and better lighting - electric lights. Notice it took 60 to 80 years before mass applications happened.

Mass Transportation and Power Generation: AC versus DC power

Sprague: (1886, in his 20s)

- American
- Constant speed DC motor for trolleys and elevators. He is one of the original inventors to replace horse draw and steam power carriages with quiet and non-polluting electric power.
- DC motors have more torque at lower spin rates than AC motors, when use an electromagnet as the stator magnet. The DC motors are called a 'traction' motors.

DC motor using iron and coils and carbon brushes, for light rail and elevators

Sprague learned about the state of the art for electric power while helping out at a world fair in Paris in 1881, while in the Navy. He developed some ideas for DC motors at that time, and recommended them to the navy. Sprague then went to work for Edison for just 1 year before leaving. He was asked to work on his DC motor designs, which would just be handing over his ideas to the Edison company, so Sprague left.

Sprague changed 'light rail' (people trains within cities, or metro) from steam power engines pulling train cars to electric trains, where each train car has electric motors. Electric light rail has the advantage over coal and steam that there is no pollution in tunnels, electricity is quiet, and that motors can be installed on all the cars so a longer train during rush hour has just as much pep as a short train.

Sprague also changed or modernized elevators using the same electric motors and electricity. With electric motors, we remove the steam engines and the loud and stinky burning coal, wood, or oil. .

In practice, Sprague and his team had to figure out how to stop the brushes from quickly wearing out, by using carbon brushes, and he had to protect the brushes from the weather.

Sprague also had to have different currents applied at a start from a stand-still, because the DC motors tended to draw a lot of current at startup when not spinning and would burn out. At that time DC current was controlled by switching in lossy resistors, which wastes power. Nowadays, DC current and average power can be controlled by pulsing the DC power, which does not waste power.

Kettering: (1911, in his 30s)

- American
- Electric starter motor for cars

Kettering with his starter motor for cars and trucks

The car starter motor enabled larger car engines.

Even modern small engines can be started by a pull cord, but those are small engines on lawn mowers or chain saws. Similarly, kick starters can turn over a small motorcycle. Modern cars have engines more than 10 times larger, and a human can not pull a cable or crank a lever to turn the engine over.

Kettering's novel approach was to gear down the starter motor by putting gear teeth on the flywheel, to gain mechanical advantage. Now the starter motor did not need to be so large.

The gear, the pinion, for the starter motor swings out when the motor spins, and the gear engages with the flywheel sprockets.

Kettering also developed an ignition system and developed the air conditioner coolant Freon.

<u>Electric motors power the motion of robots</u>: Robots are already here in different forms. Robots are already assembling our cars. Who knows what the future brings, but human-looking robots seems likely where movement is enabled by many small motors. Human-looking android robots are under development but are limited by power supply. In science fiction android robots already exist, with infinite battery power to walk and talk forever without ever saying 'please plug me in' after 20 minutes of draining their battery.

So robots have the challenges of power supply and of course they need sensors and programming, maybe even artificial intelligence.

Will androids use many powerful, high torque motors and many linear motors like relays to pull on wires? We need strong permanent magnets to get the high torque.

Or will android use materials that contract when a voltage is applied, like human muscles?

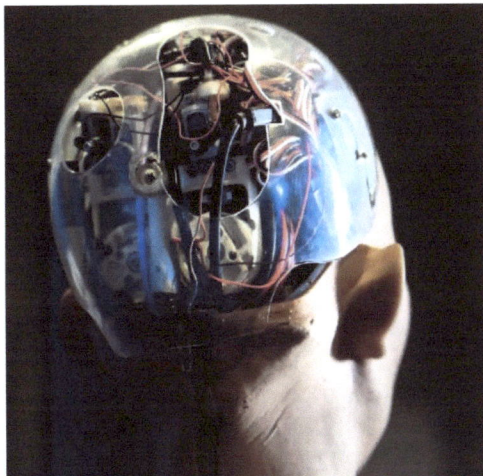

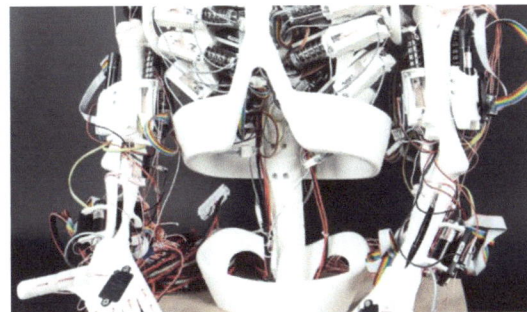

Motors are the tendons in the robot.

"Androids need to power 10s of motors, and also to power the computer chips. Batteries will run dry."

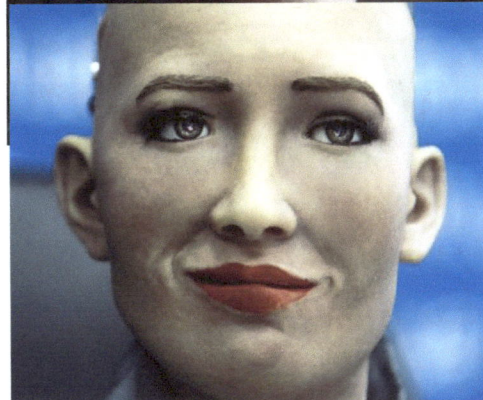

Sophia the AI robot:
Like-imitating facial muscles using motors

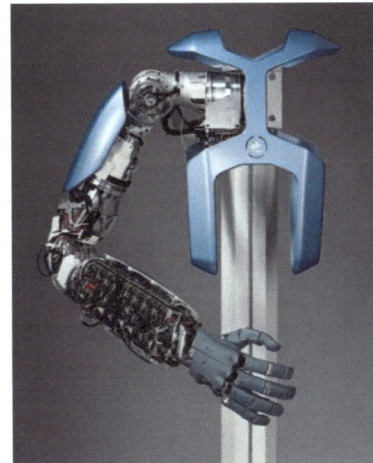

Motors and pressure sensors in the forearm.
Motors have 'tendon' wires that go to move the finger.

Robots are typically assembling cars and plugged into the wall power to keep moving.

Electric Robots on the Production Line

Electric motors power of motion of robots: Robots are here in different forms. Robots are already assembling our cars, and motorized arms are grabbing satellites in space. Toy robot dogs can move on your command.
Robots also need power for the computer chips for artificial intelligence.

Jobs and Robots?

Will robots help us, or just take away our jobs? How do people use robots to improve our life?

People and Robots work together:

One key is creating new products and new markets, where people use the robots to help them. People have much more common sense than a robot.
As a car company found out, letting people work in parallel with robots is the most adaptive and effective. Limiting factors for robots are the software and cameras necessary to get the motors to move adaptively and correctly. Robots also have no muscle memory, have limited ability to recognize objects, are not adaptable, and have no smart brain. These are basic skills for humans.

Maintenance:

Another key is that most factory lines for cars these days use robots. People need to maintain them and improve them.

Building robots:

Someone needs to design and build the robots, with motors, sensors, and software. That involves many fields in engineering.

Factories and assembly using robots.

Each joint of the robot has an electric motor.

Robot welding: robot needs optical sensors

Many robots assembling cars
These manufacturing robots are tethered to wall power.

Any human joint can be a robot's joint.

Dancing robots (Boston Dynamics)
These indoor robots need to be quiet and walk around, which means batteries. When the dancing is over, the re-charging starts.

BigDog robot
This robot uses gasoline and a loud combustion energy (not a battery) to get long hiking time, like a car.

"Electric motors and control software are being used to help manufacturing, and for future sci-fi androids."

Human helpers:
Currently there are not enough younger people to take care of people in their old age. Robots can help.

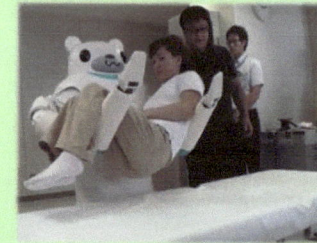

Robots for elderly care:
Batteries need to get more energy to keep these robots running untethered to a power cord.

Now → **Future?**

Outer Space

Canada's and NASA's robotic arm:
Here is a precision use of motors in an extremely cold environment, outer space.

Electric Robots Need Power Backpack

Human-looking android robots are under development but are limited by power supply. In science fiction android robots already exist, with infinite battery power to walk and talk forever without ever saying 'please plug me in' after 20 minutes of draining their battery.

So robots have the challenges of power supply and of course they need sensors and programming, maybe even artificial intelligence.

Energy and Power of human size robots:

The movies have had humanoid robots since the 1950s. What gives them super strength and super long energy? Maybe current batteries could work for 1 hour or so. The movies assume nuclear power modules, or 'AllSpark' from another world.

Robots need to 'think' or process image data or voice commands, calculate motor commands for motion, and power the motors. Thinking actually takes a lot of energy. A human uses a sizable amount of their energy just running their brain. For example, most of a person's heat loss is through their head.

"Androids need to power 10s of motors, and also to power the computer chips. Batteries will run dry."

The bottom line is that with current power technology robots need to be plugged in and are great for assembling cars. The robots can use some limited A.I. and cameras to identify parts and move them around with motors on their limbs.

Some robots are untethered and can move around for a long time. Those robots use gasoline for power, and gasoline has a lot more energy per volume than a battery. Unfortunately, these combustion engines are also very loud. Fuel cells with hydrogen might be the way to go, which is also why some car manufacturers are investing in fuel cells. Robots, electric airplanes, and cars all have the same issues. Now, if the robot is mostly a rolling battery, then maybe we have some endurance.

Energy requirements of human size robots, back of envelop estimates of size and weight of batteries:

Let's look at and compare the energy requirements for a computer, a robot and for a human. An active computer uses over 100 Watts. In comparison, humans burn about 2000 Calories (kcal) of energy per day, or about 500 watts steady power throughout the 24 hour day. That human energy is equivalent to about 10 kiloJoules, or 10 kW-hours. That is why just one person is a good heater for a room. All that energy, when all is said and done, goes into heat. When you think, you heat the room. Thinking actually takes a lot of energy. When you move, you heat the room.

Transformer 'Bumblebee'.
Now, he uses 'AllSpark' from another world, so he does not need real world gasoline or batteries or nuclear power or solar power.

Now, a human is a mammal and warm-blooded and is at a constant temperature. We know that reptiles use a lot less energy, because they don't keep their body temperature stable. Let's make the robot like a reptile, so we could assume 100 Watts for the robot, and let its temperature vary with the room. This 100 Watts is more equivalent to a computer.

How many batteries are needed to provide 100 Watts? Let's compare this reptile power draw to the energy in batteries. One D cell battery has about up to 0.03 kWh of energy. We'd need 4 D cell batteries to get our 100 Watts for 1 hour, just to think and not move. That means we recharge every hour, and we carry a lot of weight in batteries.

Recharging the batteries:

Maybe robots can have many different power supplies, to charge the batteries. One could be a plug in socket, just like any electronics. Another could be solar panels, maybe sewed into clothes.

Another power supply could be some radioactive disk that trickle charges some rechargeable batteries. Nuclear power can last for decades, but we would not want to design it to generate a lot of power at any one instant. Instead we would trickle charge a battery, basically with the radiation passing through a solar panel. The battery can discharge quickly to do physical motion for short term walking or moving.

Human looking robots are a total system and need power supplies, sensors like the human 5 senses, and powerful motors.

Chapter 6: Generators:

A simple toy motor can also be a generator, just spin the shaft by hand, or by hot steam, or by flowing water. Various toy generators, shown in this chapter, introduce all the power plants around the country.

What heats the steam? Coal, natural gas, nuclear heat all can heat water to form steam and turn a turbine and generator. 'Green' alternatives also exist like hydroelectric, geothermal, wind mills, and solar thermal. These also spin a generator. Another 'green' generator is solar cells, seen on roof tops. Solar cells are the outlier. They don't spin a generator, and instead directly convert light into voltage.

Generators, Magnets flipping near coils, enabling quality of life as we know it

- Section A: Generators, Motors can Work Both Ways
- Section B: Examples of Generators for Power Plants using Magnetic Fields
- Section C: Imagine Life Without Magnetic Fields to Create Electric Generators and Power Plants
- Section D: History of Power Plants

"There's a whole power plant and power distribution network around the world. In the USA, there are a few power plants per state. There are also local power sources like car engines and solar panels!"

Before generators, with local sources of power

Current-day generators, with electrically distributed power from burning coal or natural gas or nuclear heat

Heat to boil steam

Fans inside of a steam turbine

Generator: Make me spin, and I'll give you electricity

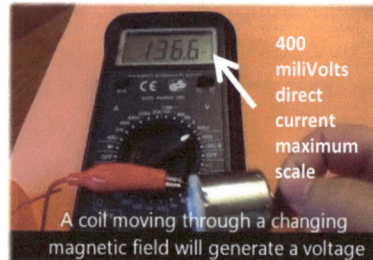

Water Power, use direct local mechanical power

Capstan to lift anchor or sails

136.6

400 miliVolts direct current maximum scale

A coil moving through a changing magnetic field will generate a voltage

Electric voltage generation by spinning shaft of motor
Spin the motor shaft by hand, to get magnet moving through coils inside motor.

The 'grid' makes electricity mostly from heat, but sometimes from gravity or sunlight, and brings it to your house.

Generators: Break Down of Chapter Sections

There's a lot to discuss regarding generators. There are prototypes, there are heat sources and magnetic generators, and there are direct sunlight-voltage conversion like solar panels.

"We all expect the power grid to just be there."

Section A: Generators, Motors can Work Both Ways

We can demonstrate voltage generation using a toy motor.

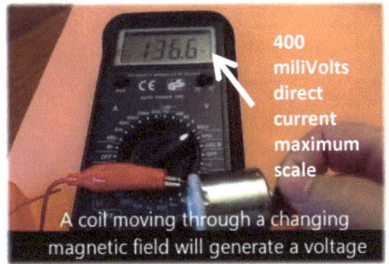

400 miliVolts direct current maximum scale

A coil moving through a changing magnetic field will generate a voltage

Electric voltage generation by spinning shaft of a dc motor
Spin the motor shaft by hand, to get magnet moving through coils inside motor.

Section B: Examples of Generators for Power Plants, using Magnetic Fields

We can demonstrate voltage generation using a toy motor.

Make steam using heat

Shovel in the coal in olden days

Nuclear fuel rods

Use gravity and flowing water, or wind

Spin turbine using steam or flowing water

Fans inside of a steam turbine

Spin magnet in coils* to get electricity

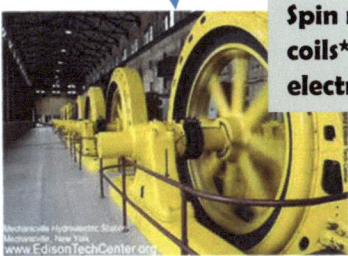

Section C: Imagine Life Without Magnetic Fields to Create Electric Generators and Power Plants

There are direct voltage generation approaches like solar panels, but heat and magnetic fields are much more common.

Sun

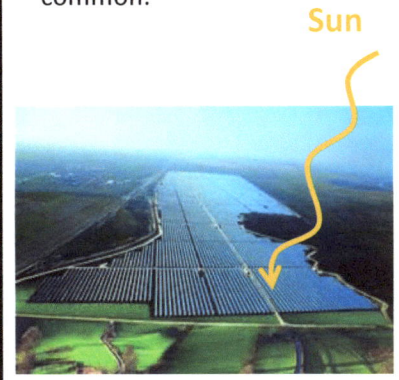

Solar Panels:
Direct electrical conversion from sun to electric, without coils

Natural Gas / Oxygen Fuel Cell direct electrical conversion:

Section D: History of Power Plants

Only in the last 150 years have we had the power grid.

1880

Hydroelectric power

Niagara falls

1890

Coal power

1960

Nuclear power

Generators: Motors can Work Both Ways, Make Electricity or Make Mechanical Motion

What brings power to electric motors? Generators, the reverse of motors, powered by steam turbines or combustion engines!

This chapter describes how electric power is generated by mechanically spinning a motor shaft to get electricity, instead of the other way around.   You depend on these generators to power your house, your business, and your schools.

Here is our power system for our country, with billions and billions of dollars of infrastructure, including power plants, transformers, and power lines. For fuel, there are coal deliveries and natural gas pipelines around the country to generate heat and steam to spin turbines for electric power.

"Dude, this just blows my mind. How simple and elegant. Motors, electricity, you are so everywhere and successful. I'm so sorry that I have taken you for granted all my life. Besides, your electricity from huge traditional power plants is cheap and plentiful, at least in states with cheap electricity."

"Let's drill down and really understand power generation!"

INPUT

Generator: Mechanical spin → Electrical energy from Generator at Power Plants

Mechanical Energy IN | Turbines spin coil in a magnetic field and get voltage | **Electrical Energy OUT**

OUTPUT

Motor: Electrical → Mechanical energy from Motors

Electrical Energy IN | Electricity creates a magnetic field and magnetic torque spins shaft | **Mechanical Energy OUT**

Various power plants:
- Heat for steam (coal, nuclear, oil)
- Wind, Water damns
- Boil water
- Solar panels
- Spin turbine
- Spin generator
- Electricity
- Spin drill / motors
- Wall outlet
- Power lines

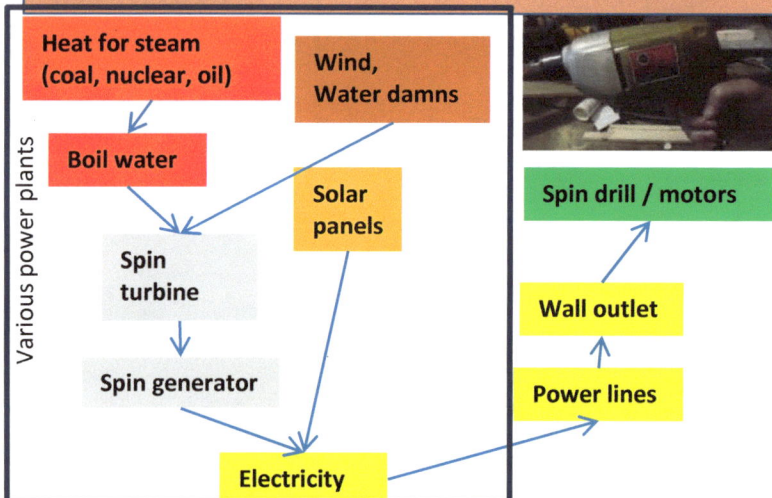

The power grid with generators and motors:
Here we have the beautiful two-way nature of motors. Motors can be used to generate mechanical motion, like spinning when current is applied. Conversely, motors can be used to generate electricity, when the shaft is spun mechanically, like at power plants, or the portable generator in your house when a tree falls on a power line.

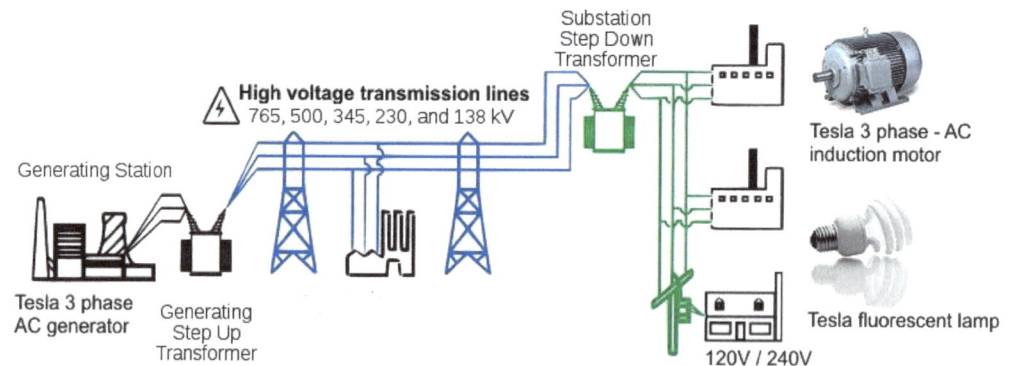

High voltage transmission lines 765, 500, 345, 230, and 138 kV

Generating Station

Tesla 3 phase AC generator

Generating Step Up Transformer

Substation Step Down Transformer

Tesla 3 phase - AC induction motor

Tesla fluorescent lamp

120V / 240V

Path from generator to your wall power outlet.

The 'grid' makes electricity mostly from heat (hot steam), but sometimes from gravity (hydroelectric) or sunlight (solar panels), and brings it to your house.

Generators: Voltage Induced in Spinning Coil

Why is there an induced voltage in a spinning coil, you ask? Lets look at the sideways orientation of the spinning loop below. This figure shows there is a force on the electrons getting dragged with the wire coil in a uniform external magnetic field, when the coil or magnet is sideways, all due to magnet force on each part of the current. The electric current is the free charges in the wire getting dragged sideways with the wire. Here is the concept for voltage generation. We can either understand the voltage generation based on forces on currents, or on change in magnetic flux versus time.

"Turn fast to pick up voltage, Padawan."

Maximum forces on free charge in the wire, because the wire is moving sideways to the external magnetic field at maximum speed.

No torque on horizontal loop or magnet, neither force on opposite sides of loop causes spin about shaft

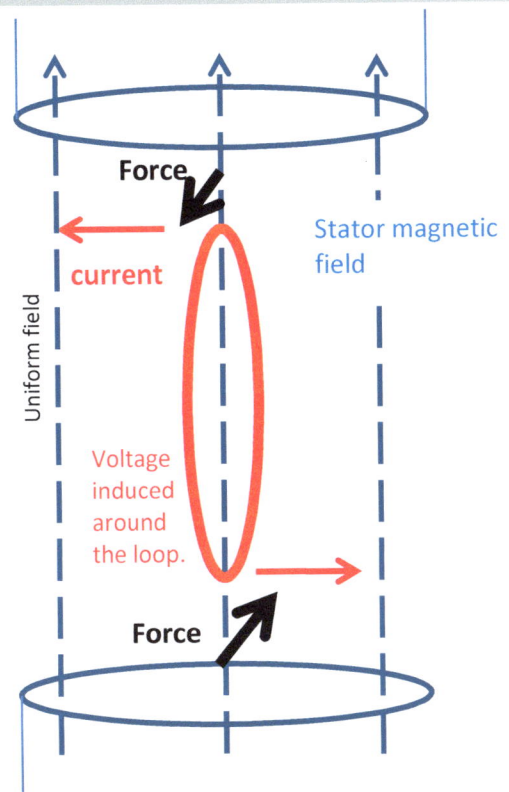

Force

current

Stator magnetic field

Uniform field

Voltage induced around the loop.

Force

Loop rotor magnetic field

Stator magnetic field

No Force

No Force

current

Rotor current loop

Force = I x B

Current

Field

Force

Fleming's Right Hand Rule

Dragged free charges:

The current is actually the wire going sideways, and dragging the free charges with it. This dragged charge is then forced along the wire due to the magnetic field force.

An external magnetic field causes opposite forces on dragged free changes (current) going in opposite directions; that is, opposite sides of the wire loop. These opposite forces act to generate a current or voltage along the loop.

Faraday's law:

The forces on electrons are just for fundamental understanding.   In practice, the voltage is a generator is predicted using Faradays's law of magnetic flux rate of change.

- When the coil is sideways, the coil gets net maximum force on free charges when current loop is moving fastest sideways to external magnetic field.

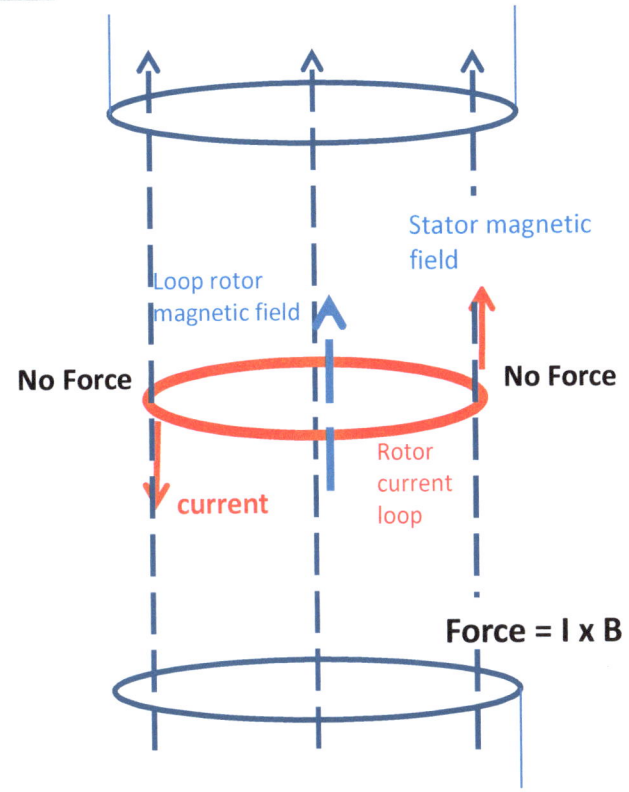

- When the coil is vertical, the coil has no force on the free charges because the free charges are getting dragged parallel to the external magnetic field.

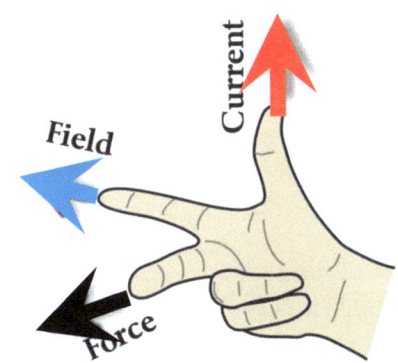

Diagram of forces (black) on dragged free charges (red) at different angles, for a spinning coil inside a magnetic field

A spinning sideways loop in a magnetic field gets forces on the free charges. The forces are along the wire, so there is an induced voltage. These forces are based on the magnetic forces on currents using the 'Right Hand Rule', where current is the dragged direction of the wire.

151

Yes, you too can make electricity! Directly spin the shaft of a toy dc motor and you are generating electricity.

In general, spin a coil in a magnetic field and you generate a voltage across the coil, which you can use to power stuff. The shaft is spun from some mechanical power source, like a steam power turbine, a waterwheel, or a combustion engine.

How is a voltage generated? The little electrons in the moving wire are also moving or getting dragged sideways to the magnetic field inside the wire and experience a force along the wire.

"Dude, he's getting a lot of mileage out of that one toy motor. Now it's a generator. Give some long overdue homage to that wall socket, using that one toy motor."

Electric charging from a shake flashlight, with a magnet sliding through a coil.

Generator discovery and prototypes:

The first generator was probably dropping a magnet through a coil, with a current meter attached. Then people realized they could spin the magnet and get oscillating current. The first motor had a battery attached to a coil and commutator in a magnetic field, the stator.

Generators and Motors are two sides of the same coin:

So, which came first, a generator or a motor? Probably a generator because scientists moved a magnetic field near a coil and saw voltage across the coil. This takes less design effort than a motor. The original motor was a straight wire spinning around another wire. That does not generate much power, but it demonstrates the concept of coupling of magnetic fields and electric currents causing a mechanical force.

Motors and generators are the same phenomenon. There is a force on a moving charge or current sideways in a magnetic field. If the charges are moving sideways with a wire that is getting spun by some mechanical means, then the charge feels a force along the wire. This generates electricity. If the charges are moving along the wire due to an applied voltage from a battery, then this sideways force is actually sideways to the wire. This sideways force is a push sideways, which spins the motor shaft.

A coil moving through a changing magnetic field will generate a voltage

Spin a dc motor by hand to generate electrical power

Generator examples:

You can demonstrate a generator. Just spin a toy dc motor shaft by hand, and put a voltmeter on the motor connector pads. The motor is now a generator. If you spin the shaft one way, you get a positive voltage. If you spin the shaft the other way, you get a negative voltage which is just a voltage in the other direction.

Every time you get in a car, you are near a generator or alternator. The battery needs to stay charged, and the car needs electrical power for the spark plugs, the fuel pump, the starter motor, the headlights, the dashboard, fuel vapor sensors, and all the other computer chips.

Even microphones are generators. A speaker can be used as a speaker or a microphone. When the cone electro coil moves in the magnetic field from the permanent magnet, a voltage is generated. Yes, that is electrical power. That is conversion of mechanical power into electrical power, caused by the energy in sound.

Spin a 2-pole demonstration kit to power a diode LED.

Generate Electricity Using DC Motor, Think Backwards

Again, spin the shaft of a toy dc motor mechanically. Voltage is generated on the electrical inputs. Voltage is generator by a changing magnetic field through a coil, which is the rotor. The voltage is largest at the angle where the coil is sideways to the stator magnetic field, because the speed of the wire is going sideways.
DC generators use the split ring as a commutator. Because of the commutator, the voltage is close to dc.
Edison patented a dc generator. The construction commutator is just like a motor, where a DC voltage will spin the rotor. In this case, a spinning rotor will generate a DC voltage.

"The DC generator is just a DC motor where the shaft is spun mechanically."

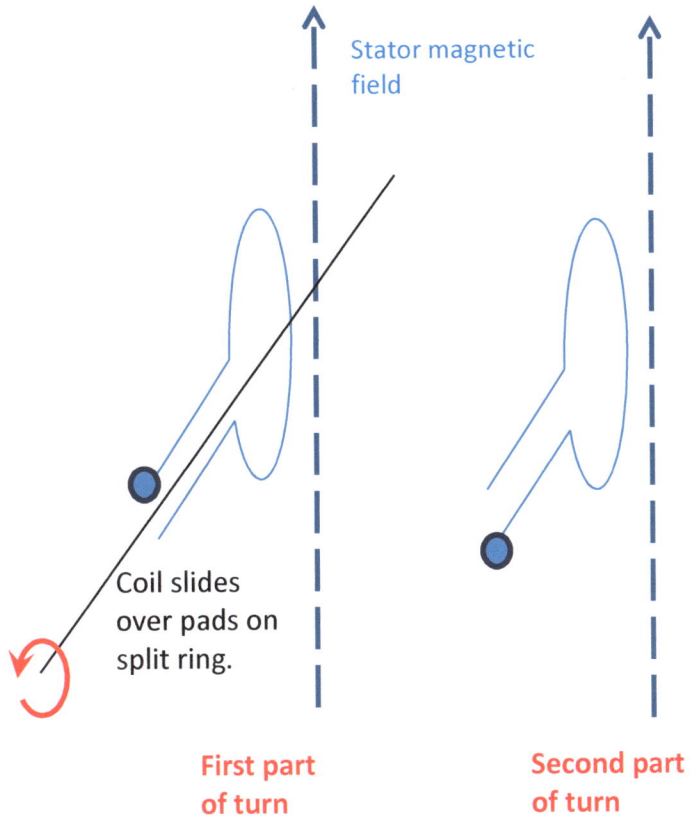

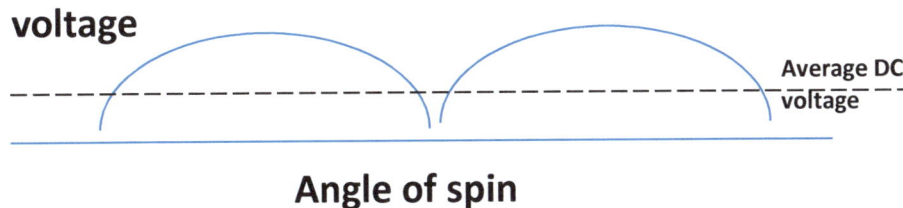

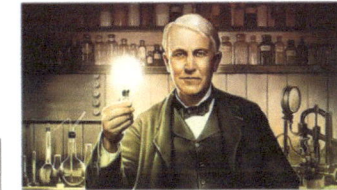

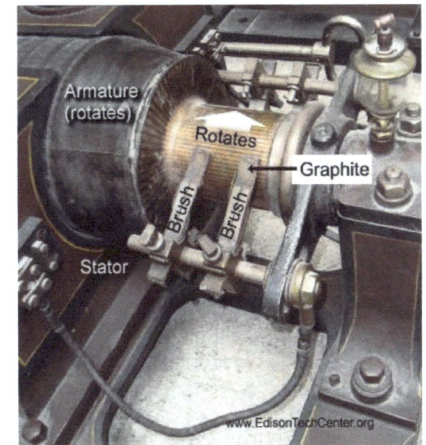

Stator magnetic field

Coil slides over pads on split ring.

First part of turn

Second part of turn

voltage

Average DC voltage

Angle of spin

DC power generator, 'Long legged Mary Ann'
Rotate coils in a large dc magnetic field to generate dc power.
Large generators require an electromagnet to create the large dc fields.

Edison and DC generator 'Long Legged Mary Anne'.
Edison patented a dc generator. The construction commutator is just like a motor.

See Appendix F for comparison of motors and generators

Generate Electricity Using DC Motor, Think Backwards

Spin a coil in a magnetic field and you generate a voltage across the coil.
How is a voltage generated by the spinning coil? The little electrons in the moving wire are also moving or getting dragged sideways to the magnetic field inside the wire and, by getting pushed by the wire in the magnetic field, the electrons experience a force along the wire to make a voltage and a current.
This voltage can also be described by the change in the magnetic field through a coil or loop.

"Dude, he's getting a lot of mileage out of that one toy motor. Now it's a generator.
Give some long overdue homage to that wall socket, using that one toy motor."

Generator concept:

Spin a metal coil in a permanent magnetic field and a voltage is induced in the coil.
Electric current is forced to flow around the coil, when the shaft spins.

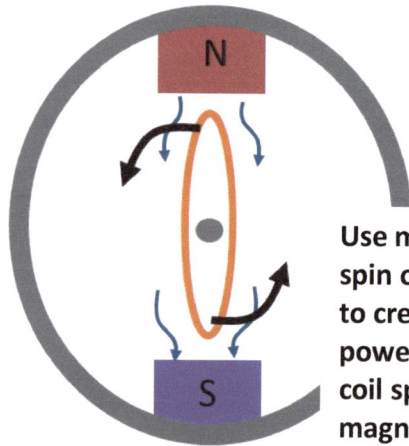

N

S

Use mechanical spin of the shaft to create electric power, from the coil spinning in magnetic field.

Generate a voltage:
Currents are induced in coil when the coil is forced to spin in a magnetic field from some mechanical force like an engine or water flow through a turbine. Electrons in moving wire are forced to move along the wire by the magnetic field, which is another way of saying current is created.

Generator concept converted to practice:

A generator! A coil on the rotor moving through a changing magnetic field will generate a voltage around the coil.

Sideways rotor:
The voltage from this sideways orientation of coils is tapped by sliding brushes, just like motors. The sideways coil in this orientation as rotor spins creates the most voltage, just like the sideways orientation provides the torque in a motor. There is the fastest sideways wire speed , or most change in magnetic field through the instantaneous sideways coil.

OUTPUT: Induced voltage from coil

INPUT: Mechanically rotated shaft or rotor, using steam turbine or combustion engine

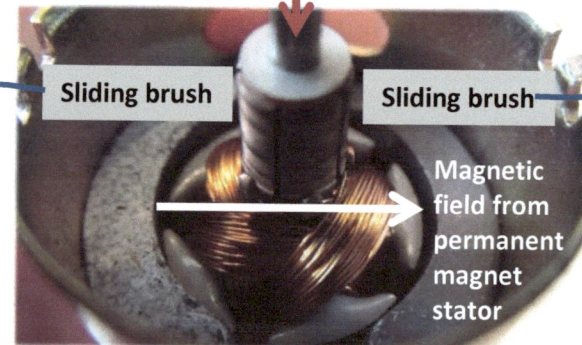

Sliding brush

Sliding brush

Magnetic field from permanent magnet stator

Generate a voltage:
Most current is generated in the coil sideways to the magnetic field, so the sliding brush commutators are always tapping into the coil at this orientation.

An AC motor will also generate electricity, with only electromagnets for stator and rotor. The stator just needs a little kick start to create some magnetic field.
A portable generator uses an AC motor.

By spinning the shaft of the rotor from some mechanical power source, you then create currents in the rotor which then generate a voltage across the coil. The voltage across the coil is used on the output of the commutator.

See Appendices F and G for forces on electrons in moving wire to create current

Fundamental Magnetic Force on Moving Charge (a Current)

For a motor, push those electrons around using a voltage. The moving electrons along the wire feel a sideways force from the magnetic field and this force spins the coil.

For a generator, move a wire mechanically with electrons in it. The moving electrons feel a sideways force, but in this case the force is along the wire, creating a current and voltage.

Either way, the electrons are moving in a magnetic field and a magnetic force is applied to these electron charges.

Coil

"All roads lead to forces on electrons."

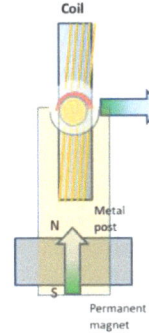

Metal post

N

Permanent magnet

S

Motor: convert electric current to wire motion

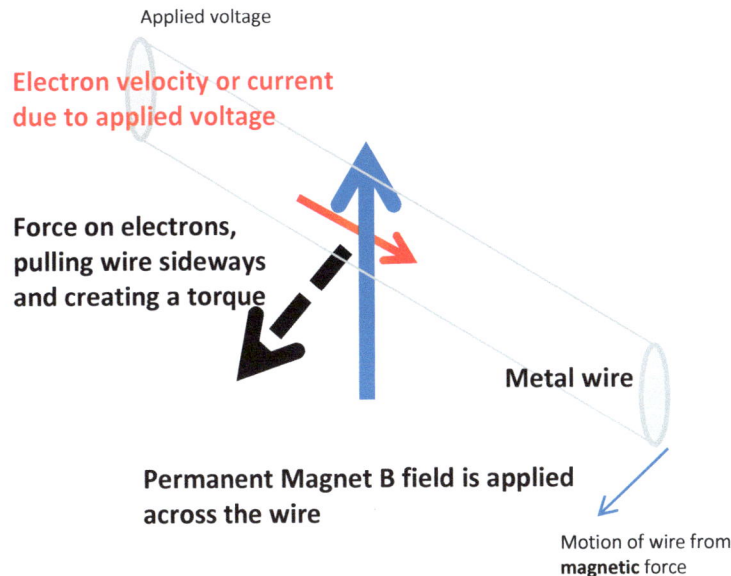

Applied voltage

Electron velocity or current due to applied voltage

Force on electrons, pulling wire sideways and creating a torque

Metal wire

Permanent Magnet B field is applied across the wire

Motion of wire from **magnetic** force

For a motor the magnetic force (black) is sideways from the wire due to a moving charge in a magnetic field

Generator: convert wire motion to electric current

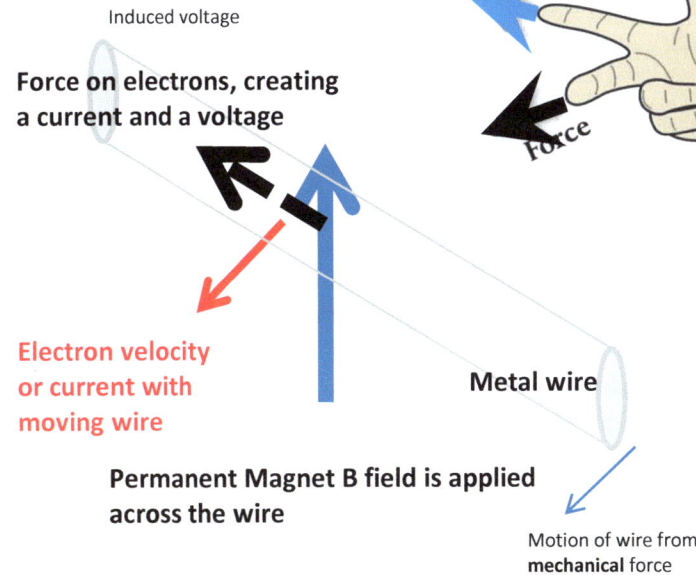

Induced voltage

Force on electrons, creating a current and a voltage

Electron velocity or current with moving wire

Metal wire

Permanent Magnet B field is applied across the wire

Motion of wire from **mechanical** force

For a generator the magnetic force (black) is along the wire due to the wire and its charges moving sideways in a magnetic field

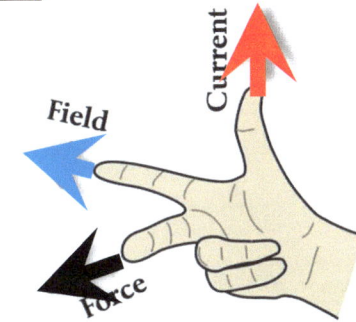

Current

Field

Force

This voltage generator validates Faradays's and Lenz's law:
The voltage generated by the coil is equal to the spin rate and the inductance of the coil.

$$V = \frac{d\phi}{dt} = \omega NAB = \omega LB$$

Equation G.7

See Appendix F for forces on moving electrons to make a current.

Force on a moving electron in a magnetic field is proportional to velocity and magnetic field. Both generators and motors use the same magnetic force.

Demonstrate Electricity Generation by Manually Spinning the Shaft

Motors can work in reverse and generate electricity. Something with energy (steam, flowing water, or a drill) or someone by hand needs to mechanically spin the shaft and a voltage will appear at the power inputs to the motor.

Motors can be a motor or a generator. Motors have the reverse case of electric power and a voltage making the motor spin. Generators have a spin which creates a voltage. This dual ability is used in hybrid or electric cars for regeneration.

- Spin the shaft by hand. For example, use your toy DC motor in reverse, similar to huge 20 foot wide power plant generators, to generate electricity from mechanical power.
- Attach a volt meter to the two leads of the motor. Just spin the motor, with a volt meter attached to the motor, and a voltage appears.

"Dude, what a digital geek. He should have used a meter with a needle, like its so much more visual."

Experiment 6.1: Spin shaft of motor (rotor) mechanically by hand, and measure voltage for generator.

Slower spin by hand:
0.13 Volts maximum generated

400 miliVolts direct current maximum scale

A coil moving through a changing magnetic field will generate a voltage

Lower electric voltage generation at slower spin speed:
Spin the motor shaft by hand, to get magnet moving through coils inside motor.

Faster spin by employing another motor:
1.7 Volts maximum generated

4 Volts max direct current maximum scale

Another motor to spin the first

Higher electric voltage generation at faster spin speed:
Spin the motor shaft with another motor, to get more speed or rpm of shaft.

The generator will get more voltage when spin the shaft faster. A faster moving wire or faster changing magnetic field will create more force on the moving charges and more voltage.

DIY Project Generator

Yes, you can quickly demonstrate a generator effect, not just a motor. Just physically spin the motor shaft (connected to the rotor), and a voltage will appear across the commutator (attached to the two motor leads).

Demonstrate Electricity Generation Manually with Kit

Spin the shaft by hand: use your toy DC motor in reverse to generate electricity from mechanical power, with same concept as huge 20 foot wide power plant generators.
- Spin clockwise to get a positive voltage on one side to light up one LED.
- Spin counterclockwise to get a positive voltage on the other side to light up the other LED.

"Dude, that looks tiring, like pedaling a bicycle to generate electricity to run your TV."

Experiment 6.2: Spin generator mechanically by hand, and light up the LEDs using generated voltage.

Here is a toy 2-pole generator spun using manual effort for spin.

Here is a toy 2-pole generator spun in both directions.

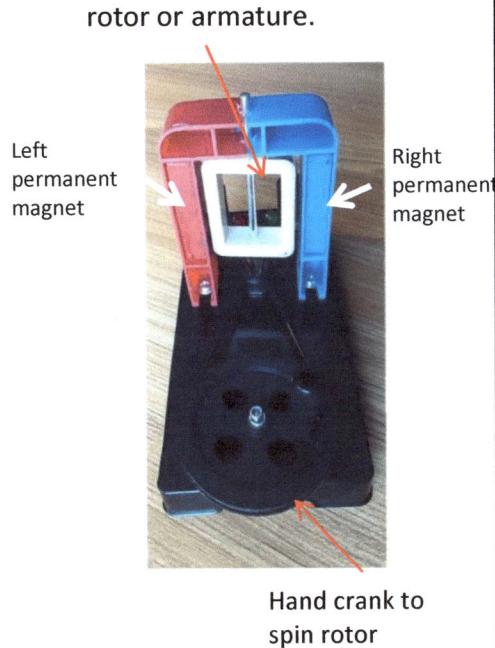

rotor or armature.

Left permanent magnet

Right permanent magnet

Hand crank to spin rotor

2-pole demonstration generator

EUDAX School DIY Dynamo Lantern
Educational STEM Building

Spin clockwise by hand: positive voltage on one side (green).

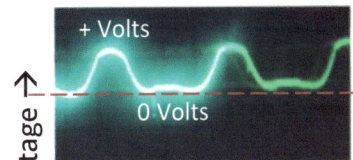

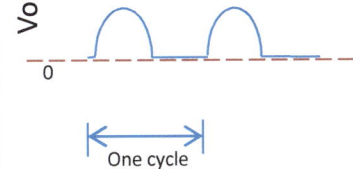

+ Volts
0 Volts

One cycle

Clockwise: Voltage drop across green diode

Spin counter-clockwise by hand: positive voltage on other side (red).

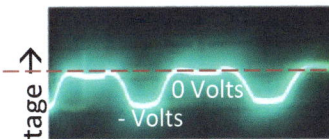

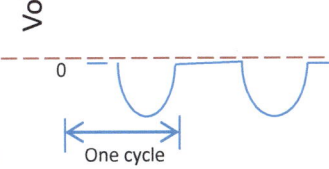

0 Volts
- Volts

One cycle

Counter-clockwise: Voltage drop across red diode

Other manual generators

Generator on bicycle wheel, for headlights

Pedal to re-charge a battery

Experiments: Get electricity by turning shaft of demonstration generator.

Induced voltage when spin rotor or armature lever by hand:
- A light emitting diode (green or red) only lights up when the voltage is positive on that diode, determined by clockwise or counter-clockwise spin.
- Commutator keeps the voltage all on one side.

Yes, you can quickly demonstrate a generator effect, not just a motor. Just spin the motor shaft (connected to the rotor), and a voltage will appear across the commutator (attached to the two motor leads).

Generators in Cars and Emergency Generators at Home

With standard combustion engine cars, your car is a mobile generator (alternator) and electric motor (starter), all rolled into one, all powered by the gasoline you pump into your car. Most people don't think of their car as a generator, but cars need to generate electrical power (12 Volts) for the spark plugs to ignite the gasoline vapor, re-charge the battery, and power the fuel pumps to bring gas to the engine, and also power purely electric things like the headlights, ABS / airbag / pollution sensors, and radio.

For a house, backup generators are also common for houses that are off the power grid, like cabins in the woods. Also, backup generators are convenient for houses even near cities that have un-reliable power due to frequent storm damage on overhead power lines from falling tree branches.

"Burn, baby burn, to get electricity to keep spark plugs sparking, headlights on and re-charge battery."

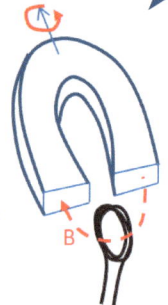

Alternating Voltage induced in Coil when magnet or coil spins

Experiment 6.3: Find the alternator in your car.

Car alternator or generator to power spark plugs and recharge battery: create the 12 Volts

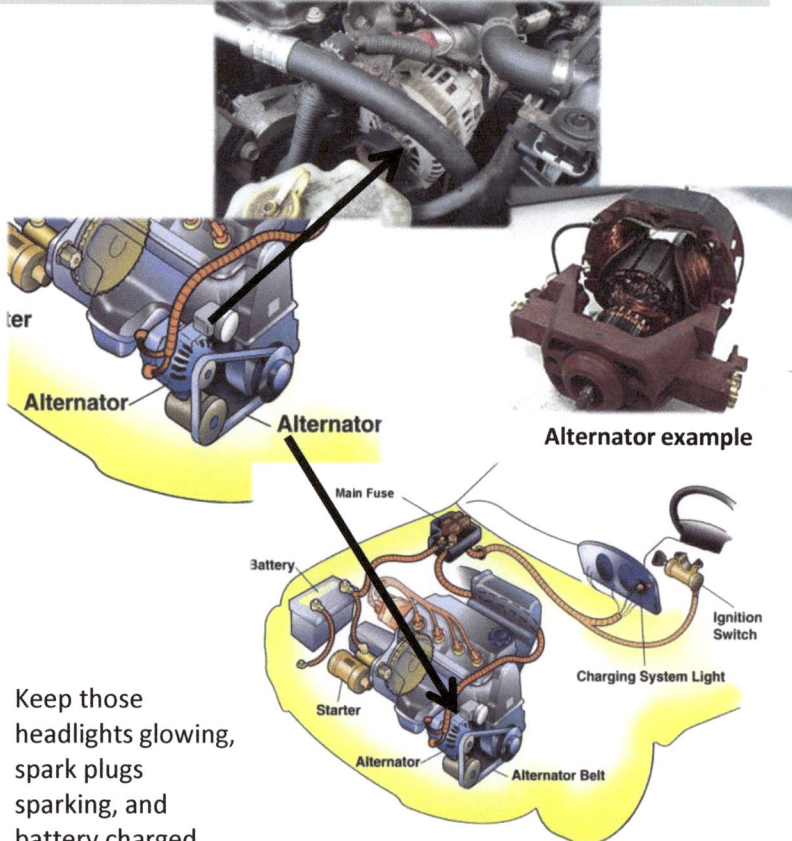

Alternator

Alternator

Alternator example

Keep those headlights glowing, spark plugs sparking, and battery charged.

Backup generators for emergency power outages

5 kW generator, runs on gasoline

During black outs, Mrs. Generator saves your refrigerated food and keeps the heat on by pumping coil or using electric heat, and may keep the water flowing if the house uses a well water pump.

Mrs. Generator is good when your power goes out for 1 week due to a falling tree on power lines in a storm, or a squirrel gets fried in a transformer.

Portable gas generators are useful.
Cars of course need generators, to be able to create sparks and re-charge the battery. Remote cabins way out in the woods might also need portable generators because there are no power lines. However, portable generators are expensive to run:

- Town power is cheap, at about 30 cents per kiloWatt-hour. The efficiency of large power plants, and their relative low cost, make home-grown power generators a challenge to justify based on cost alone.
- Small emergency generators are expensive to operate, based on the price of gas. A gallon of gas will provide up to 10 kiloWatt-hour of power, and it costs about $4 for that gallon. So there is a 1.5 times cost difference between emergency generators and the utilities.

Experiments: Find the alternator and belt and pulley that turns it near the engine

You've got generators at home. For example, every car has one.

Microphones (Speakers) and Guitar Pickups are Generators

Not everything about generators is a power plant, like spinning turbines and generators at coal plants. There are other ways that voltage generation is used. Magnetic fields are still changing through a coil.

Consider microphones, electric guitars, and exercise bikes with magnetic drag. We have microphones due to currents generated when a speaker cone vibrates with the sound. We have electric guitars because of little magnetic generators at the pickup coils. We have breaking for cars without contact friction due to induced currents and drag in rails due to magnets.

Both microphones and guitar pickups are examples of voltage generation due to changing magnetic fields, just like generators. These are all generators, baby!

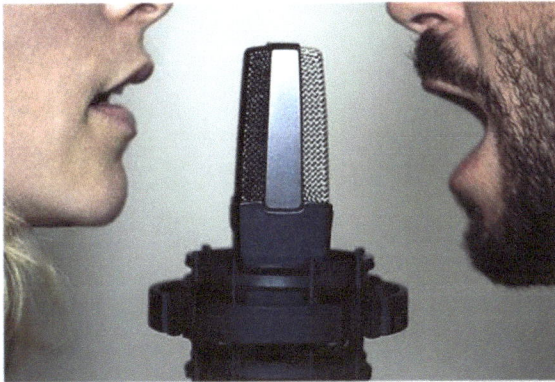

Generator of voltage: microphone

People talking into a microphone, and causing induced voltages and currents due to magnetic fields.
The microphone cone vibrates at the frequency of the sound.

Generator of voltage: guitar pickup

A magnetized string vibrates over a coil to create currents in the coil, at the frequency of the vibrating string. The strings must be magnetic, like iron or nickel, but not nylon.

Generator of magnetic drag: heavy metal disk

Exercise bikes have a magnet getting closer to a spinning large metal disk to cause currents in the disk and drag. The magnets are heating up the disk.
The distance of the magnets from the disk and the spin speed will determine the amount of current and drag.

Speaker and microphones are also reverse effects, just like motor and generators are reverse effects. Both effects are due to voltage generation when a magnetic field changes in a coil.

Microphones (Speakers) and Guitar Pickups are Generators

Both microphones and guitar pickups are examples of voltage generation due to changing magnetic fields, just like generators. These are all generators, baby!

When it comes to electric currents representing sound waves, we don't care about huge currents. We just care about a clean signal with little noise and high dynamic range (quiet sounds are still present and are not distorted by loud sounds) and no distortion or saturation.

Long established microphone, the reverse effect of a speaker

You can use a speaker as a microphone. When the coil on the speaker cone is vibrating near the permanent magnet, a voltage is created, just like a generator. From the point of view of the wire coil in the cone, the magnetic field is changing through it.

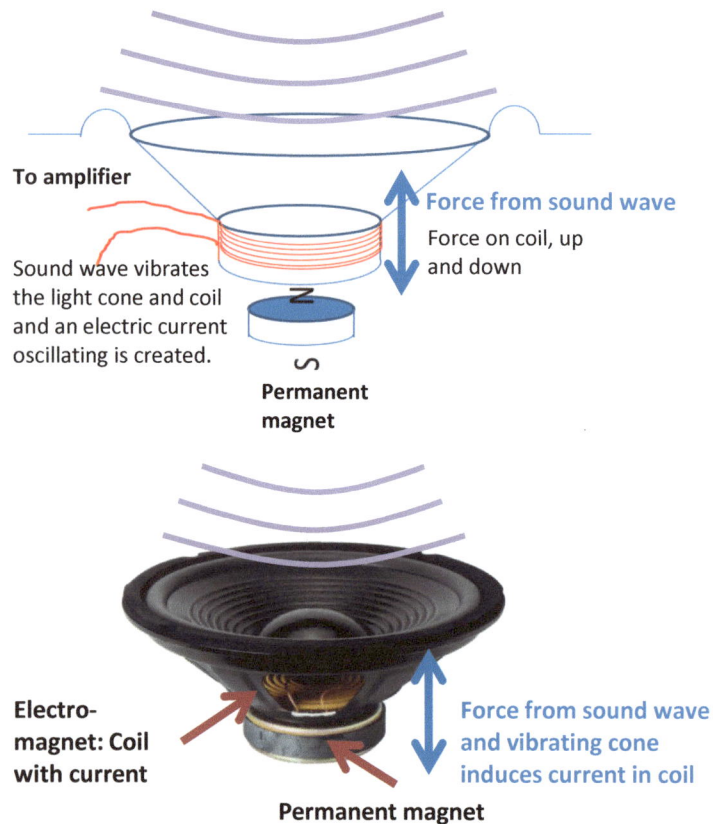

To amplifier

Sound wave vibrates the light cone and coil and an electric current oscillating is created.

Force from sound wave

Force on coil, up and down

N
S
Permanent magnet

Electro-magnet: Coil with current

Permanent magnet

Force from sound wave and vibrating cone induces current in coil

Speaker can both create sound when a current is applied to the coil, and create current as a microphone when the cone vibrates from outside sound.

Electric guitar pickup, the reverse effect of a speaker

A guitar pickup is another example of voltage generation due to a vibrating magnetic field. The iron guitar string is magnetized, and it vibrates over a coil. This coil is wrapped around a permanent magnet below the iron guitar string. The guitar string gets magnetized, and has its own magnetic field. As the guitar string vibrates, this magnetic field from the string is changing the magnetic field in the coil and there is a voltage generated.

Nylon strings or aluminum strings will not work for electric guitars, because they can not be magnetized. Iron (steel) and nickel strings work because those elements are magnetic.

In an electric guitar, there is one pickup coil and permanent magnet below each string to, first, magnetize the string, and, second, then measure the vibrating magnetic field.

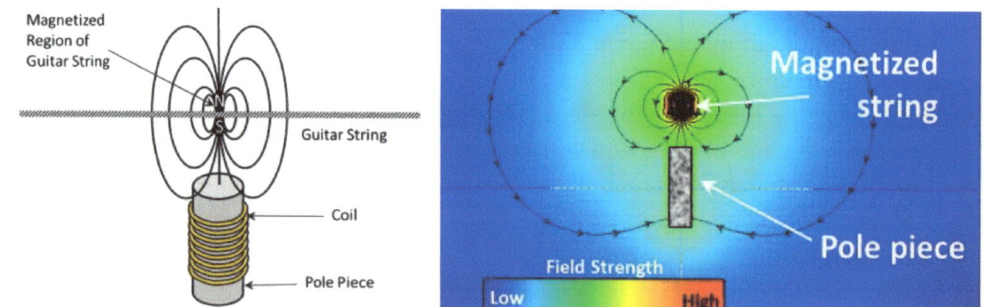

Magnetized Region of Guitar String

Guitar String

Coil

Pole Piece

Magnetized string

Pole piece

Field Strength
Low   High

Electric guitars need a permanent magnet below the string in order to magnetize the iron string. The coil is wrapped around the permanent magnet because that insures that the iron string has the most magnetization right at the coil for the largest signal.

Speaker and microphones are also reverse effects, just like motor and generators are reverse effects. Both effects are due to voltage generation when a magnetic field changes in a coil.

Magnetic Brakes for Trains, Wet Tracks Don't Matter: Drag from Fields in Metal, Generating Current and Heat ★

Magnets create currents and drag when moving over conductive metal. This can be useful for braking trains and exercise bicycles. The eddy currents induced in the metal are the same currents induced in a generator, except the currents are not harnessed to get power but instead just generate a reverse magnetic field and heat.

Experiment 6.4: Drop a penny and a permanent rare earth magnet on a conductive aluminum sheet, like below. Which is slower, and does the magnet feel the drag from eddy currents?

Magnets create currents and drag when moving over conductive metal. The mechanism is the same as generators. We are using the induced currents to create drag instead of power for electronics.

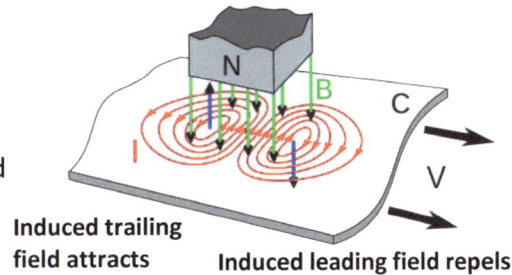

Induced trailing field attracts

Induced leading field repels

Eddy current braking of magnet over metal aluminum:

- Magnet slides down the metal ramp at a much slower speed, from the drag of eddy currents induced in aluminum.

- Aluminum has no magnetic attraction, just large electrical currents due to its low resistance when a magnet moves over it.

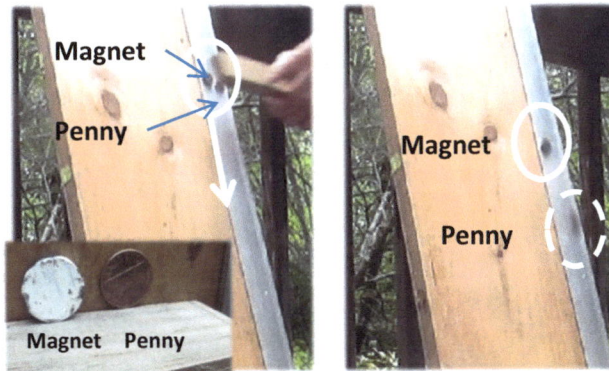

Eddy current drag on aluminum metal: Magnet and penny start at same time and magnet falls more slowly.

Braking using eddy current drag of moving magnet over a conductive metal:
- Bring a moving magnet close to a thick piece of metal, or raise a metal plate close to the moving magnet.
  - ✓ Instead of a coil, use a fat piece of aluminum (non-magnetic) metal moving through a magnet, or a magnet moving over aluminum, and all the induced currents just heat up the aluminum, and provide drag.
  - ✓ We're creating a voltage and electric power, like a generator, but, un-like a generator, we are immediately shorting the currents all out in the block of aluminum, creating heat and drag.
- The induced currents in the metal create their own magnetic field, which is delayed behind the moving magnet and keeps attracting itself to the magnet. This is a braking or backwards force.

Fun Facts: Applications of magnetic eddy current braking:

Magnetic eddy current brake for roller coaster

- Eddy current brakes don't care about rain or ice: they work just as well. This reliability is important for safety.
- There is no wear and tear.

Magnets cause drag on metal plates.

Eddy current brakes on the roller coaster Goliath

Japanese Bullet train: This train used to use eddy current braking, but has upgrades now using regenerative braking, like hybrid cars.

Eddy current drag for winter exercise bike.

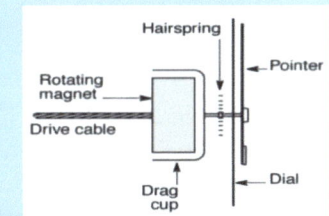

Eddy current tachometer in cars, to tell speed with a needle.

Experiments: Show eddy current drag on falling magnet over metal

The same currents to generate electricity can also generate drag, because it takes energy to create currents.

Summary of Generator Experiments

Here are the experiments to show what a generator can do and where they are, already explained in more detail in the preceding pages.

Go ahead, make a copy. Then check off the list.

Experiment 6.1: Spin a dc motor mechanically by hand, and measure voltage for generator.

- You could also spin a motor with a waterwheel to get consistent power, or use a drill to spin the shaft.
- A coil spinning in a dc magnetic field sees a changing magnetic field as seen by the coil. That generates of voltage.
- For an alternative explanation, the electrons in the wire feel a sideways force as the electrons and wire move sideways in the dc magnetic field.

A coil moving through a changing magnetic field will generate a voltage

Spin shaft of motor by hand to get electricity

Experiment 6.2: Spin a toy generator mechanically by hand, and light up the LEDs using generated voltage

- You might find is takes a tremendous amount of effort to keep a lightbulb going. Even a steady 10 Watts is exhausting for a human.
- Even long distance runners are only using 10s of Watts during the run. The health benefit comes from the increased metabolism during the entire day from the run.

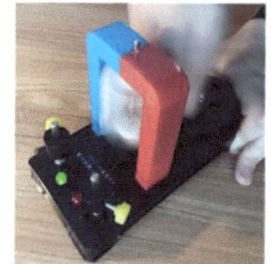

Spin shaft of 2-pole generator by hand

Experiment 6.3: Find the alternator in your car.

- Some belts and pulleys from the crankshaft should wrap around the alternator and spin it.
- The alternator makes ac voltage. The ac voltage goes to a regulator to make dc voltage, typically a little over 12 volts in a car, which goes to the battery to keep the battery charged.

Alternator/generator in car to get electricity, for spark plugs and lights
Combustion engine spins the generator shaft.

Experiment 6.4: Slide a penny and a permanent magnet down a metal aluminum sheet, like the picture on the right.

- We have non-magnetic pennies (just a weight) and a permanent magnet. Which is slower sliding and falling down the aluminum surface, and does the magnet feel the drag from eddy currents? If both are slid down wood, there should be no difference.
- Here are explanations of eddy current drag. The charges or electrons in the metal are effectively moving past the magnet and feel a sideways force. These forces create little swirls of current, which creates a delayed magnetic field behind the magnet. This delayed magnetic field attracts the magnet and slows it down.

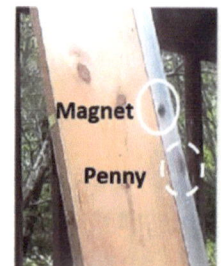

Magnet

Penny

Eddy current braking with magnets moving over metal

Use some mechanical force to spin the shaft of a motor, and electricity is generated.

Edison: DC Electricity Development ... a Little History

Thomas Edison had a fantastic impact on modern life, with electrical power to homes and lighting.

Mass Transportation and Power Generation: AC versus DC power

Edison: (1878, in his 30s)
- American
- Worked with lots of helpers
- Believed in huge effort, with many trials and failures before getting it right
- Nasty business competitor. Portrayed ac power as a way to kill people with high voltages, but then bought out the AC generator patents when AC current was proven better for long distance power transmission.

Inventions:
- DC Power plants (with help from others) using dc generator, 'Long legged Mary Ann'
- Long life light bulb, using carbon dust on cotton filament to last 10 hours. Modern incandescent (hot) lightbulbs use a meandered tungsten filament and last 1000s of hours. Also, for perspective, incandescent light bulbs are getting replaced nowadays by LEDs and fluorescent bulbs, which are more efficient.
- Sound recording using phonograph, using wax and plastic disks.

Net worth at death:
- Super Rich

Background:
Thomas Edison started work young by running away and being a telegraph operator. He knew the Morse code so well that he could instantly translate signals.

How did he work?
Edison worked as a team leader, and apparently back in the day he could claim credit for any and all ideas that originated from his laboratory, about 1000.
He and his employees tried 100s of different solutions in the quest to solve an engineering need. So he was a master at trial and error.

Did he ask for help?
Edison demanded hard working employees.

What did he invent?
Edison built a version of a DC generator, called the 'Long legged Mary Ann'. He also invented a more practical version of the light bulb using a longer lifespan filament, the play piano using paper with holes as the memory, and a mechanical version of a movie projector.
Edison developed DC power about 10 years before AC power took hold. Edison went to dramatic lengths to preserve his DC power market by accusing AC power of killing people due to the high voltages that AC power can generate. He scared people by saying AC power is used to execute convicted felons, with the high voltage. He even had dogs electrocuted to prove the point, and created negative publicity.
It is precisely these higher voltages that enable lower loss transmission on power lines, so an AC power plant can be more than a few miles away from the city. DC power needs fat wires, and needs a power plant within a few miles, or the transmission losses are too high. Hence AC power wins.

How was he known to the world?
'The Wizard of Menlo Park', and 'The Father of Electricity'

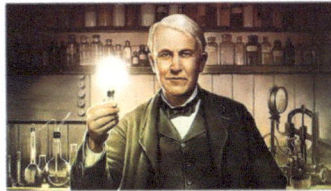

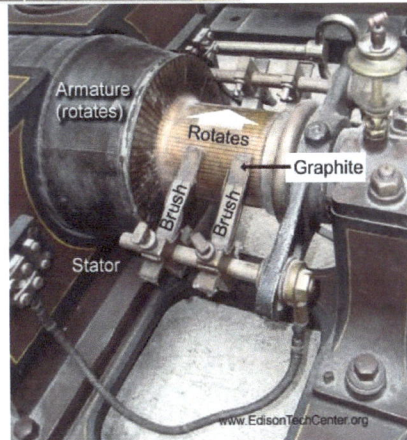

DC power generator, 'Long legged Mary Ann'
Rotate coils in a large dc magnetic field to generate dc power.
Large generators require an electromagnet to create the large dc fields.

Thomas Edison is mostly known for DC power and the light bulb.

Tesla: AC Electricity Development ... a Little History

Nikola Tesla made the long distance power grid possible, by creating AC generators and high voltage low loss transmission of the power along transmission lines.

AC power development in the United States: AC versus DC power

Tesla: (1887, in his 20s)
- Serbian, but worked in America
- Worked alone, at night
- Dreamer, always had a next technology revolution in mind
- No business sense, because he ripped up the contract for royalties from the AC motor and Westinghouse Electric (General Electric) when his AC invention was too successful and too much money was coming to him.
- Knew many languages

Inventions:
- AC motor and generator, for long distance power transmission, powered by hydroelectric power
- Tesla coil

Net worth at death:
- Zero

Tesla started the mass development of AC power in the USA.

Alternating electric current generator, March 10, 1891.

Background:
In his late 20s, Nikola Tesla came to the USA, New York, to work for Thomas Edison. Tesla had already developed ideas for an AC motor in Serbia, while at a technical university. AC power was already well known in Europe.

Tesla came to the US and worked for Edison's company. However, Tesla's ideas and patents for AC motors and generators were getting ignored, so Tesla left after only 6 months. The Edison company apparently did not want to build a competitive technology to its DC power grid.

After leaving Edison after 6 months, Tesla worked as a laborer and struggled to get by for 1 year, in complete misery. Then Tesla received support from Westinghouse to develop his patents, where he more fully developed the AC motor and generator. His generator was first used at the large hydroelectric power plant at Niagara Falls.

How did he work?
He worked alone, typically at night with the least distractions.

Did he ask for help?
He did work with technicians to build his motors, and he did license his ideas. However, his inventions were typically done alone.

What did he invent?
Tesla is best known for inventing versions of AC motors and AC generators, and a wireless controlled boat.
His generators and motors use poly-phase current to create a rotating stator magnetic field, instead of using a commutator. The rotor needs to rotate at the same frequency as the alternating current to keep getting a torque in the same consistent direction.
Tesla also invented the Tesla coil, where his goal was to wirelessly power everything. He thought that wires were not needed, and that everything could be powered using pickup coils. Unfortunately, that has not proved to be practical or efficient. At 60 Hz, everything else around also feels the magnetic field and absorbs it with losses. 60Hz has a super long wavelength.

How was he known to the world?
He was partly a showman, and he performed stunts at the World Fair and allowed electric current to flow through his body to power his various types of lighting.

Nikola Tesla is mostly known for AC power for a practical long distance power transmission system, and fantastic dreams.

Generators, Headlights, and Lights in Your House, Q and A

Everyone needs electricity to have a modern lifestyle of lights, cars, and computers.

Question 1: When did generators first start getting used? Over a century ago, electricity and generators replace whale oil and gasoline for lighting.

- In 1879, electric generators powered arc lamps for street lighting.
- In 1882, Edison used generators to light up a house with electric bulbs, not burning candles, using DC current. To see the house, visit Hearthstone near New York city, at the Appleton house museum.

Question 2: What were the first applications of generators, for local power plants for cities?

- Powering street lights,
- Powering electric trollies to reduce congestion and horses on city streets.

Question 3: Does your car depend on a generator? Yes, electricity puts a spark in your engines.

- Yes, the battery needs to be constantly charged by the onboard generator, called the 'alternator'. This 12V battery is used to power the lightning sparks that ignite the gas, and this charge is used for all other electrical needs, like headlights, the sensors (fuel level, oxygen exhaust to adjust air-fuel ratio, carbon dioxide emissions, engine temperature), and the dashboard.   Basically, the car is a self contained power plant, with motors and generators.
- For spark plugs, the 12 V is up-converted to a 1000 V spark using a coil or transformer.

Question 4: Does power to your house depend on a generator? Yes, we don't get electricity without a generator somewhere.

- Absolutely, some hot steam somewhere (probably within 100 miles) is turning a large generator rotor to make voltage and electricity. The steam is created from burning coal, burning natural gas, or heat from nuclear fission.
- More 'green' power sources with generators are taking the energy of water flowing down hill and turning a hydro-electric turbine and generator, or windmills turning a generator on hilltops or open ocean.   Solar power could be solar thermal, where sun-heated fluid turns a generator. In contrast, solar power can also be solar panels, which generator a voltage directly from sunlight, without turning a coil.

Question 5: How much of the electricity on the grid comes from solar panels? The percentage is growing, but coal, natural gas, and nuclear power still dominate.

- Only a small part of the electricity to your house comes from solar panels.   Power plants using solar panels are becoming more common, so you could say solar power is the faster growing source of power.
- About 3% of the USA power grid comes from solar panels and solar-thermal power.   Solar panels use semi-conductors, which exploit modern silicon processing.   Solar thermal is something completely different, which focuses sunlight to heat a liquid and turn a generator.

Question 6: Are large power plants more efficient than portable generators? Yes, the temperature of combustion is higher.

- Absolutely, larger power plants are more efficient using higher temperature turbines.   A large power plant is about 60% efficient using the high explosion temperature in a turbine engine, and a portable gasoline generator is only about 30% efficient using the lower explosion temperature and lower pressure in a combustion engine.   Try running your house using a portable generator, always going to the gas station to get more gas, and compare the monthly energy bill of the portable generator to the bill from a utility company using a large power plant.
  - Let's compare portable generator and city electricity costs. A portable generator which is on for about 10 hours a day will use a least 5 gallons of gas, which is about $15.   One month will add up to $450. In comparison, a typical electrical bill at a house per month from the utility power grid is $100, without any drive to the gas stations and with a full 24 hours a day service.

Arc lamp for streets

Spark plug, with high voltage sparks snapping away to ignite gasoline vapor in engine cylinders

Headlights, with many amps of current

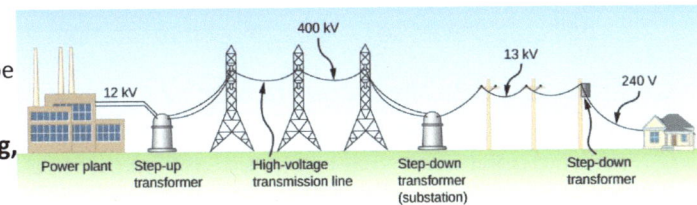

Generator at power plant to house, with power transmission over 100s of miles, thanks to voltage transformers and AC power to up convert and down convert.

During transmission line repair and power outages after a storm, portable generators keep your food refrigerated and your heat on. There are many trips to gas station to fill containers and fill the tank of the portable generator

150 years ago, steam engines ruled the rails all across the country on train engines. These engines powered the wheels directly, without the go-between of generating electricity.

Do we still use steam engines? Yes. Now, instead of moving trains, steam turbines spin generators and create electricity for all of us.

"Dude, this toy steam engine is fun, a very real benefit to people …Steam engines are the core of power plants... Go get one of these toys."

Generators powered by burning something to create hot steam:

How do we know that steam generators have reached the big time? Well, there are toys of steam powered turbines or flywheels. Seriously, steam generators realized the big time over 100 years ago, but, still, there are toys.

Coal is plentiful in the USA and can heat the steam. Hydro power is also plentiful, but most suitable rivers have already been used.

Steam was a natural way to spin turbines and generators. People in the 1800s already were powering trains and steam ships using steam so they just used the steam to spin a turbine and generator.

Nuclear power was invented in the 1950s and 60s. Nuclear power is just another way to heat steam and turn turbines and generators.

Oil was also available. Oil and diesel were so available that factories did not need to be near rivers for waterwheels and hydroelectric power, so factories moved to locations with cheaper labor.

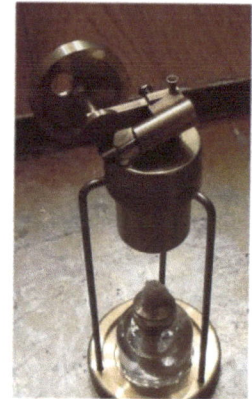

Toy steam engine

Generators that use alternatives to burning something: Hydro-electric water flow, windmills and solar panels:

Large rivers and hydroelectric power have been used for 100 years, and already create 6% of the power in the US.

Now wind power and solar cell power are becoming effective. Both are not on 100% of the time, but they can fill in peak periods from other power plants.

Wind power can happen during most days with moderate wind. The winds over the ocean can be very stable, with no land structures to block the wind. Also, the wind going up the side of a mountain is strong, as the air flow needs to get over the mountain.

Solar power happens mostly during the day and mostly during the summer. Some climates are just asking for solar power. Consider the summer in Texas or Arizona, when in mid day there is peak use of air conditioners. This is exactly when solar panels will be generating the most power.

Toy hot gas engine, lighting up an LED light

Power distribution from the generator power plant to people's homes:

How does the energy at power plants get distributed to cities and towns? Well, there is a huge power line infrastructure around the country and around the world. There are transformers to raise the voltage and avoid some resistive losses over 100 miles of transmission lines. Near the consumer, there are down-converting transformers to get to the typical 120 volts at the businesses and homes.

Fans inside of a steam turbine

Make a toy steam engine spin, using a flame and water. The toy is only a few inches high, but will turn the flywheel for a few minutes, until the water boils off and disappears, in an open water cycle. Hot steam has more force than hot air.

Examples of Generators for Power Plants:
Powering a Generator Using Steam: Flame →Toy Steam Engine

Get a toy model of a steam engine, and see first hand the spinning flywheel. The flywheel is powered from escaping steam from boiling water. A large spinning flywheel, in a small or huge electric power plant, could turn an electric generator or power a train.

"Dude, this toy steam engine is fun, a very real benefit to people ...Steam engines are the core of power plants... Go get one of these toys."

Experiment 6.5: Start a toy steam engine with a kit and flame. Flame, steam, and a piston push a flywheel.

Steam engine demo kit 1, just a spinning flywheel

Steam engine cutaway, with piston and wheel

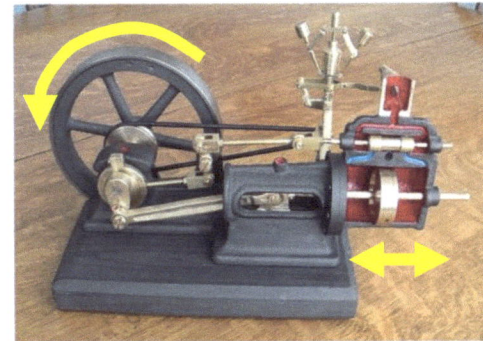

I need my electricity!

Before heat and water: Flywheel will be turned with the piston, when flame starts

With heat and water: The piston is powered by steam, which spins flywheel

Cutaway of toy steam engine with piston and wheel

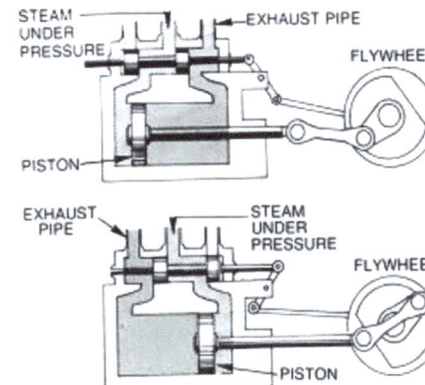

I'm not strong enough to run a treadmill generator all day, and generate maybe only 20 Watts! TVs take over 100 Watts.

Reservoir for boiling water

Flame for heat

Flame and escaping steam: Water boils off and the high pressure steam pushes a piston. The water is gone, and needs to be refilled. This is called 'open cycle'.

*Sunnytech Mini Hot Live Steam Engine

STEAM UNDER PRESSURE — EXHAUST PIPE — FLYWHEEL — PISTON

EXHAUST PIPE — STEAM UNDER PRESSURE — FLYWHEEL — PISTON

Steam engine and piston: Just like a combustion engine, the piston needs a valve to let steam enter, and a valve to let steam escape.

Experiments: Rev up a toy steam engine

Make a toy steam engine spin, using a flame and water. The toy is only a few inches high, but will turn the flywheel for a few minutes, until the water boils off and disappears, in an open water cycle. Hot steam has more force than hot air.

Powering a Generator Using Hot Air?
Flame→Air Engine→Generator

Here is a toy hot-air engine, single cylinder. It will run until the alcohol runs out for the wick and flame. As a toy, hot air is used to make the flywheel spin, but air does not have much density. The piston concept is shown, but a hot air engine is not a practical engine to spin a generator.

More practical generators use steam or hot flowing liquid with higher energy and more force, but air makes the point.

"Burn, baby burn, to get electricity."

Experiment 6.6: Start a toy hot air engine.

Ok, this air engine demonstrates the concept, but hot air, just slightly hot, won't make enough electricity.

Both the hot air or steam are heated by burning coal or wood or oil. Steam just has more energy.

Ahh, yes, now I know why steam, not hot air, from burning coal and wood became more useful and popular ... Steam just has more energy than air.

3. Piston turns flywheel
- Heavy steel disk smooths out the pushes

2. Hot air engine: Piston pushed out by expanding hot air

Flame from Alcohol

4. Electric generator spins, turns on light

1: Air from outside cool room comes in here

Air piston: Room cooler air is sucked in as the piston comes back in, heated up with the flame, causing the air to expand and push the piston back out again.

Tabletop demonstration of this hot air generator makes you think

*Sunnytech Hot Air Stirling Engine

Experiments: Rev up a hot air engine and attached generator

A hot air piston does not carry the same 'umph' as a steam piston, but it demonstrates the generator concept.

Old Fashion Steam Engines: Flame → Steam Engine

For generators, steam engines do, and did, create electricity for all of us. To spin an electric generator, something hot needs to create steam, and push pistons (in the old days) or turbines (current days), which spin the shaft of a generator.

For transportation 150 years ago, steam engines ruled the rails across the country on train engines. Instead of generating electricity, steam directly powered the wheels using pistons. Water was boiled in the engine using coal or wood, from the tender or coal-car behind the engine. Steam and pistons also powered huge ocean liners like the Titanic, using coal for heat.

"Dude, steam powers generators. Also, steam ruled the rail, just like 'Thomas the Train', or the 'Little Engine That Could'."

Good old days and modern days of steam powered generator

Massive 1907 steam generator

Good old days of steam travel

Steam engines opened the West!
People could cross the United States quickly and directly. Also, cattle could be shipped back to the East coast in cattle cars.

Before steam engines, travel across the country used horse and wagon for 60 days, or used a ship going around South America or crossed the narrower Central America using lakes.

For locomotives, steam engines which don't use electricity at all gradually got replaced with diesel engines.

Water refill:
The locomotive water is boiled off as steam, and needs to be refilled every 50 miles or so.

Steam engines are not so good for cars:
- Can't turn off heat during time in grocery store for ½ hour.
- The startup time is a long ½ hour to get the steam.
Trains on the other hand have steady use for hours, so the long distance train application does not care about the idle time to turn off. For trains the steam engines are very appropriate.

Steam engine train:
Steam engine trains started in the 1820s, burning coal or wood. No electricity or liquid fuel is used.

Titanic ocean liner:
Steady steam and 10 day steady speed.
- Ocean liners are perfect for steam power.

Steam engines are still used to power large generators, although more up to date turbines are used instead of pistons.

Steam Power Plants: Heat-Source and Electric Generators

Heat Power: Any fuel or heat source can be used to heat water, then generate steam, which then turns the generator shaft and generates electricity.

Experiment 6.7: Don't believe that steam is used in power plants? Go and see the hot water or steam exhaust of your local power plant, be it powered by coal or nuclear or natural gas (methane) or oil.

The main practical questions for fuel or heat sources are:

- how expensive?
- how reliable the supply?
- how polluting?

Burning Coal heats water

- Coal has been around for over 200 years, dug out of the ground, from the start of the industrial revolution.

Nuclear Reaction heats water

- Nuclear power sounds high tech compared to coal, but nuclear fuel and its heat is just another heat source.

Man, it's hot in here.

Burning Natural Gas heats water

- Easy distribution using pipes for the gas (yes, a gas state, not a liquid)

Burning Oil heats water

- Oil or petroleum is not that common for power plants recently, because oil is better to use as gasoline for transportation cars and fuel for airplanes, or material for plastics.

Most power plants:

Make steam using heat

Shovel in the coal in olden days

Nuclear fuel rods

Pop culture

Flames from natural gas (methane)

Oil

"Dude, everything has issues. Coal is dirty, nuclear waste is hard to store for over 1000 years, and bio oil (made from corn) for cars just competes with food.
But, man, we have the cleanest air we've had since the 1940s. Something must keep improving."

Spin turbine using steam

Heat makes steam, which turns a turbine

Fans inside of a steam turbine

Spin magnet in coils* to get electricity

Turbine spins magnets in coils which makes electricity, which is the power going down transmission lines.

Generator: Make me spin, and I'll give you electricity

The amount of steam allowed into the turbine is regulated by the power draw in the city. When the city all turns on their air conditioners all at once, then the turbine gets harder to turn and the power plant releases more steam through the turbine to generate more power and keep the same 60 Hz spin rate.

*Generators can spin magnets in coils or spin coils in magnets. Both generate a voltage.

Experiments: Go see the steam come out of a power plant

Here are three basic ways to heat water to make steam and push a generator around: Coal, Nuclear, and Oil.

Various Power Plant Examples in New Hampshire

Let's look at the many generating power plants, using magnetic generators, in just one small state: New Hampshire. New Hampshire uses mostly natural gas, nuclear, and imported electricity from hydroelectric plants in Canada.

Other states could use more coal, if coal is local and inexpensive. Other states could have and use more wind, if have large open plains, or open water like oceans and lakes, or mountain ridges.

Here are generating stations in New Hampshire. There are three large generating power plants, each with more than 400 MegaWatts output. There are many small generating stations, with a few MegaWatts output.

New Hampshire has ~ 1.5 million people, with ~3000 MegaWatt capacity max, so there is about 2 kW power capacity per person. When you consider that a typical house uses 0.5 kW on average, and there are workplaces (shops, factories), this power capacity makes sense.

Q: Why did the gardener plant a light bulb?
A: He thought he would get a power plant!

"Some big plants, some small plants, harvest energy where ye may."

Utility	Plant Name	City	State	MegaWatt
Algonquin-Cambrian Pacific GenLLC	Four Hills Nashua Landfill	Hillsborough	NH	3
Algonquin Power Systems Inc	Lochmere Hydroelectric Plant	Belknap	NH	1.2

Nuclear Power Plant: Seabrook
~1200 Megawatts
(1,200,000 kilowatts)

Lumber Co	Lumber	Sullivan	NH	6.1
Errol Hydroelectric Co LLC	Errol Hydroelectric Project	Coos	NH	3
Foss Manufacturing Co Inc	Hampton Facility	Rockingham	NH	11.9
FPL Energy Seabrook LLC	Seabrook	Rockingham	NH	1242
Franklin Industrial Complex Inc	Franklin Industrial Complex	Merrimack	NH	1.9
Fraser NH LLC	Berlin Gorham	Coos	NH	40

Natural Gas Power Plant: Newington Power Facility with a design capacity of 605.5 Megawatts (605,500 kilowatts)
- Natural gas (methane gas) is delivered in pipelines from Canada and Maine.

Co	Newfound Hydroelectric	Carroll	NH	1.4
Newington Energy LLC	Newington Power Facility	Rockingham	NH	605.5
Pembroke Hydro Associates	Pembroke Hydro	Merrimack	NH	2.7
Pinetree Power Inc	Pinetree Power	Grafton	NH	17.5
Pinetree Power Tamworth Inc	Pinetree Power Tamworth	Carroll	NH	25
Plymouth Cogeneration LP	Plymouth State College Cogeneration	Grafton	NH	2.8

Coal Plant: Merrimack Power Station Plant
~500 Megawatts
(500,000 kilowatts)
- Coal plant is only used for over load, typically in winter.

Public Service Co of NH	Lost Nation	Coos	NH	16
Public Service Co of NH	Merrimack	Merrimack	NH	496.4
Public Service Co of NH	Schiller	Rockingham	NH	171.2
Public Service Co of NH	Smith	Coos	NH	15
Public Service Co of NH	White Lake	Carroll	NH	18.6
Public Service Co of NH	Newington	Rockingham	NH	414

Comment: If desire to visit a power plant, please respect the private property rights of the facility. Generally a request for a tour needs to be made, if tours are even available.

http://www.powerplantjobs.com/ppj.nsf/powerplants1?openform&cat=nh&Count=500

Small hydro electric power plant: Mine Falls ~4 MW

Even a small state like New Hampshire has the full spectrum of power plant energy sources, using standard nuclear, natural gas, and coal plants to spin turbines.

Coal Power Plant

Coal Power: Here's an outside look at a coal power plant. This coal plant actually hasn't been used for more than 2 years because coal is expensive compared to natural gas and nuclear in New Hampshire. Tours were available before 2015, but not anymore.

Other states with a more local source of coal, such as Pennsylvania or West Virginia, would have more competitive rates for coal plants.

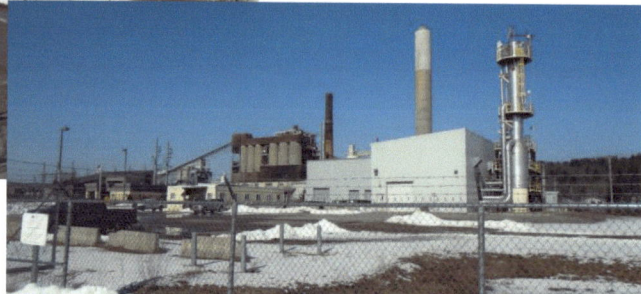

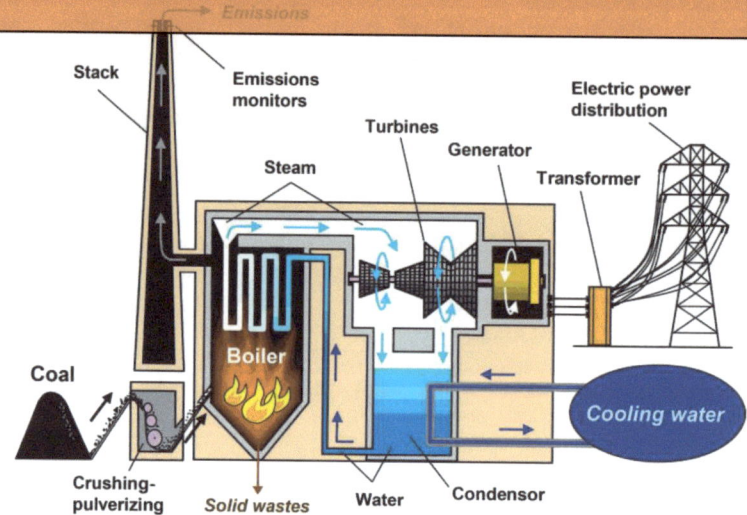

Schematic of plant:
Steam from the boiled water turns the turbine.

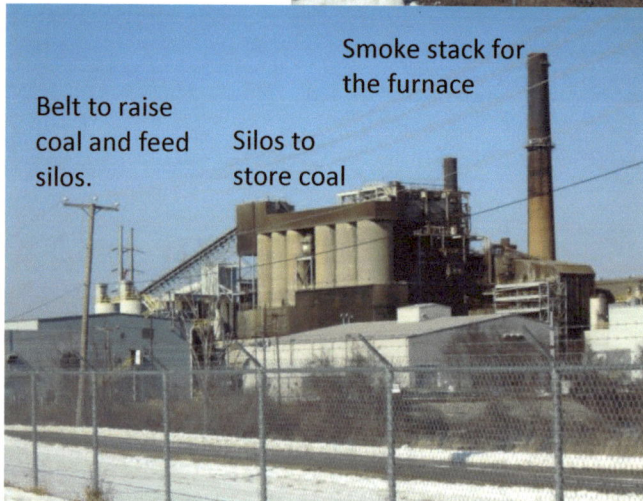

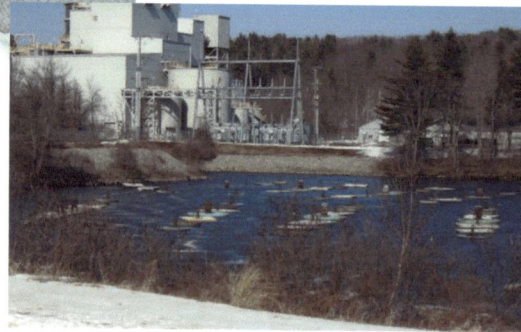

Smoke stack for the furnace

Belt to raise coal and feed silos.

Silos to store coal

Water feed from river:
River water or 'cooling water' is used to cool and condense the steam power for the turbine and to cool the parts of the furnace.
The water intake system uses fine mesh screens to remove fish, and a sluice to return trapped fish to the river.

Merrimack Station Coal plant, NH:
This coal station is turned on only during overload conditions, typically in the winter. New Hampshire in New England has no local sources of coal. Coal takes more manpower to handle and is more expensive to run than natural gas, or renewable sources like wind and solar which require no fuel at all.

The coal plant in New Hampshire was heavily used up until 2010, but then other power plants – natural gas and nuclear – have proven less expensive to operate. So now this coal plant handles high demand periods, but not base power levels (levels during non-peak typical demand).

Hydroelectric Power Plants: Cold Flowing Water and Huge Turbines

<u>Water Power</u>: Hydroelectric power plants are already on major rivers across the country. The turbines spin due to flowing water, usually at the bottom of a dam.

In general, turbines linked to generators at power plants convert mechanical spinning into electric energy by moving magnets through coils. In the turbine, the shaft moves due to hot steam, turning the generator. In the generator, magnets move past coils and electricity is generated in the coils, like a car alternator. This is the reverse of an electric motor.

"These must be my big and little brothers - Large hydroelectric dams, and a tiny water flow from a stream."

Experiment 6.8: Look for the water run-off down stream of the local hydroelectric dam.
- Water needs to flow out with some speed (not zero) because water needs to leave the turbine and not back up the incoming water. This is why turbines can never capture all of the gravitational energy of the water, because there needs to be energy for the exhaust water to get out of the way.
- Some hydroelectric dams have fish ladders, which are cascaded pools going up the side of the dam. During the spawning season, visitors can see fish through glass windows swimming up stream and fighting the fish ladder.

Gigantic water power at Hoover dam

Water spins the shaft on these generators at Hoover dam

Largest hydro-electric power plants in US

Go on a 'nerd' trip, and search for power plants. Maybe you can get a tour of a hydro-electric power plant. Historically, the first hydro-electric generator was at Niagara Falls, using a DC motor.

Rotor under construction

These 500 MW power plants have huge generators, with huge rotors. These rotors needs to be perfectly balanced, or the bearing will melt down. If these huge rotors go out of balance, they vibrate and self destruct.

Small water power

Small stream can provide power
- Water going downhill will also turn a turbine and generator. This small size water flow can probably provide a few watts to charge a battery.
- For this stream, there are probably no salmon swimming up stream, so there are no wildlife concerns.

Experiments: Drive across a dam, see the run-off stream

Water flowing down rivers has huge energy. Let's use it partly for electricity.

Wind Power Plants, Up and Coming

Wind Power: A generator can also be powered by spinning a turbine using wind power, instead of hot steam.

Large windmills are installed on the ocean, where the air currents are strong. As sailors discovered long ago, the wind can blow strong and steady over the ocean. There are no trees or buildings on the ocean to break up the wind.

Large windmills are also installed on top of hills, where consistent wind blows because air from below is forced over the hill.

Q: Why is wind power popular?
A: Because it has a lot of fans!

"Windmills, I'm free and non-polluting, but only provide 1% of the power in the world."

Experiment 6.9: Find a windmill near you, that simple. It could be on a farm, or in a coastline bay, or on a mountain top, or along the open plains, wherever there is steady wind.

wind

Wind power, with 3 blades, 100 foot each, for high winds

Large wind turbines notably found in Denmark and California and Texas.

Wind power with helix shape, for moderate winds

Smaller wind turbines in less wind on household roof tops.

Toy windmill with turbine generator, experiment kit

red LED light.

Toy windmill with air from fan, from Kosmos

Wind turbine is powering the red LED light.

...must find wind!

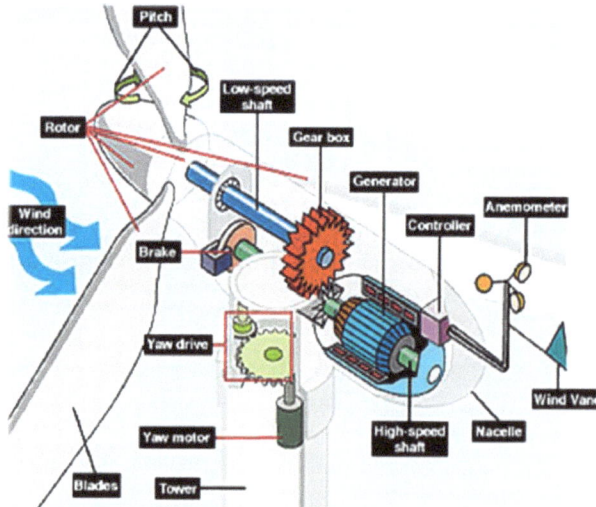

Pitch
Low-speed shaft
Rotor
Gear box
Generator
Wind direction
Controller
Anemometer
Brake
Yaw drive
Wind Vane
Yaw motor
High-speed shaft
Nacelle
Blades
Tower

Electric generator inside pod or 'Nacelle'

Note: Windmills need the right location with just the right wind. Some older windmills were placed in locations without enough wind, or even with too much wind that would damage the blades.

Windmills, besides electricity, have other purposes too. If you live in the Netherlands, then some windmills do not generate electricity but directly turn a water pump to irrigate the fields or pump ocean water back over the dikes.

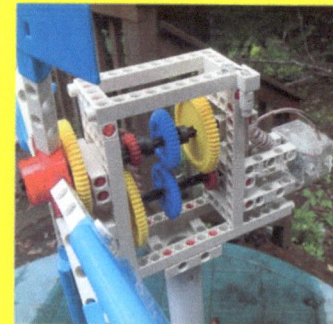

Spin a little generator. The spin rate is increased from windmill to generator using gears.

Gears increase spin for generator

Experiments: See if there are any windmills near you

Let nature, wind, geo-thermal, water dams, generate the power instead of burning fuel. These are 'green' because there is no burning fuel, and they do little damage to the environment.

Wind Power Plants, Up and Coming

Wind Power: A generator can also be powered by spinning a turbine using wind power, instead of hot steam.
Even though the blades spin rather slowly, there is a lot of power or torque behind them. The shaft speed is geared up to provide more power from the generator.

Q: Why is wind power popular?
A: Because it has a lot of fans!

"Windmills, I'm free and non-polluting, but only provide 1% of the power in the world."

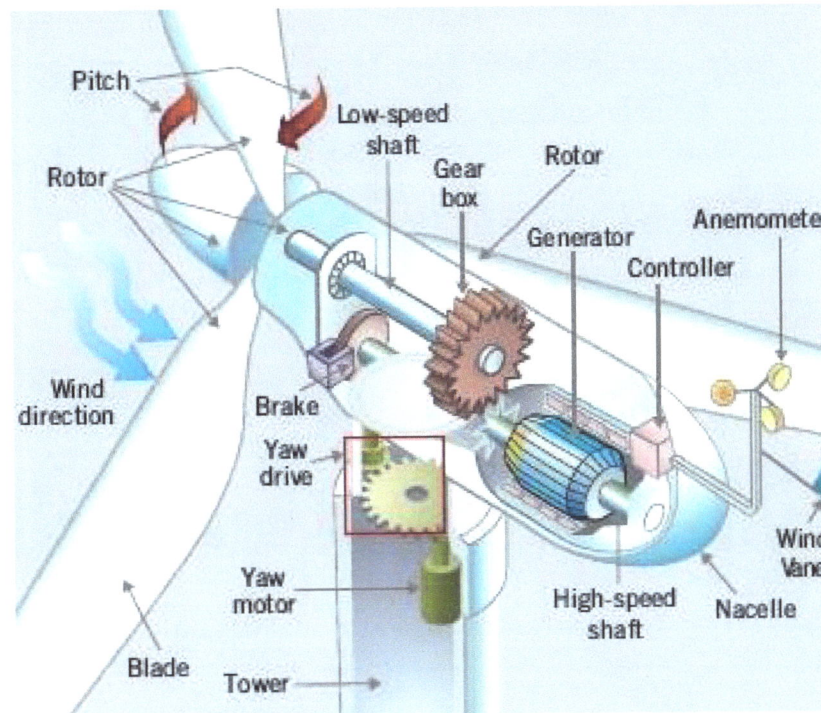

Pitch, Rotor, Low-speed shaft, Gear box, Rotor, Generator, Anemometer, Controller, Wind direction, Brake, Yaw drive, Yaw motor, Blade, Tower, High-speed shaft, Nacelle, Wind Vane

The generator is similar to a DC motor with a permanent magnet.

The low speed shaft is connected to the blades. The low speed shaft of the blades is geared up to much faster spin rates using a gear box.
The blades are pitched to provide the most power for a particular wind speed. Typically more pitch means more speed and power.
When there is a hurricane, then the blades are angled all the way back so there is no force and torque on the windmill, for protection against spinning too fast.
The transmission gear ratio should be designed so that all the power possible is extracted from the wind, and the wind blades are feeling the drag of the torque on the generator. If the blades just spin with no generator drag, then we are not extracting energy.
High wind speeds can tolerate a lot of generator drag, so the gear box should really be in high gear where the generator is spinning very fast.
Low wind speeds can not tolerate a lot of generator drag, so the gear box should be in low gear where the generator is only spinning slowly.

Let nature, wind, geo-thermal, water dams, generate the power instead of burning fuel. These are 'green' because there is no burning fuel, and they do little damage to the environment.

Power Plants: Hot Steam Turns an Electric Generator

On the vast national power grid, traditional heat sources – coal, nuclear, natural gas – are providing the vast majority of electrical power to the country. More environmentally friendly 'Green' sources – water, wind, and solar – are less than 10%.

The power needs of cities and countries are huge. The power industry is huge. It is difficult and limiting to drop any source of energy. We all love our refrigerators, heating systems, air conditioners, and TVs, and let's remember we need power for businesses and factories.

As a backup, a small personal home generator is great when your town electricity goes out during a winter storm and you want the heating furnace to run (pump fuel, sparks to start the flame of the oil or gas), water to flush the toilet, water pipes to not freeze, or summer hurricane and you want the refrigerator with an electric motor to keep the food cold.

Q: What did the light bulb say to the electric generator?
A: "You spark up my life!"

How does the country generate its electricity, and what source of electricity is most common?

- There are heat sources for generating steam and pushing a generator, which power most of the electrical grid.
- There are also non-heat generators, like wind power, hydro-electric, and solar panels.

If the town power goes out, then start an electric generator spun by a gasoline combustion engine.
Gasoline typically is used in cars, not at power plants.

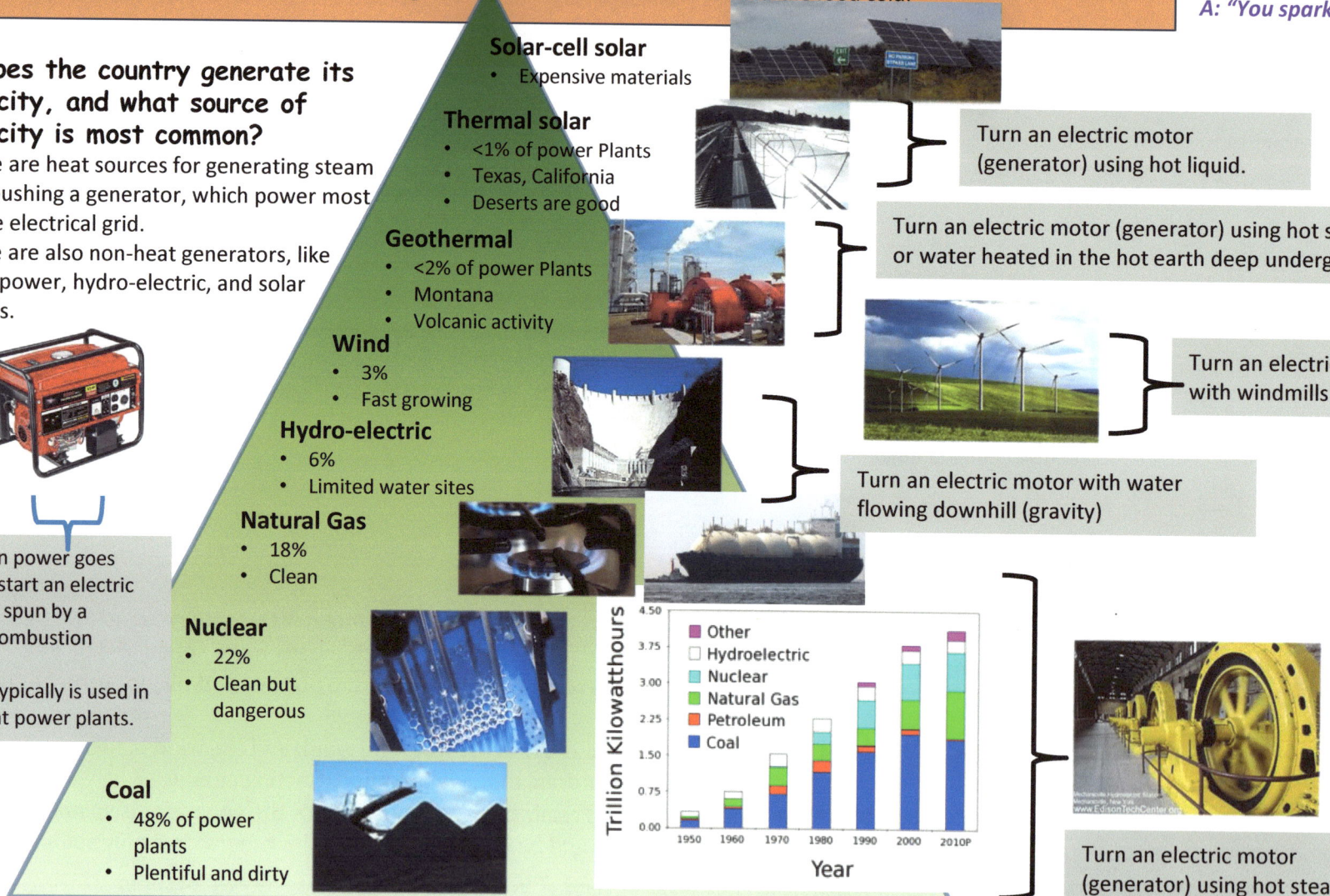

Solar-cell solar
- Expensive materials

Thermal solar
- <1% of power Plants
- Texas, California
- Deserts are good

Geothermal
- <2% of power Plants
- Montana
- Volcanic activity

Wind
- 3%
- Fast growing

Hydro-electric
- 6%
- Limited water sites

Natural Gas
- 18%
- Clean

Nuclear
- 22%
- Clean but dangerous

Coal
- 48% of power plants
- Plentiful and dirty

Turn an electric motor (generator) using hot liquid.

Turn an electric motor (generator) using hot steam or water heated in the hot earth deep underground.

Turn an electric motor with windmills

Turn an electric motor with water flowing downhill (gravity)

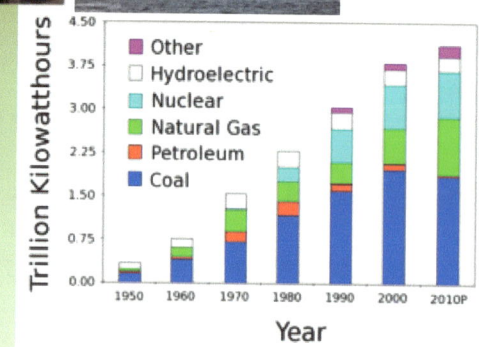

Turn an electric motor (generator) using hot steam.

Chart: Trillion Kilowatthours by Year (1950–2010P); legend: Other, Hydroelectric, Nuclear, Natural Gas, Petroleum, Coal.

There are a diversity of energy sources for our electricity, and most power plants rely on heat and steam and magnetic generators.

Non-Renewable and Renewable Energy Resources

The USA has huge land area and huge natural resources, both non-renewable like coal, petroleum, and shale, and renewable like solar, wind, and hydroelectric.
The USA is an exporter of energy, so the country is lucky.

Q: What did the light bulb say to the electric generator?
A: "You spark up my life!"

Inventory of Domestic Energy Resources

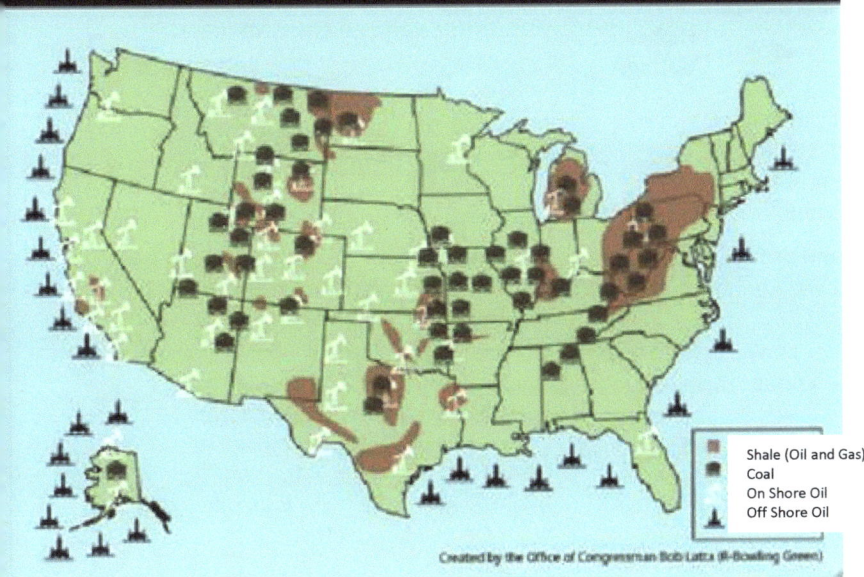

Shale (Oil and Gas)
Coal
On Shore Oil
Off Shore Oil

Created by the Office of Congressman Bob Latta (R-Bowling Green)

A Simplified Look at
Renewable Energy Resource Abundance
in the Conterminous United States

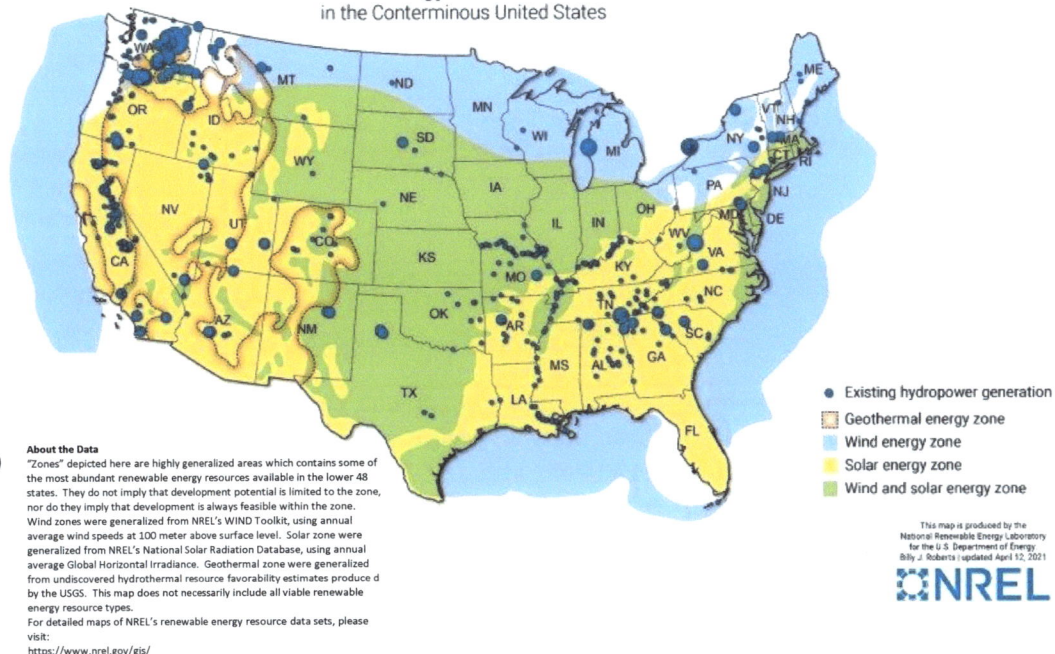

About the Data
"Zones" depicted here are highly generalized areas which contains some of the most abundant renewable energy resources available in the lower 48 states. They do not imply that development potential is limited to the zone, nor do they imply that development is always feasible within the zone. Wind zones were generalized from NREL's WIND Toolkit, using annual average wind speeds at 100 meter above surface level. Solar zone were generalized from NREL's National Solar Radiation Database, using annual average Global Horizontal Irradiance. Geothermal zone were generalized from undiscovered hydrothermal resource favorability estimates produced by the USGS. This map does not necessarily include all viable renewable energy resource types.
For detailed maps of NREL's renewable energy resource data sets, please visit:
https://www.nrel.gov/gis/

- Existing hydropower generation
- Geothermal energy zone
- Wind energy zone
- Solar energy zone
- Wind and solar energy zone

This map is produced by the National Renewable Energy Laboratory for the U.S. Department of Energy
Billy J. Roberts | updated April 12, 2021

NREL

Fossil Fuels, Non-renewable Energy:
Coal is good near the western Rockies, and the eastern Green Mountains.
Oil is good all over the central states, and on the West Coast.

Renewable Energy:
Wind is good in the Great Plains.
Solar is good in the south, especially around Arizona.
Hydropower is good near many rivers.

The USA is fortunate to have huge fossil fuel natural resources, and to have huge potential for renewable energy resources.
We have mountain chains which typically have lots of coal. We have lots of petroleum (oil) in the western half, and especially off the coasts.
We have open plains at the Great Plains which have lots of wind. We have deserts with lots of solar sunshine in the Southwest.

Island countries like Japan, or smaller countries in Europe don't have access to all the resources.

There are a diversity of energy sources for our electricity, and most power plants rely on heat and steam and magnetic generators.

Power Lines and Voltage Conversion using Magnetic Fields

Transmission lines: You need to get the power to the people somehow, to bring electric power from remote power plants at the energy source to your house, apartment, office, store, or factory.

These AC power lines were cutting edge technology around 1900, with a standoff between AC and DC power. The best way, with lowest resistive loss, is in high voltage AC power lines, which use magnetic coils and AC current to convert voltages up and down between high (>1000 V) and household voltages (120 V).

In the future, using solar panels, fuel cells, and wind, maybe power will be generated locally at the house.

Q: Why do transformers hum?
A: They don't know the words!
(actually the vibration from the iron magnetization changing direction.)

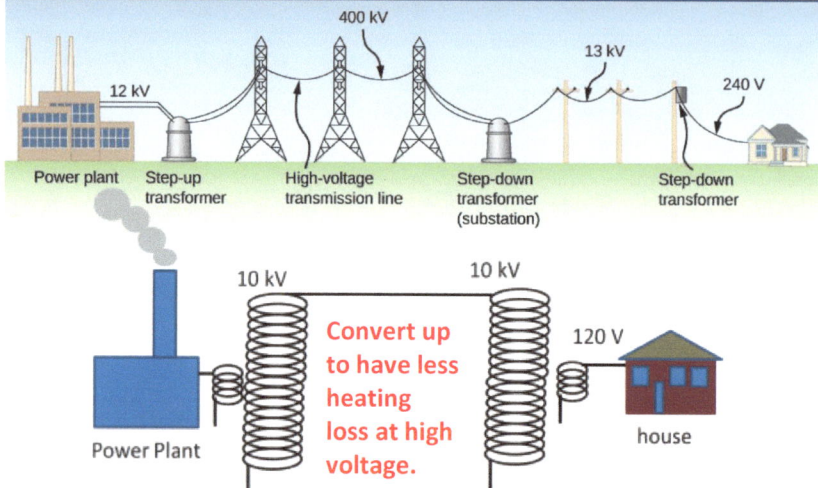

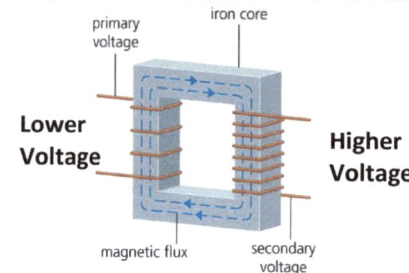

400 kV
12 kV
13 kV
240 V

Power plant | Step-up transformer | High-voltage transmission line | Step-down transformer (substation) | Step-down transformer

primary voltage — iron core
Lower Voltage — **Higher Voltage**
magnetic flux — secondary voltage

Coils with different number of turns for voltage transformer.

10 kV | 10 kV
120 V

Convert up to have less heating loss at high voltage.

Power Plant | house

Power distribution first converts from low voltage to 10,000 Volts and higher at power plant, for long distance transmission. Near the point of use, transformers convert back down to 120 Volts at house, for household use.

Two coils are wound around each other, with the magnetic field enhanced by an iron core.

Voltage transformer converts from one voltage to another voltage, using magnetic coils.

- If there are more wire turns on the output coil, then there is more output voltage.
- If there are less wire turns on the output coil, then there is less output voltage.

Fun other voltage transformers:

Fluorescent bulbs   **Plasma Lamp**

Practically, fluorescent bulbs use higher voltage to excite the gas inside. On the fun side, a plasma lamp is a silly high voltage conversion example. A plasma lamp converts from household voltage to something much higher, to create lightning in the low pressure bulb. It uses two coils wound around each other too.

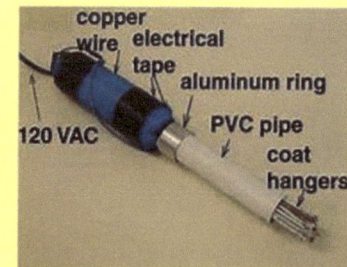

10,000 V
120 V

Step up transformer before distribution: a lower voltage at the power plant can convert to a higher voltage for the line

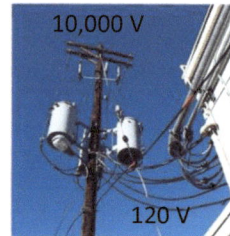

Outdoor voltage transformer near point of use, down conversion

In-door transformers: Convert 120 Volts household power down to 9 Volts, for electronics.

copper wire — electrical tape
aluminum ring
PVC pipe
120 VAC
coat hangers

Electro-magnetic gun, firing conductive metal ring:
Current induced in aluminum ring from magnetization of the iron coat hangers, and the ring is shot away (opposite poles for magnets).

Fun Fact of competition between AC and DC city power:
In the beginning of electric power, around 1890, Edison, using DC power, and Westinghouse, using Tesla's AC power, had a duel between AC power and DC power. The DC power could not step up the voltage to get low losses on long power lines, so AC power won. However, Edison's company General Electric won in the end because it brought the legal weight of lawsuits to share patents and it bought another company that owned AC technology.
Nowadays, with local power generation like solar panels, maybe DC power will make a resurgence without AC/DC converters.

AC Power lines crisscross the entire country. Lines bring power from power plants 100 miles away to the cities.

Summary of Generator Experiments

Here are the experiments to show what a generator can do and where they are, explained in more detail in the preceding pages.

Go ahead, make a copy. Then check off the list.

Experiment 6.5: Start a toy steam engine.
* To get steam, heat a water reservoir with a candle or a burning wick fed with alcohol.

Experiment 6.6: Start a toy hot air engine.
* Show that hot air does not have the same energy as hot steam. Hot air has less energy than steam because there is less mass.

Experiment 6.7: Don't believe that steam is used in power plants? Go and see the hot water or steam exhaust of your local power plant, be it powered by coal or nuclear or natural gas.
* River water could be used to cool down the hot steam in a closed system, or to directly spin the turbine in an open system.

Experiment 6.8: Look for the water run-off downstream of the local hydroelectric dam.
* Some water turbines, with twisting tubes with huge water flow, are large enough to walk through.
* Fish can survive the trip through the fast water, and swim up stream in the fish ladder.

Experiment 6.9: Find a windmill near you.
* It could be on a farm, or in an ocean coastline bay, or on a mountain top, or along the open plains, wherever there is steady wind.
* If the wind is too fast, the windmill is turned off or it will break. If the wind is too slow, the windmill will not spin.

Steam engine to spin flywheel

Hot air engine to spin flywheel and generator to get electricity

Huge flowing water to spin generator to get electricity (Hoover dam)

Windmill to spin generator to get electricity

Use some mechanical force to spin the shaft of a motor, and electricity is generated.

Electric Power Plants, and Whole National Economies, Q and A

Typically anything that burns can be used to make hot steam which turns a turbine and generator. The issues, or choice of heat, are ease of local access to the fuel, cost of fuel and cost to build the plant, and pollution.

Question 1: What makes a good fuel for a large power plant?   Anything that burns.
- A good fuel is whatever is low cost and available and burns, with extra points if environmentally friendly.

Question 2: Why haven't we all gone solar, either photo-voltaic panels or solar thermal, when the sun is free and these power plants seems like such a good idea?
- Solar panels are photo-voltaic, with direct conversion of sunlight into voltage, but solar panels need to be produced cheaply and cleaned frequently.   More efficient solar panels also cost more.
- Solar thermal, or heating of steam or other liquid to turn a generator, needs bright sunlight. Solar thermal is more efficient than solar panels. So locations closer to the equator are good, like the deserts of New Mexico.

Question 3: Why is burning natural gas good? Natural gas is available and it burns relatively clean.
- Natural gas (mostly methane) is relatively clean burning, compared to coal. However the methane in the air has more greenhouse effect than carbon dioxide, even though it can not be seen with the naked eye.
- Methane is produced by decomposing organic material, and methane is found near coal seams. Cattle release methane from stinky digestion.
- Natural gas has a high energy density

Question 4: Why is burning coal good? Coal is very available.
- Coal is very abundant and available in the ground.   Supplies in the USA could last many 100s of years.
- Coal is compressed dead stuff (plants) from 100 million years ago.
- Unfortunately, digging up coal tends to strip the ground, and burning coal tends to emit lots of soot and pollution, unless lots of money is spent on special smokestacks that clean the exhaust.

Question 5: Why is nuclear power good? Nuclear power has provided lots of power and very little radiation issues for the last 50 years.
- Nuclear power has no pollution when the nuclear fuel is well protected with plenty of backup coolant systems. Of course, if there is a radiation leak, then there is uninhabitable land with toxic waste for a 1000 years.
- Unfortunately, we have not solved the challenge of how to store the spent radioactive fuel rods for a 1000 years. Many spent rods from 50 years of nuclear power are now sitting in large water pools around the country near the power plants.   Nobody wants them stored nearby, or traveling across their land on highways or trains to even get to a permanent storage site, due to the possibility of accidents.
- Interestingly, a more radioactive source means more heat during use, and also less storage time after the rod is spent because the half life is shorter.

Question 6: Why has oil for power plants been rejected?
- Burning oil is dirty
- Oil is more expensive than other fuels, and is less reliable on the world production.
- Oil or petroleum power plants would compete with transportation car fuel, and that seems unnecessary. Power plants can use other fuel, and cars generally don't.

Solar panels
- **Expensive semiconductor materials**
- **<20% efficiency, although the sun is free**

Solar thermal
- **Heat from sunlight heats liquid to turn generators**
- **More efficient**

Natural gas
- **Very plentiful**

Coal
- **Very plentiful**

Nuclear fuel rods
- **Used still-radioactive rods are stored for 1000 years**

Imagine the ridiculous 'what-if' scenario where magnetic fields could not produce electricity in spinning coils.

What? Who turned off and denied physics?

"Ha … we could ban me, but we'd need major changes … solar panel, fuel cells, more use of oil or natural gas to spin things or heat homes."

What are the alternatives to generators that depend on magnetic fields?

New technology to generate electricity has been discovered since the 1880s, back at a time when generators using spinning magnetic fields were the next big thing. Some of these newer technologies without spinning magnetic fields are solar panels and fuel cells with hydrogen gas.

Alternatives to generators that use magnetic fields:

Natural gas for heat:

Distribute nature gas in pipes to each house. Natural gas pipes are already present in densely populated cities.

Solar panels:

Solar panels used to be super expensive because it requires highly purified silicon. It is essentially a computer circuit material. Now solar panels have gotten less expensive and more efficient. Now there are power plants made of acres of solar panels. In the 1970s and 80s, NASA helped develop solar panels to provide steady and long term power to satellites and to the International Space Station.

Solar panels large power plant

Fuel cells:

People know that hydrogen gas and oxygen gas have a lot of energy when they combine. These gases can also be combined through a membrane, where an electron goes around a circuit to generate power.

Not a candidate alternatives: these energy sources still use magnetic fields:

Hydroelectric and solar thermal still use spinning magnetic fields:

If we really don't want to use magnetic fields, then hydroelectric power is off the table because the water turbine just spins a magnetic generator.

Also, if we really don't want to use magnet fields, then solar thermal is off the table because that just focuses light to heat a liquid that spins a turbine and generator.

Solar panels for local power on house roof

Without electricity, we scrap long distance power distribution using electricity in power lines. Instead, we have long distance power distribution of natural gas, for local steam engines (generate pressurized gas throughout the house) or fuel cells (hydrogen gas combines with oxygen to create electricity directly) at the home. Many cities have natural gas pipes to each house. There are ways to have a completely different energy economy.

Imagine Life without Electric Generators: Big Changes in Power Plants

Imagine the ridiculous 'what-if' scenario where magnetic fields could not produce electricity in spinning coils.

Without getting voltages from magnetic fields, there is no longer a purpose to turn the rotor of an electric generator, by heating water to spin a turbine or exploding fuel to push a piston.
Can we still get electric power? Are there other ways? Yes, there are. Large power plants may disappear in favor of solar panels and hydrogen fuel cells.

Local de-centralized power from alternative sources:

Maybe the biggest technological hurdle of no magnetic generators would be different types of power plants, and how to generate large voltages to send around the country. Natural gas can be piped everywhere to local combustion engines at your house, which then pressurize air tanks.

Some greener energy sources would still work without generators, like solar panels, fuel cells, and thermal electric materials. These sources do not need electric motor generators, but they are much less efficient than steam-generated electricity from heat sources. Of course, sunlight is free, so efficiency is not the main point.

What? Who turned off and denied physics?

Rejected generators because magnetic fields not allowed	Allowed alternatives: direct conversion to electricity
Steam from Coal	Solar panels
Steam from Oil	Fuel cells
Steam from Nuclear	Thermo-electric
Hydro-electric	
Wind	
Thermal-solar	

No magnetic field, so no mechanical to electric energy conversion

"Ha ... we could ban me, but we'd need major changes ... solar panel, fuel cells, more use of oil or natural gas to spin things or heat homes."

Solar Cells, converting sunlight to electricity:
1. Free energy from the sun, but less than 30% efficient even on sunny days.
2. Need sunny days.

Sun

Solar Panels:
Direct electrical conversion from sun to electric, without coils

---

Fuel Cells, combining hydrogen gas and oxygen gas for electricity
1. Fuel is expensive
2. Fuel is currently from non-renewable natural gas.

Natural Gas / Oxygen Fuel Cell direct electrical conversion:
Combine hydrogen (H) gas from natural gas (methane) and O2 gas (air) to get electricity.
Currently used for :
- Backup power for technology companies
- Already at 5 MegaWatt capability.

Thermo-electric ceramics, a voltage from a thermal gradient across a ceramic
1. Less than 5% efficient.

Thermal-Electric materials:
Direct electrical conversion: a temperature gradient creates an electric voltage across a crystal.
Interestingly, the reverse works too: electric current will create a temperature gradient for cooling.

Used for :
- Spacecraft (temperature gradient from hot radioactive source → provide decades of electric power),
- A camp stove (temperature gradient from fire → 5 V output to charge cell phone)
- In reverse, small refrigerators (applied current → temperature gradient, with cool end near the food).

Fire for heat   Voltage out

Without electricity, we scrap long distance power distribution using electricity in power lines. Instead, we have long distance power distribution of natural gas, for local steam engines (generate pressurized gas throughout the house) or fuel cells (hydrogen gas combines with oxygen to create electricity directly) at the home. Many cities have natural gas pipes to each house. There are ways to have a completely different energy economy.

Substitutes for Magnetic-Based Electric Generators (that is, Magnets or Electro-Magnets do Not Exist)

Substitutes for generators and motors? People did not have a power grid and electric motors in their house 100 years ago, so we could find a way to survive if we went back to the 1900 style of living, but with some improvements! This has been a TV movie script.

<u>Generators</u>: Electric generators do have 'also-ran' modern solid state replacements (solar panels and thermo-electric ceramics), which are less efficient. There are also fuel cells, if we can get hydrogen.

<u>Motors</u>: Some electric motor replacements can just use a different non-magnetic engine: combustion, or steam engine.

- **What are utility power replacements, instead of spinning generators using coal, oil, nuclear or hydro-electric?**
    - Solar panels replace utility power, because, again, we are assuming we can't spin a magnet to generate a voltage in a coil.
    - Fuel Cells can convert natural gas directly into electricity, but expensive.
    - Get oil from algae, to power Fuel Cells?

- **What can keep car batteries charged and headlights on, without an alternator?**
    - Solar panel charging (but not enough juice with small panels on a car)
    - Thermo-electric voltage generation (but not efficient enough, about 1%)
    - Oil burning lamps for head-lights? For example, acetylene gas lamp below, used around 1900.

- **What will create sparks for a car combustion engine?**
    - Cars could still run on diesel, because heat ignites the gas, not a high voltage spark.
    - Compressing a crystal (piezo-electric crystal) quickly can generate a high voltage and can ignite regular gasoline.

- **Be your own walking power source:**
    - Solar panels on clothes, connected to a re-chargeable battery
    - Piezoelectric charging by walking, by compressing a crystal with each foot step.
    - A spring on your foot that pushes a magnet through a coil on each step.

Solar panels beside a highway

Plenty of sun and open spaces:
People may want to move to Arizona to get cheap electricity from solar panels!

Fuel cells converting natural gas to electricity directly

Solar? Fuel cells? Hey, what are these new gizmos? Nobody told me there are innovations.

Oil burning headlight on cars, forget using electricity

Still can have solar panels and fuel cells to generate electricity. For cars, without a magnetic generator for the spark, we could use diesel instead.

Power plants became a big deal in the 1880s, with the invention of generators, motors, and electric lights and trolleys. Power plants were built to power home and street lights, and power trams.

In general, for coal powered power plants, the cheapest power came from locations near coal or a river. Nowadays, high voltage power lines bring power from far away, and natural gas and nuclear power plants can be built anywhere.   Coal power plants generally stay near mountains where the coal is dug up. An ocean, lake, or river is convenient for cooling the steam.

To supply hydro-electric power plants a steady flow of water, a river or reservoir is dammed up in a valley. These power plants have been around from the beginning of generators in 1880.

"Pay respect to my great-great-great- grandfather Niagara. He demonstrated power for the world."

Hydroelectric and coal power:

Say it is the 1880s. New electric lights are replacing gas lights and need to be powered in cities. Light rail for moving around in cities are getting converted to electric motors, replacing steam engines and horses.

What energy would you use to generate electric power? If there is moving water, then use hydroelectric. Maybe create a reservoir using a dam for more steady power. If you are near mountains, then coal is probably a good power source, like in Virginia USA.

Flowing water has always been useful. Look at the New England fabric mills back when they used water power, like in Lowell, MA, along all the rivers. Look at the huge hydroelectric power construction in the 1930s at the Tennessee Valley USA and the Hoover Dam on the West USA.

Some older hydroelectric dams built 100 years ago are being torn down, which is a good thing for nature. Due to age, the cement dams are a safety risk and show cracks. Also, these original dams did not include fish ladders to allow fish like salmon to go back upstream to reproduce.   Dams without fish ladders in general can severely cripple populations of fish, and dams also can reduce flow and downstream sediment, which impacts all animals.   So the natural negative impact of dams – fish reproduction and fresh sediment - needs to be reduced as much as possible.

Oil and nuclear:

Petroleum was available from drilled wells in Pennsylvania starting from the 1860s, but it was mostly used to make kerosene for oil lamps instead of burning whale oil.

Also, from the 1950s, oil and nuclear power started getting used for power plants.

Let's look at oil. Oil can run in diesel engine power plants, after conversion to kerosene or gasoline. The fuel is expensive and price depends on the roller coaster price of gas, but the diesel engines can start and stop quickly, so diesel engine power plants can be used to handle the extra power required for peak demand.

Let's look at nuclear power. Nuclear power was supposed to be the dream power supply, but issues with safety derailed it.

In the first generation nuclear power plants, there is the fear of meltdown if no coolant is flowing. Look at the reactor that melted down and released radiation in Ukraine (Chernobyl) due to relaxed safety standards. Look at the reactors that melted down and released radiation in Japan (Fukushima) due to inadequate design to protect against huge waves (a tsunami from an earthquake). Fortunately, there are next generation nuclear power plant designs these days that will not melt down if coolant fails. The nuclear reaction simply stops naturally.

Spent nuclear waste raises the issue of where to store the radioactive waster for the next 1000 years. The waste is currently sitting in deep pools of water at the nuclear power plants. No one wants the radioactive fuel to travel through their city on a train, in case of a derailment or accident, on their way to a permanent storage site. Chicago, where most trains converge, would be most vulnerable to accidents and contamination, or deliberate attacks.

Power plants originally uses the local energy sources available, be it flowing water or coal. Later on oil and nuclear power can be placed anywhere.

History of Power Plants:
Niagara Falls Hydro-Electricity: 1886 First Large Power Plant

Here are some major dams and reservoirs that have been built to supply hydroelectric power plants with a reliable supply of water. Most major rivers have already been dammed up. Niagara was the first in the 1880s, directed by Mr. Tesla. The Tennessee valley in the 1930s was used to power military production during WW2.

"Pay respect to my great-great-great- grandfather Niagara. He demonstrated power for the world."

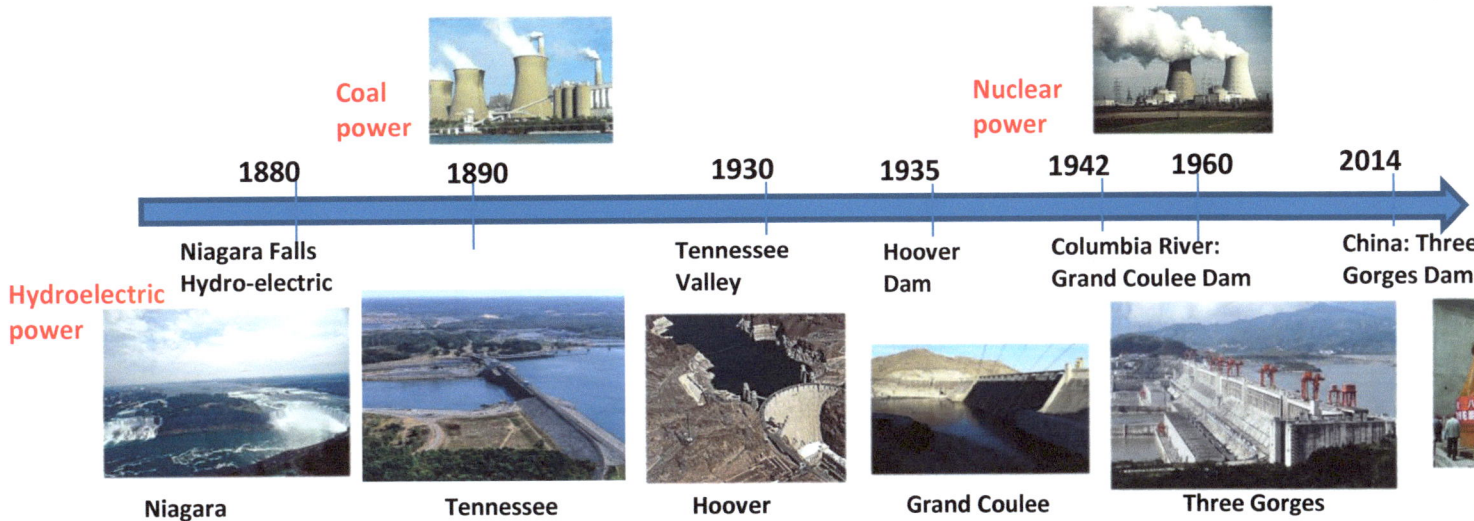

Coal power

Nuclear power

1880	1890	1930	1935	1942	1960	2014

Hydroelectric power

Niagara Falls Hydro-electric

Tennessee Valley

Hoover Dam

Columbia River: Grand Coulee Dam

China: Three Gorges Dam

Niagara

Tennessee

Hoover

Grand Coulee

Three Gorges

Other prime players with great water resources:
- China
- Canada

Experiments: Find hydroelectric power plants near you

- Water flowing downhill can push water turbines
- At a waterfall, a water tube can be made alongside to turn the turbine.

FIG. 385.—Cast-steel Casing for 70,000 h.p. Niagara Falls Power Company units 19 and 20, showing how sections of casing are fitted together, also arrangement of guide vane levers, links, and shifting ring. (I. P. Morris.)
Turbine

Set of generators

Stator and rotor of generator

1880: Niagara Falls, hydro-electricity from the inventor Tesla

Each turbine contains a giant waterwheel attached to a shaft; as the water rushes through, the strong water flow spins the wheel, the shaft, and all the equipment attached to the shaft

http://spiff.rit.edu/classes/phys213/lectures/niagara/niagara.html

There is always a first: Niagara Falls was the first, using free energy from the flowing water.

Power Plant Examples:
Hydroelectric: 1930

"Here's a water powered power plant in Tennessee that powered the industrial might of the US during the WW2."

These hydroelectric dams in large river valleys were built in the 1930s, to put people to work in the Great Depression era and the 'New Deal'. They have been humming along ever since.

Public works projects are a favorite way for government to get the economy and employment going again during recessions. In this public works case, that employment to build huge generator power plants was extremely successful. The generators on the rivers even powered the factories in WW2.

The flow rate can be adjusted during the day to match demand for electricity.

Tennessee Valley, Tennessee

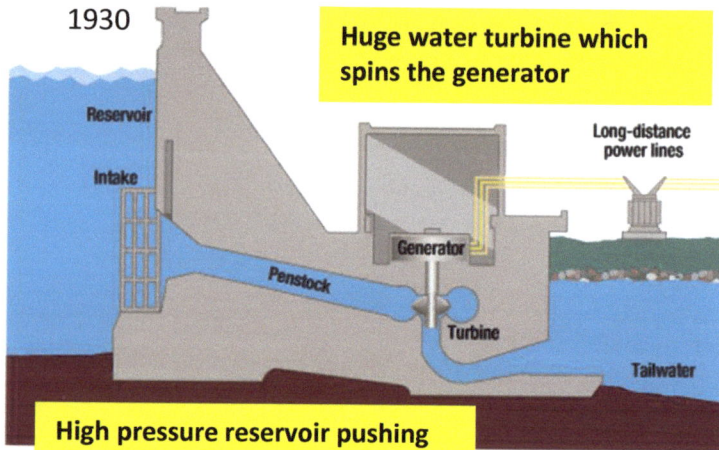

1930

Huge water turbine which spins the generator

High pressure reservoir pushing high speed water

Below are the hydro-electric power plants along a river opened after the depression, just in time to power factories for WWII.

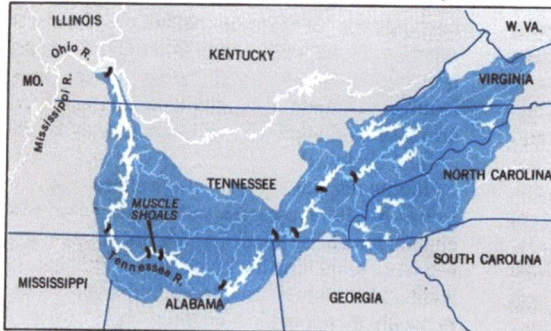

Tennessee valley with many dams for hydroelectric power:
A whole series of dams were built along the Tennessee valley.
http://www.tva.gov/power/hydro.htm

Grand Coulee Dam, Washington State

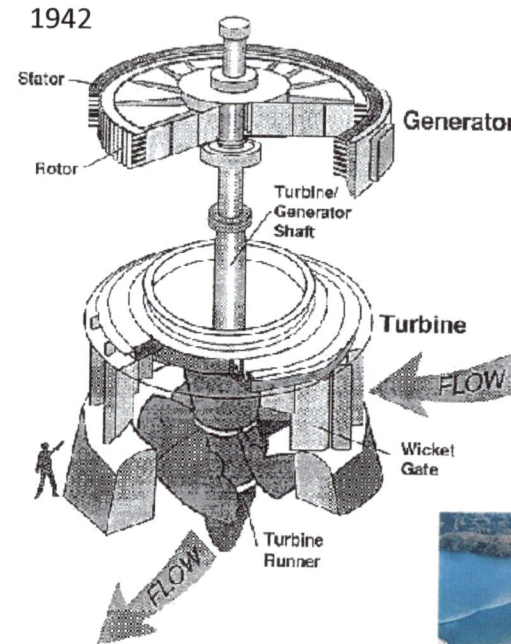

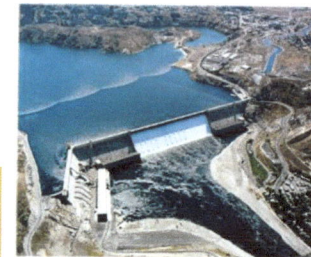

1942

Huge water flow, from the free energy of mother nature

Generators inside the dams, spun using flowing water through helix tubes of a turbine.
http://users.owt.com/chubbard/gcdam/html/hydro.html

Experiments: Look up hydro electric micro-generators, estimate if a stream near you would generate much electricity.

A political movement to employ people actually got some huge power plants built in the 1930s.

Chapter 7: Summary of Magnets, Motors, and Generators

For powering our modern world, the existence and dependability of all magnetic devices like permanent magnets, electro-magnets, motors, and generators are a big deal. You depend on them every day.

Permanent magnets stick notes on refrigerators, but also power DC motors and brushless motors, which are critical to many products in society.

- Magnets can lift iron objects
- Magnets twist to align their magnetic fields, or 'opposites attract'

'Opposite poles attract'

Electro-magnets allow electric motors, speakers, and computer memory to work. The fact that someone long ago in 1820 (Mr. Faraday) showed that electric currents in coils produce a magnetic field is huge. Electro-magnets are critical for every motor in production. Those magnetic fields from the rotor make the rotor want to twist. Without electromagnets and current switching, two permanent magnets would just line up and stop spinning.

- Electro-magnets allow motors to spin.
  - ✓ Commutators keep the rotor magnetic field a steady sideways direction with maximum torque.
  - ✓ Spinning electric motors power our industries, our home tools.
  - ✓ Spinning electric motors are car starter motors.
  - ✓ Spinning electric motors more commonly these days are powering hybrid and EV cars.
- Electro-magnets cause speakers to vibrate
  - ✓ Speaker cone moves and pushes air and sound.
- Electro-magnets write data onto spinning hard drives in computers
  - ✓ Magnetize a spot on the film one way or the other.

'Electro-magnetics allow motors and speakers'

Generators need magnetic fields in order to generate a voltage and a current, in other words generate electric power. The basic mechanism is that changing magnetic fields in a coil will generate a current.

Generators are very similar to but opposite to a motor. In a generator, a wire is mechanically moved by a shaft, and the electrons in the moving wire feel a magnetic force which generates a voltage and a current. In a motor, the force on an electric current in a wire in a stator magnetic field gets a magnetic force on the wire, which pushes sideways on the wire and turns the rotor. A DC motor can operate as both a motor or a generator.

- Generators provide the power in your wall outlets.
- Generators are how microphones pick up sound.

'Let me give you instant power in your wall outlets.'

'Power lines and transformers for low loss power distribution'

For powering memory storage for computers, magnetic films or tapes are the backbone of information storage.

'Magnetic disks provide memory on computers'

For saving our planet's atmosphere from solar wind, which otherwise would sweep the atmosphere away and cause us cancer, the Earth's magnetic field is a necessity.

'Northern lights from Earth's magnetic field and sun's radiation'

Magnetic fields and magnets are part of nature and are an everyday part of your life, with motors, electrical power, and computers.

Magnet Field Physics

Magnetic fields cause forces for attraction. Magnetic fields cause rotation for motor, to turn a shaft. Magnetic fields cause a voltage, for generators when a wire is moving in a magnetic field.

Permanent magnets stick notes on refrigerators, but also power DC motors and brushless motors, which are critical to many products in society.

Electro-magnets allow electric motors, speakers, and computer memory to work.

Generators need magnetic fields in order to generate a voltage and a current, in other words generate electric power.

Lift forces

These two magnets are attracted to each other because of the magnetic field gradient that pulls them together.
Magnetic forces are easily stronger than gravity, end to end
The energy is lower when the magnetic fields are aligned and close together.

Magnets feel a force due to an external magnetic field gradient.

Motors

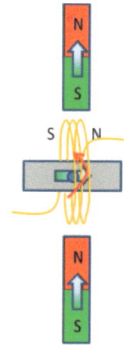

Magnets want to rotate to align their magnetic fields.   Permanent magnets will align and stay there.
Electromagnets can keep switching their magnetic direction and keep rotating.
The energy is lower when the magnetic fields are aligned.

<u>Motor</u>: convert the need for magnetic fields to align to a spinning shaft.

An electric current feels a force from an external magnetic field, which enables motors.

Generators

Induced voltage

Force on electrons, creating a current and a voltage

Electron velocity or current with moving wire

Permanent Magnet B field is applied across the wire

Metal wire

Motion of wire from **mechanical** force

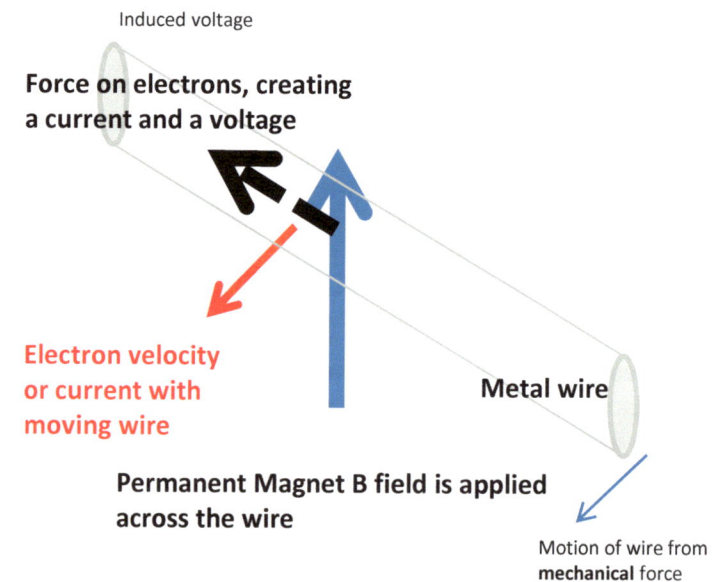

<u>Generator</u>: convert wire motion to electric current

A wire moving through a magnetic field will generate a current in the wire.

Summary of Demonstrations

There are demonstrations for all applications of magnetic fields: static forces for attraction and repulsion, twisting torque to get magnetic fields to align which enables motors, and electricity generation by moving a coil near a magnetic field.

Forces and attraction, for lifting and speakers converting electric current to sound:

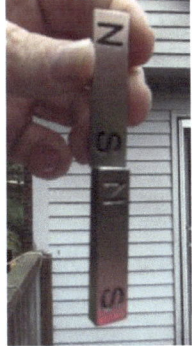

Magnetic forces are easily stronger than gravity, end to end with opposites attract

Attraction: Strong magnetic field controls magnetic compass needle

Magnet near iron, which magnetizes and gets attracted

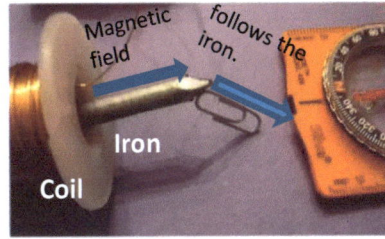

Magnetic field follows the iron. **Iron** **Coil**

Magnetize a steel paper clip, and get attraction.

Electro magnet on a tractor arm: pick up large heavy sheet of iron

Light electro-magnet on cone: Coil with current   Heavy permanent magnet

Motors and spinning, doing work when magnetic fields want to align:

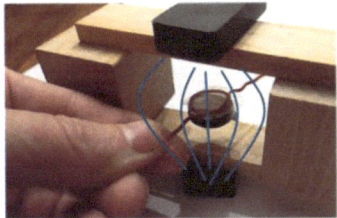

Alignment of middle magnet with 'stator' magnetic field

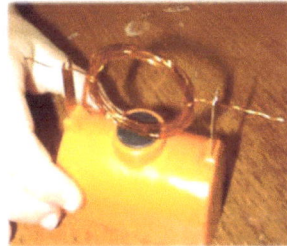

One-pole motor demonstration, with current less than ½ rotation.

Two-pole motor demonstration, with current most the rotation

Toy DC motor parts: electromagnet rotor, permanent magnet stator, and commutator

AC electric motor inside portable drill

Generators and power from the wall outlet, from spinning coils in magnetic fields:

Two-pole generator demonstration

Hot air powers a piston which spins a generator

136.6

A coil moving through a changing magnetic field will generate a voltage

Spinning rotor magnet to get voltage: Spin rotor or shaft mechanically to convert from mechanical energy to electrical energy.

Emergency generator Use combustion engine to spin the generator

Industrial generators in hydroelectric power dam

Bigger size, higher performance

Past, Current, and Future of Applications of Magnetic Fields

Electricity will probably be with us for a long time in the future, doing new and better things. Maybe we'll get the electricity from 'green' power like the sun, wind, or ocean tides instead of coal, but we'll still need electricity.

Past, pre-electric

The pre-electric era involves a lot of manual work, with kings, knights, wars, and small farms.

Steam engine trains, without electricity
Manually load wood or coal from the tender

Tilling the soil without a tractor
manual effort for both you and the ox.

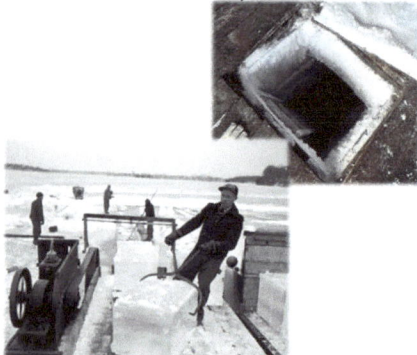

Ice from the lake and ice cellar to freeze your deer meat.

Present, industrial and information age

Heat and steam turn generators, which power our refrigerators, computers, and lights. Life is more convenient and interesting than 300 years ago for regular people, as well as providing a much longer life span.

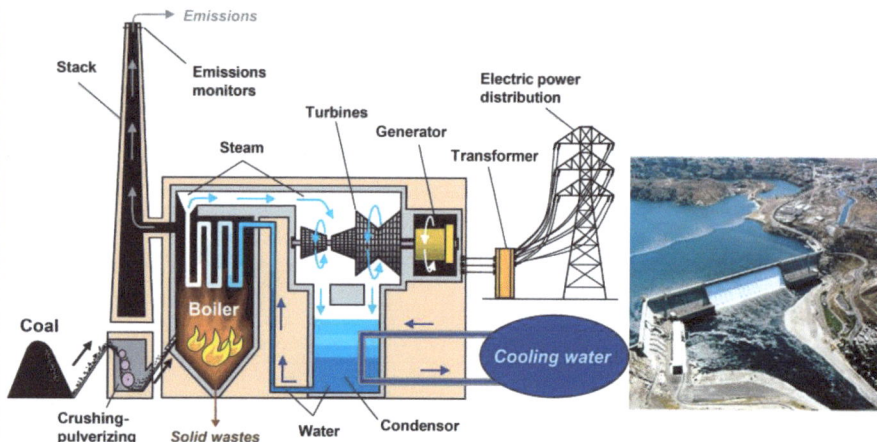

Schematic of generator powered by heat and steam
Decayed plants from millions of years ago are burnt to power our lights. Power plants have been around from 1880s.

Huge water flow spinning the turbine, free energy from mother nature, and the water cycle. Thanks storms!

Refrigerator
Refrigerators have electric pumps to circulate the coolant. Refrigerators have been around since the 1920s.

Computers and the internet
Electricity powers your electronics. Electronics makes a stable voltage where a few volts turns transistors on and off.

Electric car charging station
Need credit card and bank of car batteries to store energy for over 200 miles.

Future, distributed energy and renewables

The future can include levitating fast trains. New high energy density batteries can give electric cars longer range, and electric cars can be charged by many different fuels at power plants instead of gasoline.

Common high speed levitating trains

Hydrogen fuel-cell powered electric motors for airplanes
Hydrogen has more energy density than batteries.

More large wind turbines notably found in Denmark and California and Texas.

More solar panels:
Direct electrical conversion from sun to electric, without coils

People: Electricity Development

First discovery and early science had to happen, such as electrostatic forces, creation of magnetic fields, the explanations, DC and AC current, and prototype motors and generators. And then the ideas get applied by industrious engineers who see a need, like trolleys and electric lights. Notice it took 60 to 80 years before mass applications happened from the science.

Early Science

Oersted: (1820, in his 40s)
- Danish
- Demonstrated a magnetic field around a current in a wire

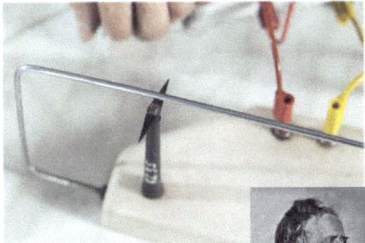

Magnetic field around a current

Jedlik: (1828, in his 20s)
- Hungarian
- First DC motor, with stator, rotor, commutator
- Motor powered skate board

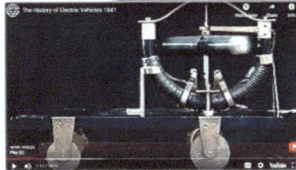

Demonstration of motor powered skateboard

Mass Transportation and Power Generation: AC versus DC power

Edison: (1878, in his 30s)
- American
- Worked with lots of helpers
- Believed in huge effort, with many trials and failures before getting it right
- Nasty business competitor. Portrayed ac power as a way to kill people with high voltages, but then bought out the AC generator patents when AC current was proven better for long distance power transmission.

Inventions:
- DC Power plants (with help from others)
- Long life light bulb
- Sound recording using phonograph

Net worth at death:
- Super Rich

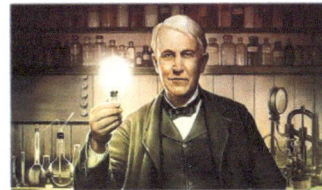

DC power generator, 'Long legged Mary Ann'

Tesla: (1887, in his 20s)
- Serbian, but worked in America
- Worked alone, at night
- Dreamer, always had a next revolution in mind
- No business sense, because he ripped up the contract for royalties from the AC motor and GE when his AC invention was too successful and too much money was coming to him.
- Knew many languages

Inventions:
- AC motor and generator, for long distance power transmission
- Hydroelectric power
- Tesla coil

Net worth at death:
- Zero

Alternating electric current generator, March 10, 1891.

1820　　　　**Science Demonstrations**　　　　**1890**　　　　**Mass Development**

Ampere: (1820, in his 40s)
- French
- Two wires carrying current attract
- Solenoid

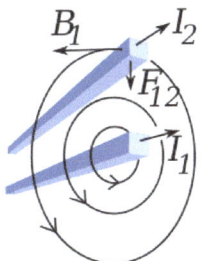

$$B_1 \quad I_2$$
$$F_{12}$$
$$I_1$$

Magnetic field and force around a current

Sprague: (1886, in his 20s)
- American
- Constant speed DC motor for trolleys and elevators
- DC motors have more torque at lower spin rates than AC motors, when use electromagnet as the stator magnet called a 'traction' motor.

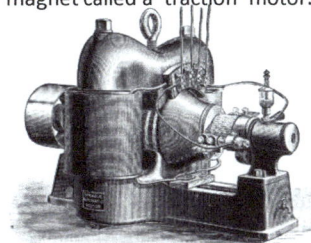

DC motor using iron and coils and carbon brushes

Kettering: (1911, in his 30s)
- American
- Electric starter motor for cars

Did it or Didn't it Depend on Electric Motors and Generators?

Not everything in this world depends on magnetic fields, but a lot does. Can you find which of the items below owe a debt of gratitude to electric motors and generators, and which don't?

Flushing a toilet?
Yes, electric water pumps are in wells and pumps to town water tower to keep the water pressure.

Hugs and kisses?
No, people and love have been around a long time. Of course, the extra free time from the convenience of motors helps.

New e-bikes?
Yes, a small battery and a small electric motor power the wheel. Riders can select whether to get large battery assist, or just a little extra push. Thank lithium ion batteries, which have lighter weight and higher energy density.

Chocolate chip cookie?
Yes, use an electric oven, and use farm tools to grow the wheat.

Commercial fishing?
Maybe not, boats pull in the fish net with a combustion engine, not electric. Boats typically need gasoline because boats are isolated from land power and need lots of energy over many days. Batteries will drain, but gasoline can be stocked up.

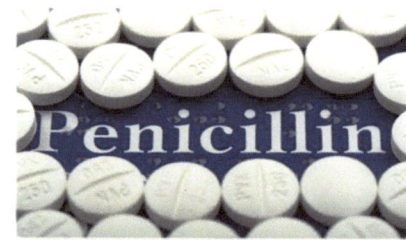

Health and penicillin
Maybe not, unless motors are used to separate chemicals.

Cities and skyscrapers?
Yes, elevators use electric motors so people don't object to 40 flights of stairs and enable skyscrapers. Steel I-beams also enable skyscrapers by holding up more stories than brittle cement or brick.

Electric Cars?
Yes, large electric motors power the wheels.
Standard combustion cars?
Yes, a smaller electric motor is the starter motor.

Computers and information, or radios and TV?
Yes, for electric power. Also, magnetic hard drives spin using electric motors.

192

Magnetic Fields in the Motors, and in Solar System

Local magnetic fields from magnets in their small volume can be a larger magnitude than planet fields, but planet fields have a huge volume. These local strong magnetic fields enable motors.

1) Rotor with 5 pads or poles or coils

2) Stator permanent magnets
- Balanced B field across rotor, with no gradient.

3) Brushes or electric sliding contacts to rotor, called a commutator

Rare earth permanent magnet:
There is about 500 Gauss on surface. The magnet shape matters. There is a larger magnetic field when stack up disks.

Motor:
There is about 100 Gauss from stator, using ferrite magnets (iron oxide) instead of stronger rare earth magnets.

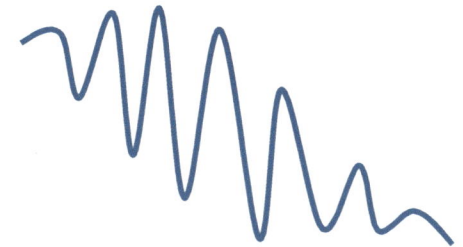

Photon of visible light:
Photons have less than about 1 Gauss, for a light photon pulse. Magnetic and electric fields are oscillating as the photon moves along. A photon is a packet of oscillating magnetic and electric fields.

The magnetic fields of solar system planets and other objects in the galaxy have a huge range of magnitudes.

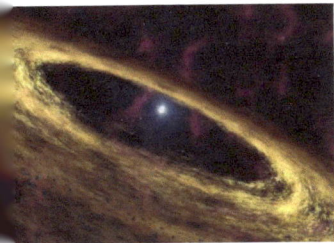

Neutron star surface: astronomically huge field
Magnetic field is an unbelievably huge 10^{13} Gauss, more than a million million times larger than the field on Earth.
- How do we know? X-ray radiation of particles confirmed it.
- The magnetic field source is not fully known, but is either huge currents inside or aligned of neutron spins.

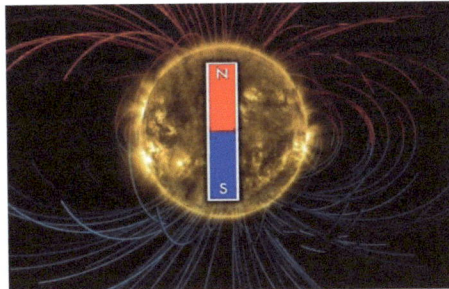

Sun: similar to Earth's field
Magnetic field is 1 Gauss on surface.
Every 11 years the magnet field switches, causing the sun spot cycle.
- The magnetic field may be due to the flow of ionized gasses in the convection zone due to the heat.

Venus: no field structure
Magnetic field on surface is about 0.001 Gauss, with no structure.
- Gravity: 1 earth
- Huge atmospheric pressure from gravity, so solar wind not strong enough to sweep away atmosphere due to larger gravity.
- No magnetic field implies no conductive gasses or no iron core.

Earth: 1 Gauss
Magnetic field is 0.5 Gauss on surface.
- Field from iron core and moving magna with charges.

Mars: no field structure
Magnetic field on surface has no structure.
- Gravity: 0.37 earth
- The no magnetic field and the low gravity means the atmosphere can be swept away by the solar wind (electrons, protons, alpha particles, gamma rays).

Jupiter: a little more than Earth's field
Magnetic field is 4 Gauss on surface.
- The large magnetic field may be due to a rapidly spinning metallic hydrogen core.

Glossary of Magnets, Motors, and Generators

If you are going to understand magnets, motors, and generators, you need to do a few hands-on experiments to 'walk the walk'. But here are names, or lingo, that also help you 'talk the talk'.

Magnets:

- **Bar magnet**: a magnetized piece of iron typically, or rare earth magnet.
- **Magnetic field**: a field that aligns magnets along its direction.
- **Magnetic field B**: sum of the external field H and internal magnetized material M
- **Magnetic field H**: applied field, external.
- **Magnetic moment**: strength of a bar magnet (dipole moment)
- **Rare earth elements**: a row of the periodic table that has great magnetic properties, with the property that some of the electrons are in wide orbits, or can couple strongly with the nucleus, aligning their magnetic spins or magnetic orbits.

Motors:

- **AC**: Alternating Current from a wall socket and power lines
- **DC**: Direct Current, from AAA batteries, or car batteries
- **Linear motor**: a motor that pushes magnets sideways on a rotor to get spin.
- **Motor commutator**: The ends of the rotor that connect with the sliding electrical contacts to run current through only the sideways coil.
- **Motor rotor**: The shaft with coils that spins inside the motor.
- **Motor stator**: For a DC motor, the fixed magnet around the rotor and shaft. For an AC motor, coils excite the magnetic field.
- **Synchronous motor**: a motor where the stator and rotor fields both rotate together.
- **Torque**: force that wants to twist something, instead of move it sideways or to a different location.

Generators:

- **Alternator**: a generator in a car
- **Transformer**: converting voltage up or down to reduce resistive heating losses in the national power grid.
- **Turbine**: shaft with blades that spins due to steam, flowing water, or even windmills.

As stated in the introduction to this book, math is included throughout the book, but the main points are the pictures. Please skip over any math you want. The most detailed math is reserved for the appendices, for those people with more math interest and background.

The appendices describe the energy interpretation of forces and torques, and the math for the magnetic forces and twisting torques for motors, along with the fundamental description of the forces on moving charges. The appendices also describe the more subtle observations of 1-pole and 2-pole demonstration motors, flipping rotor coils and flipping stator magnets. In addition, the appendices describe electric circuits and instruments to measure the current flowing through these demonstration motors.

Symbols of level or difficulty of math (with a nod to ski slopes):

Easy : 4th grade

Algebra : 8th grade

Trigonometry: 10th grade

Calculus: 12th grade

At a very high level we can talk about energy, like a Hindu disciple. All forces and torques are the result of the magnets and rotors finding a lower energy state, like rolling down a hill. Although any understanding of magnets and motors needs to include magnetic fields and forces, these magnetic field interactions can be described as an energy.

A magnet has an orientation (physical direction of the magnet) and a different orientation has a different force and a different energy level near another magnet.

Magnets have their fields roll off with farther distance. This gradient causes a different energy if the magnet moves farther or closer. The magnet feels how much the energy would change by moving, and more energy change means more force.

Lift forces to go to a lower energy

"Magnets will repel or attract, to reduce energy, Padawan."

Field

Force

Opposites poles attract (slam together):
- B fields aligned
- Field falls off (gradients)
- Low Energy state

Similar poles repel (twist apart):
- B fields opposite
- Field falls off (gradients)
- High Energy state

Attraction and energy:
- The magnets are in a lower energy state when all the fields are aligned in the same direction, and they move to share each other's fields, or
- Opposite poles attract

It's all about energy... Things want to move to a lower energy state.

Same-Direction Fields Have Lower Energy When Closer

Opposite poles attract. If two magnets are positioned so that North touches South, then there is attraction.

Whenever a the magnetic field of the second magnet is aligned in the same direction with the magnetic field of the first magnet, then the energy is lower and the two magnets are attracted together (that is, energy is lower). The field gradient pulls the two magnets together. This happens when opposite poles touch.

Displayed are the fields around a permanent magnet, and a top magnet which is attracted because its fields are aligned with the lower magnet, and the top magnet can get into a higher magnetic field by moving closer to the lower magnet.

"Feel your magnetic fields aligned in harmony, and feel your energy level go calmer and lower (which is a good thing for attraction)."

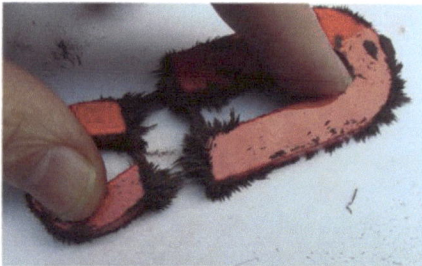

The iron filings follow the magnetic field

The iron filings collect where the magnetic field has a large gradient and is intense.

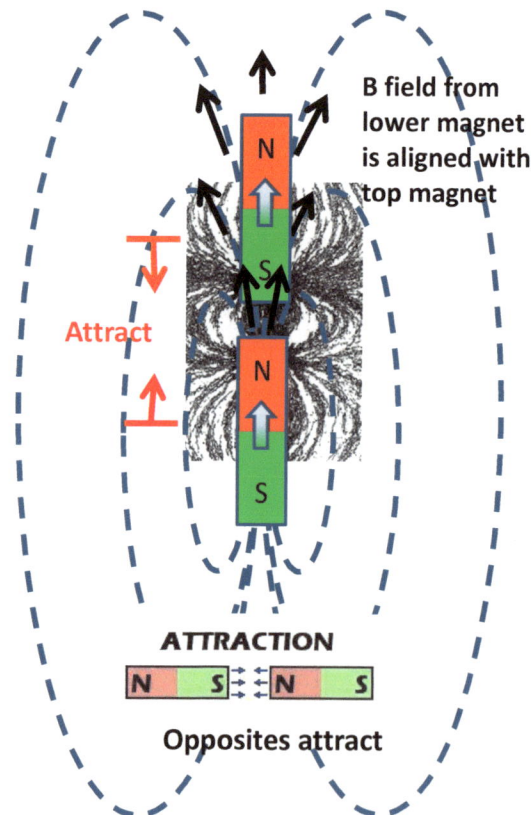

B field from lower magnet is aligned with top magnet

Attract

N
S
N
S

ATTRACTION

N   S   N   S

Opposites attract

Energy versus separation for attractive opposite poles

Find the direction to the lower energy. That's the way to know the direction of the forces.

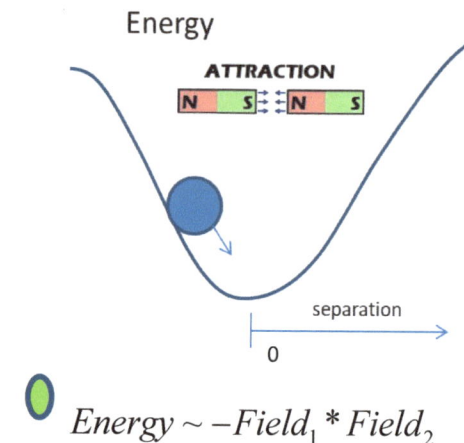

Energy

ATTRACTION

N   S   N   S

separation

0

$$Energy \sim -Field_1 * Field_2$$

Physicists have made up an energy to describe this attraction, and the slope of the energy is the force.

See Appendix G, equation 10

Behind the opposites attract rule, magnets attract when B field aligned with magnet, and where magnets want to move to get into a larger external magnetic field. Magnets get pulled in by a large field gradient.

Iron Filings Show When Fields are in Same Direction

Magnets snap together, or fly apart. If the poles are opposites and the fields are aligned, then there is lower energy when the magnets get closer together. If the poles are the same and fields not aligned, then there is higher energy closer together and, instead of attraction, the magnets want to get away from each other.

Experiment A.1: Show two magnets attracting and repulsing

- Magnets snap together, or fly apart.
- Coloring the magnetic poles with paint really makes the attraction easier to understand.
- How to tell same poles of different magnets? Snap the magnets together as a long cylinder, and then you know that the poles are alternating North-South, to align the magnetic field.

"To see the invisible fields, use iron filings, or just stick me on a refrigerator."

Opposites poles attract for magnets

The magnet fields in this opposite pole's case are in the same direction, and add together in the middle.

The magnet fields in this same pole's case are in the same direction, and cancel in the middle.

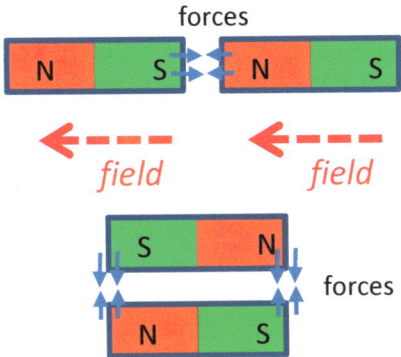

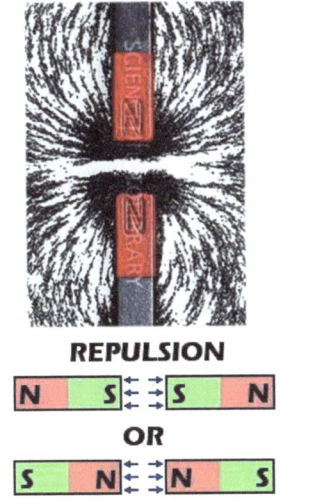

forces

field   field

forces

ATTRACTION

REPULSION

OR

field   field

field   field

Fields aligned with opposite poles

Fields opposite with same poles

The iron filings get magnetized and rotate to be parallel to the magnetic field.
The iron files also are attracted to the high magnetic fields near the pole regions. This attraction is like an iron core of a relay coil getting pulled in.

- In terms of magnetic poles, an easy way to remember how magnets behave is to remember that 'opposites attract'.
- In terms of magnetic fields and energy, whenever a second magnet is aligned with the magnetic field of the first magnet and magnets would be in a higher magnetic field by moving closer, then the energy is lower the closer the magnets get together and the two magnets are attracted together.

You can see the iron filings cross the gap, with N-S opposites, which means that the magnetic field is nicely crossing the gap and the magnetic fields of the two magnets are in the same direction.

Mantras for Energy, the Buddha Way

Energy, forces, and torques are all good to talk about, but energy is easier to analyze. If you know in what direction the energy is lower, then the magnets will go in that direction, either by rotation or sideways movement.

Energy is all around us and it wants to settle down: Heat energy goes to cold, which is less energy. Storms, caused by extra heat or energy in one spot, go to blue skies. Electric batteries get discharged. Batteries are charged by some energy source, either heat or the sun.

Practice some Zen Buddhism, and focus on your core energy, and relax and settle down.

Does it take more energy to do homework, or more energy to watch TV? Yes, there's disciplined thought required to exert yourself and do homework. So that's your excuse for watching TV instead of doing homework – You are part of nature and nature has a natural tendency to go to lower energy.

To predict magnetic behavior and attraction, you can first just think about where the lower energy is.

You could go to a deep vector analysis of current and magnetic fields, like discussed in Chapter 3, but the logic of energy is much easier.

"Zen and the art of attraction"

Energy rules:

- **Lower energy → Direction of force**
  - People don't know why this lower energy rule works, but the rule works, and it works for most everything. Lower energy, representing direction of force, is a simple math approach to describe attraction, be the attraction magnetic, chemical, electrical. Rocks want to roll down hill, skate boarders want to roll down the half pipe, rain wants to fall, chemicals want to combine, magnets want to align...all to get to a lower energy state
  - Magnets want to get sucked into higher magnetic fields for lower energy.
  - If the external magnetic field is uniform (for example, between two magnets or in the Earth's magnetic field), then the bar magnet does not get pushed anywhere. The bar magnetic does not want to move anywhere. Instead, the bar magnet wants to rotate to align the fields, but the bar magnet does not want to move position.
  - The direction of lower energy is reversed when the magnet is flipped, and changes force from attraction to repulsion, or vise versa.

- **'Opposite poles attract', and 'Like poles repel'**
  - This is the same as 'Lower energy' → 'Direction of force'.
  - When opposite poles are touching each other, the two magnetic fields are in the same direction at the other magnet's location.

In the strange world of physics description of reality:

- **Higher energy, or more positive energy**
  - Something to get rid of!

- **Lower energy, or more negative energy**
  - Something that pulls things together. If the magnets are pulling themselves together, such as actually pulling your arms so that the magnets come together, then the magnet is giving you energy. When something gives you energy, that is negative energy for that something.

If you are pulling two magnets apart, and you exert yourself by using your own muscles, then you are giving a positive energy to the magnets.

Objects go in the direction of lower energy.

Energy versus separation

ATTRACTION

N S ⇄ N S

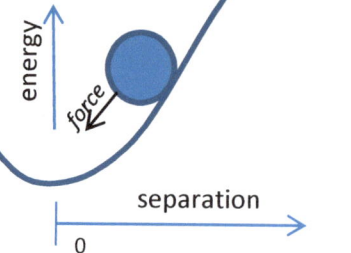

Energy of attraction for opposite poles:
Think of a ball. The ball wants to roll downhill, where the ball is in a stable and lower energy state.
Leaving that stable bottom state would require energy.

REPULSION

N S ⇄ S N

OR

S N ⇄ N S

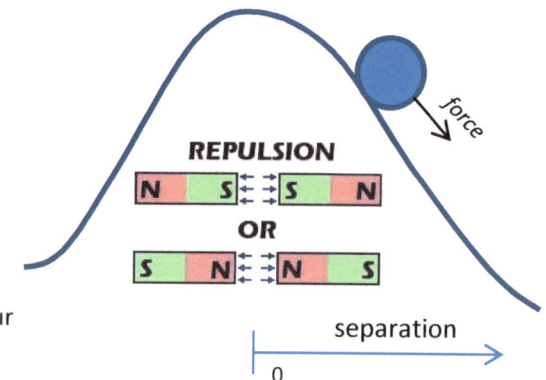

Energy of repulsion for same poles:
The ball wants to roll downhill, away from peak.

Forces and Lower Energy

Forces come from a desire for lower energy. These rules of 'opposite poles attract' and 'like poles repel' can all be described as a desire for lower energy.
Opposite magnetic poles attract means fields want to be in the same direction or aligned, where the energy is lower.

"Lower energy, things want, Padawan."

Repulsion:
Mis-aligned magnetic fields have higher energy, and repel.

Attraction:
Magnets want to be aligned to the other magnet's magnetic field, for lower energy.

Higher Energy

Lower Energy

S

N

N

S

Repel

N

S

N

S

Attract

Same Poles repel and twist apart:
*   **B Field in opposite directions**
*   **High gradient**
*   **High Energy**

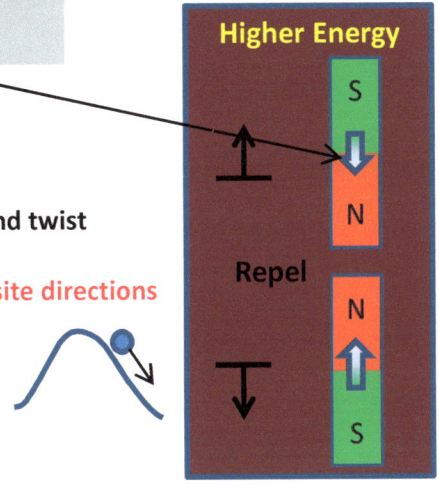

Opposite Poles attract and slam together:
*   **B Field aligned in same direction**
*   **High gradient**
*   **Low Energy**

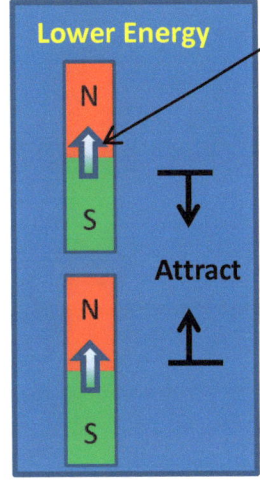

It's all about energy … Things want to move to a lower energy state*.

Examples of lower energy: rocks want to roll down hill, rain wants to fall, chemicals want to combine, magnets want to align…all to get to a lower energy state.

Repulsion and higher energy:
Higher energy is when the magnetic moment is opposite to the magnetic field of the other magnet, or when North pole is closest to North pole. The two magnets want to flip over to get to the aligned orientation, or the two magnets want to fly apart.

Attraction and lower energy:
For attraction, 'opposites attract'. The magnetic fields are then aligned, as shown by the arrows in the figure. The energy is just the magnetic moment of one magnet times the magnet field of the other magnet. When magnetic moments are aligned with the magnetic field, the energy is lowest, at

Energy = - magnetic moment * magnetic field (U = -m*B).
Higher B fields and moments will have even lower energy, which means the forces get even larger as magnetics get closer together.

* The desire for a lower energy state is also the 2nd law of thermodynamics, where heat always moves downhill from hotter to colder objects.

There is lower energy when the magnetic fields are aligned, or 'opposite poles attract'.

'Likes' Repel and 'Opposites' Attract

Do you think opposites attract like shy and bold, tall and short, beautiful and ugly? Maybe yes, maybe no, but that's psychology. For magnets, opposite poles attract, simple as that. The attraction is not so opposite as it sounds. Opposite poles attract because the magnetic fields are in the same direction.

Both repel and attract are variations of lower energy, depending on aligning the magnetic fields. Would you rather learn the phrase 'fields the same' direction, or 'opposite poles attract'?

'Like poles repel'
Higher Energy: fields in opposite directions

'Opposite poles attract'
Lower Energy: fields in same direction

"You hear about the love struck magnets?
Whenever they met face to face, they just couldn't seem to connect, however the moment one turned to walk away, they were nearly inseparable."

Magnets want to be aligned to the external magnetic field, for lower energy, NOT oppositely aligned, like in the cartoon.

Unstable:
Same pole magnets either fling apart, or flip over and slam together (to get from higher to lower energy)

Magnets want to be aligned to other magnet's magnetic field, for lower energy. With a field gradient, there is an attractive force.

Stable:
Top magnet has the same orientation as the applied B field of the other (lower) magnet, so the top magnet is attracted, and is in a stable lower energy state.

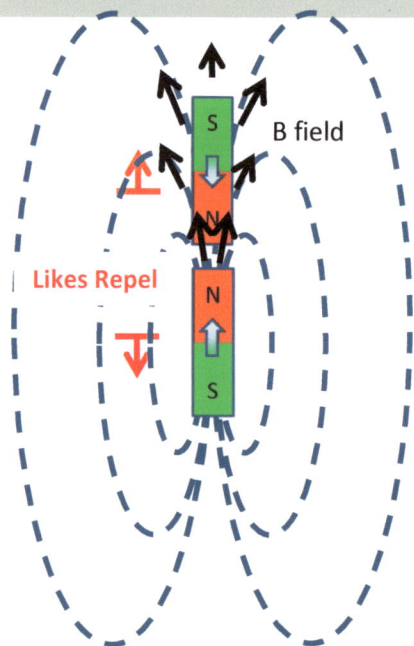

Likes Repel

B field

Same poles repel
Get two magnets to fly apart. Force two magnets to kiss the same poles, and they want to fly apart. The picture of the two magnets captured just before the end magnet flies away, due to the repulsion.

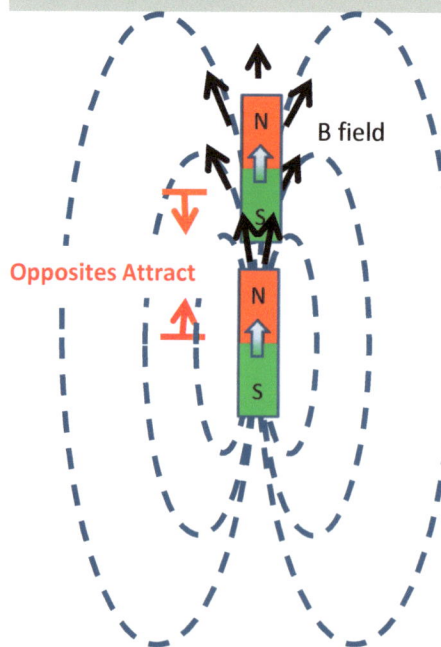

Opposites Attract

B field

White paint side has the same pole.

Opposite poles attract
Magnets on top of each other want to kiss opposite poles. Get two magnets to attract and squeeze against your finger. The finger was not crushed, but the magnetic attraction is strong enough for the magnets to stay on the finger.

Repel: Magnetic fields are opposite each other

Attract: Magnetic fields are aligned with each other

Displayed are the fields around a permanent magnet. The top magnet is repelled because its fields are opposite the lower magnet. The picture of the two magnets is un-stable, due to the repulsion.

Displayed are the fields around a permanent magnet. The top magnet is attracted because its fields are aligned with the lower magnet. The picture of the two magnets is stable, due to the attraction.

Play with two magnets and feel the attraction and repulsion, based on what sides are closest.

Attracting Alternating Magnets, Side by Side

Here is a great example of side by side magnets getting attracted to each other, beating gravity, because the fields from neighboring magnets are aligned with the magnetic moment of their neighbors. Yes, the attraction is stronger than gravity. The proof is the levitation.

Forces happen when two magnets get closer together, and forces happen, either attraction or repulsion, dependent on the magnets' orientation. The magnets can be side by side, or end to end. 'Opposite poles attract' applies to all these geometries.

"There are many ways to attract, and achieve harmony."

Experiment A.2: Make chains of magnets.

- Magnets can also be chained together side by side, instead pole to pole. The poles will reverse from neighboring poles, because that way the fields are aligned.

Side by side attraction, with opposite or alternating poles, when fields are in same direction at the magnets:
- Be stronger than gravity.
- Magnetic forces are strong.
- Alternating side by side magnets attract. Their fields are aligned through each magnet.

Magnetic forces lifting other magnets

Opposites or alternating poles attract

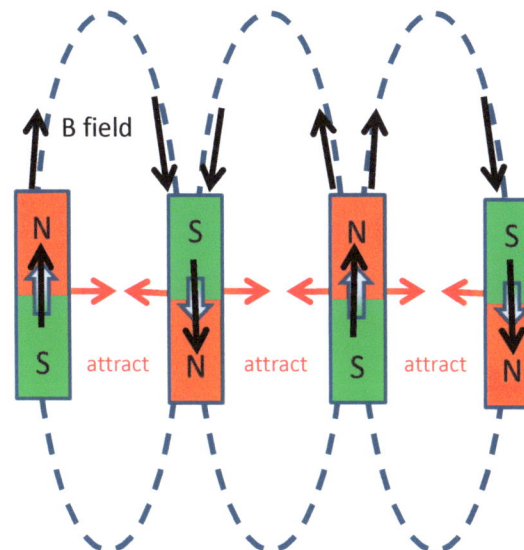

Attraction: Magnets line up with magnetic B fields of neighboring magnets, so there is a lower energy and attraction.

Experiments: Side by side attraction

Side by side magnets are also following the aligned magnet fields if they keep reversing their direction, which also follows the rule 'opposites attract'.

Attract Magnets, Side by Side, Opposite Poles

Any two close magnets - any arbitrary orientation like side by side or parallel - get attraction or repulsion, based on their field alignment. Side by side magnets still follow the rule that fields align.

"I can release a lot of energy to get to a lower energy."

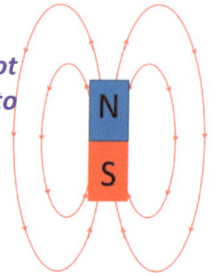

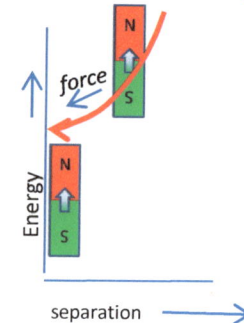

'Like poles repel'
Higher energy: unstable

Like poles repel. These magnets have oppositely aligned magnetic fields through each other, with unstable high energy.

Magnets want to be aligned to other magnet's magnetic field, for lower energy. Where ever there is a magnetic field, there will be magnetic forces. The two magnets can be held side by side. If they are both oriented with North up, then the two magnets will repel: the magnetic fields are not aligned with each other. Said another way, the magnetic field of one passing through the other is opposite.

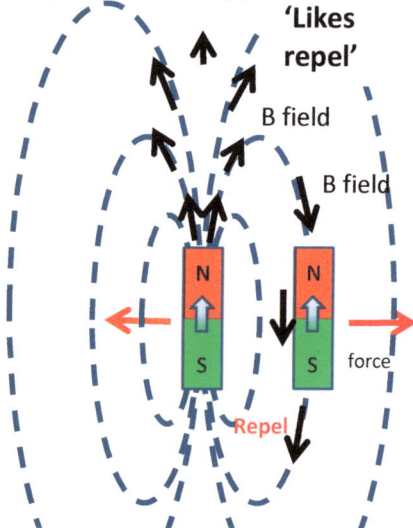

Lower energy away from other magnet. (or magnet can flip and get attracted.

Side magnet is opposite the applied B field of the other magnet, so the side magnet is repelled, and is in an unstable higher energy state.

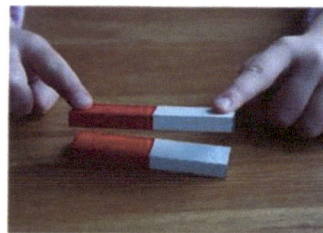

'Likes repel'
B field
B field
force
Repel

Like poles repel
Magnets can be held, levitating about another magnet, although not stably.

The same poles will repel, and the fields will be mis-aligned.

Repel: Magnetic fields opposite each other, inside each magnet

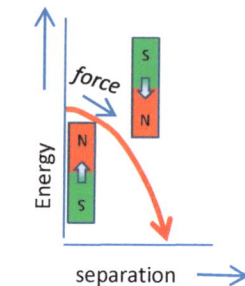

'Opposite poles attract'
Lower energy: want to get closer

Magnets want to be aligned to other magnet's magnetic field, for lower energy. Opposite poles attract.

Where ever there is a magnetic field, there will be magnetic forces. The two magnets can be side by side. If they are oriented the reverse of each other, then the two magnets will attract: the magnetic fields are aligned with each other, or the magnetic field of one passing through the other is in the same direction as the other magnetic moment.

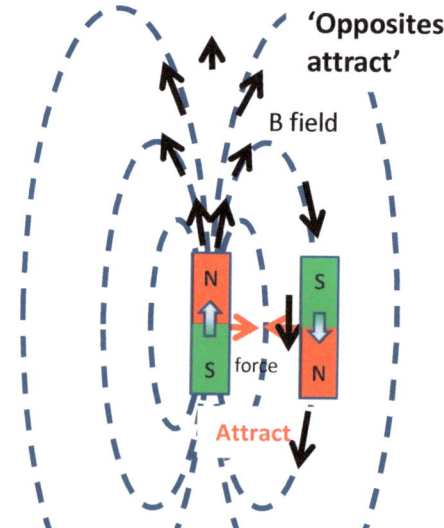

'Opposites attract'
B field
force
Attract

The opposite poles will attract, and the fields will be aligned.

Attract: Magnetic fields aligned with each other, inside each magnet

Lower energy closer to other magnet, so magnet is attracted.

When opposite poles are close, the side magnet is same orientation as the applied B field of the other magnet, so the side magnet is attracted, and is in a stable lower energy state.

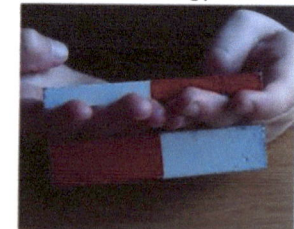

Opposite poles attract

Notice the poles are all opposites, so these two magnets attract. Magnets side by side, with opposite poles closest, have aligned magnetic fields and attract.

You can see the forces when you hold the magnets over each other, side by side.

To understand how magnets are used in motors, any understanding needs to start out looking at magnetic fields and forces, and how different alignment of these magnetic fields creates different energy. We will demonstrate forces and torques between magnets, and how those torques can be used to design a strong motor with the most torque. A magnet has orientation (physical tilt or direction of the magnet) and a different orientation has a different torque and a different energy level.

Electro-magnets have electric currents in a wire coil and produce a magnetic field, the same as a permanent magnet, except that the magnetic field switches direction when the electric current switches direction. We can be clever and switch currents to the sideways electro-magnet in the rotor, so the rotor field is always sideways to the stator magnetic field. Hence we get perpetual rotation, which is a motor.

We use the fact that the magnetic field turns off when the electric current turns off.

"Rotor turns to align the magnetic fields, Padawan."

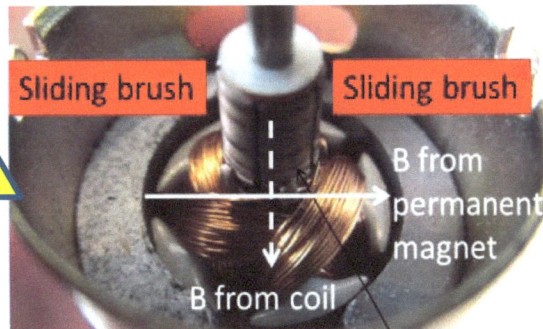

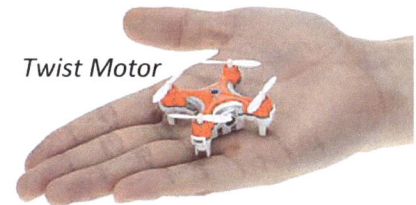

What creates the energy of the motor?
→Electric currents

Electric current creates a magnetic field and mechanical torque

S ◁ N

Twist Motor

Sliding brush   Sliding brush

B from permanent magnet

B from coil

N
↑
S

"Spinning, all the little motors allow me to control the speed and direction of flight really easily."

Sliding connector excites correct coils.

It's all about energy ... Things want to move to a lower energy state.

Rotate Magnets to Reduce Energy, using Torques

This figure shows that the top magnet wants to rotate so that 'opposite poles attract', where N and S are closest and the magnetic fields are in the same direction. This desire for lower energy is just nature, so let's relax and enjoy our cars, air conditioners, ceiling fans, DVD players...

"Turn to align the magnetic fields, Padawan."

No torque, but balancing on an unstable energy plateau

Like a ball in gravity, tottering to fall down.

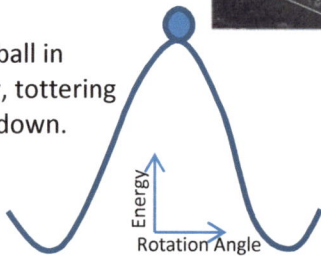

Energy

Rotation Angle

Higher Energy

Steepest change in energy versus angle, so most torque

torque

No torque, and stable energy valley

Lower Energy

Sideways magnetic orientation has most torque.
This is the geometry when the electro coils are turned on.

Opposite alignment and higher energy: Unstable, no torque when perfectly mis-aligned, but any tilt and will immediately want to flip.

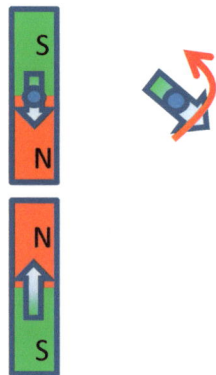

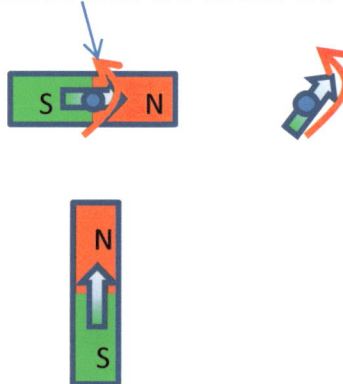

S
N

N
S

Parallel alignment and lower energy: Stable, no torque and at lowest energy.

S
N

N
S

N
S

N
S

Trouble with permanent magnet for rotor:

Permanent magnet effect, not an electro-magnetic motor: Unfortunately, when the rotation overshoots the top , a permanent magnet will just reverse the twist direction (causing an oscillation, not spin), which is why electromagnetic rotors and switches are needed instead.   Electromagnet rotors can reverse currents and keep the same twist direction, using commutators.

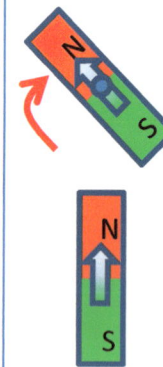

N
S

N
S

See Appendix G, equations 6 and 11

The spinning magnet wants to rotate to the lower energy direction where fields are aligned and opposite magnetic poles attract

Ideal Multi-Pole Motor Magnetic Orientation

Using an electro-magnetic, the rotor always has the sideways field, compared to the DC field of the permanent magnet stator. This rotor field direction has the largest torque. The electrical current is always going through the exact one of the many coils that aims a magnetic field sideways, in the same direction with torque in same direction. The commutator brushes guarantees this. The soft iron inside the rotor coil enhances the magnetic field, so the fields are larger and the torques are larger.

Every force has an optimum location, Padawan.

"Dude, I bet that Jedi master thinks he just froze the perfect motor diagram. But I'd like to see the real motor."

Most torque when sideways:
- Largest change in energy versus change in angle or orientation:

Just tilt to align with permanent magnet, no rotation.

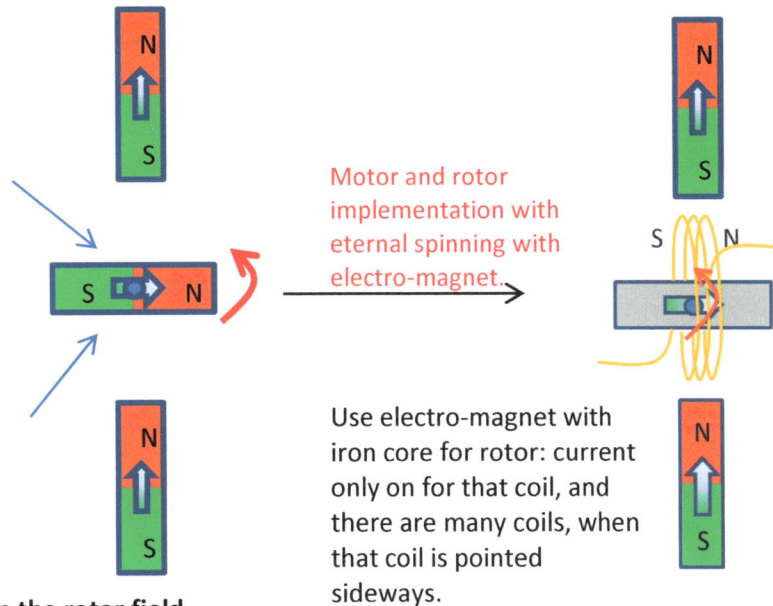

Motor and rotor implementation with eternal spinning with electro-magnet.

Use electro-magnet with iron core for rotor: current only on for that coil, and there are many coils, when that coil is pointed sideways.

Split-ring metal connectors for the brushes, attached to the coils.

An electro-magnet can always keep the rotor field pointed sideways, using the commutator.
If can keep the spinning magnet's magnetic field oriented sideways to the permanent magnet, then get the most torque out of the motor.

When looking at the world through energy, lots of physical terms involve a change in energy over a change in something, like change in angle for torque and change in location for force.
- A torque is the twist on the rotor, like pedaling bicycle pedals. Torque is the slope of the change in energy of the magnets over the change in angle of the rotor.
- Another quantity, force, is the slope of the change in energy over the change in distance.

8-pole strong motor:
- Practical motors have multiple poles, or coils in the rotor.
- For a 2 or greater pole motor, the number of poles is equal to double the number of coils (equal to the number of pads), because each coil accepts current going in one direction, and then the other direction when that coil is flipped around 180 degrees.

By using many coils in the rotor, the right one can get the electric current to keep the rotor magnetic field in the same direction, sideways.

Pull Magnets to Reduce Energy, for Linear Motor

This figure shows that the rotor magnet wants to get pulled into the stator magnet field so that 'opposite poles attract', where N and S are closest and the magnetic fields are in the same direction. This desire for lower energy is just nature, so let's relax and enjoy our cars, air conditioners, ceiling fans, DVD players...

"Turn to align the magnetic fields, Padawan."

No pulling or torque, but balancing on an unstable energy plateau

Like a ball in gravity, tottering to fall down.

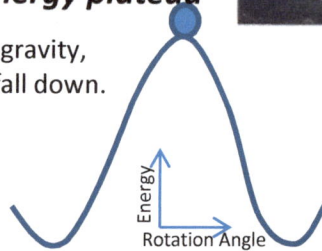

Energy / Rotation Angle

Higher Energy

Steepest change in energy, so most torque

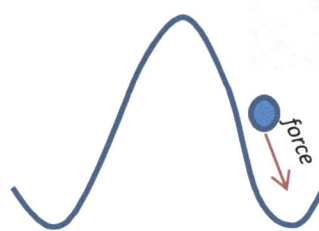

force

Most sideways force on rotor magnet, which pulls rotor disk around.
This is the geometry when the electro coils are turned on.

No pulling or torque, and stable energy valley

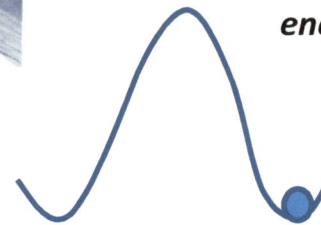

Lower Energy

Opposite alignment and higher energy: Unstable, no torque when perfectly mis-aligned, but any sideways offset and will immediately want to push away.

Rotor magnet, permanent magnet

force

Stator magnet, electro-coils

Parallel alignment and lower energy: Stable, no torque and at lowest energy. The two magnets have parallel magnetic fields and are the closest to each other.

The rotor magnet wants gets attracted to the opposite magnetic poles on the stator, and pulls the rotor disk around.

Alternative Motor Configurations Based on Energy

The idea of energy gradient can lead to different electric motor designs.
Here we show the successful standard approach of a rotor that is essentially a magnet rotating about its center.
We also show the novel approach of many electro coils around a disk, and the electro coils try get out from between two stator magnets. If we put a ring of electromagnets on a rotor disk, and place the rotor disk between a ring of stator magnets, then there is repulsion and movement and rotation of the disk.
We can also avoid spinning and just have linear reciprocal motion by pulling a magnet into a coil.

The concept of energy can provide different answers to motor design, Padawan.

"We just need two magnets to push against each other around a shaft to get a motor."

Most common energy approach:
Rotate the middle rotor about a shaft.

Specialized energy approach for Linear Motor:
High torque, low profile:
Push magnets sideways away from the stator field. This linear sideways push between magnets can be used for brushless motor or for acceleration force for levitating trains, and this sideways push is more about forces than rotation.

Energy approach specialized for linear motion:
Push magnets into a coil.

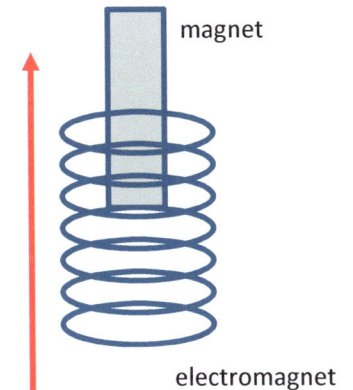

Most torque when sideways:
- Largest change in energy versus change in orientation:

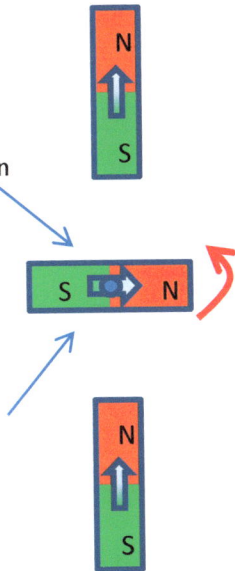

Just tilt to align with permanent magnet, no rotation.

Most sideways push when polarization is opposite and center magnet is a little offset:
- Largest change in energy versus change in offset location:

Magnet is pushed sideways, along with the rotor disk and many other magnets.

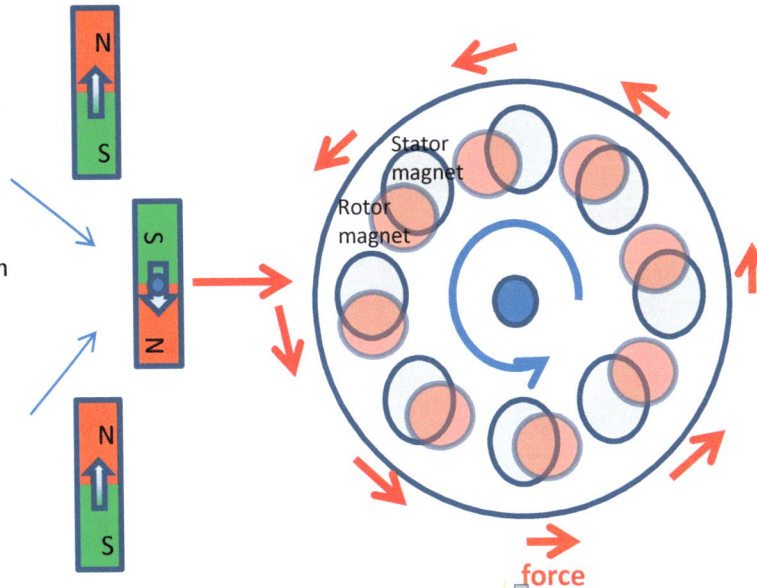

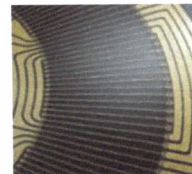

Stator magnet

Rotor magnet

force

magnet

electromagnet

Linear motion can be generated directly when pull and push a magnet into the electromagnet.

Example of etched traces to get a electro coil for the rotor on a disk if a brushed motor, or electro coil on the stator if brushless motor.

The energy idea can lead to different designs for rotating motors.

Appendix C: 1-Pole and 2-Pole Commutator:
Current versus Time Revealed with Oscilloscope

For a 1-pole motor, here are the repeating spins of the coil, over and over again. The current on/off behavior repeats continuously. The current needs to turn off for half the cycle so the rotor can coast back to the starting point and get a consistent torque direction.

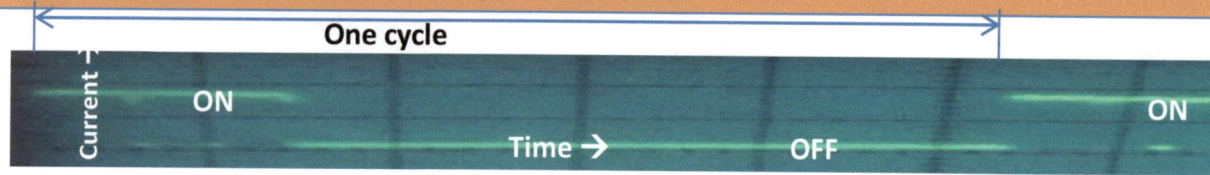

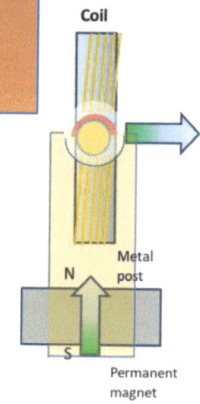

One cycle

Current ON ... Time → ... OFF ... ON

Coil

Metal post

N

S

Permanent magnet

"I'm half on."

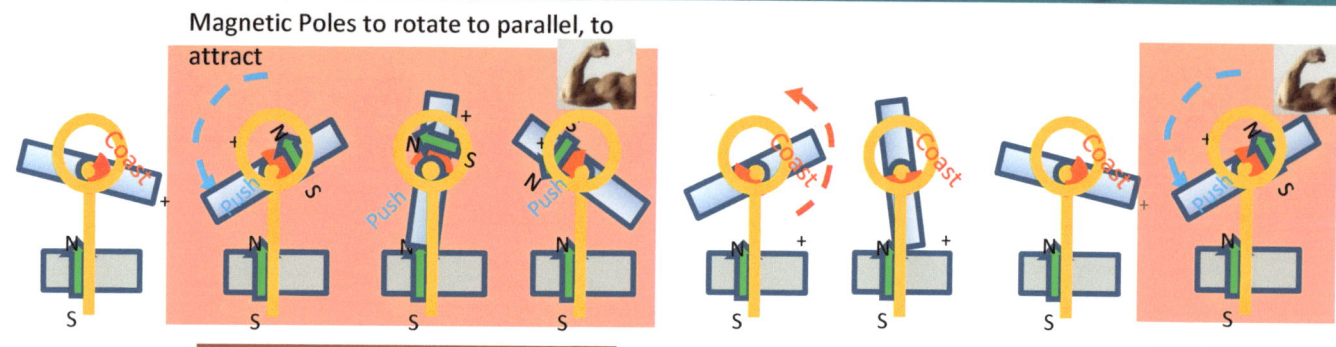

Magnetic Poles to rotate to parallel, to attract

Current ON, Torque — Get magnetic field interaction

Current OFF, No Torque, Coast through turn with inertia — No magnetic field interaction

Homemade commutator: When the bare metal of the coil slides over the bare metal of the post, then current flows and there is a rotor magnetic field, with torque. Straighten out the rotor wire so that the coil is not bouncing around as it spins, and make sure that the bare side of both ends of the coil are in the same direction.

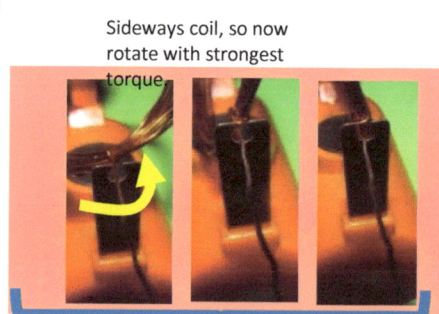

Sideways coil, so now rotate with strongest torque

Coil resting on insulating enamel, so no current or torques. Keep rotating with no torque resistance.

CURRENT and Torque: Scraped Shiny side of Copper wire DOWN, connecting with bare metal support.
- Get coil magnetic field

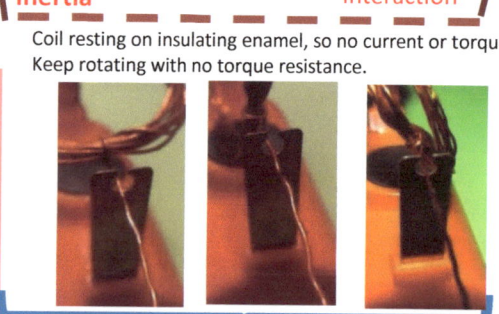

NO CURRENT and no Torque: Scraped Shiny side of Copper wire UP, not connecting with bare metal support.
- No coil magnetic field

Constant current would not spin: Remember, if kept current on all the time, in same direction, the coil would get an opposite torque back up to align the magnetic fields, just like two permanent magnets that can not change their magnetic moment. The coil would just hover around the top, instead of rotating. That is where commutators and multiple coils around the rotor save the day, and allowed a positive torque over the complete rotation, by keeping the rotor magnetic field sideways in the same direction.

Current off: Current turns off, and Magnetic Pole disappears, allowing coil to keep rotating back to North anti-parallel. Coil is resting on enamel, so current stops. Magnet wire enamel is blocking electrical contact, so there is insulation and no current and no coil magnetic field. Coil is coasting around the turn, with no torque, just inertia.

Current on: You did a good job scraping the enamel off this side of the wire, and both axle wires are centered, straight, and have the same exposed metal on the same side.

The coil will keep spinning until the battery runs dry, or your finger stops the spin, or you actually try to turn something useful. Let's be blunt, this 1-pole coil needs typical motor improvements (multi-pole commutator with more coils, iron core, tighter spacing) to be useful.

1-pole motor only has power less than half the time.

2-Pole Commutator: Current versus Time

"I'm full on."

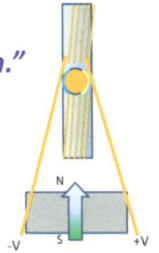

For a 2-pole motor, here are repeating spins of the coil, over and over again. The current on/off behavior repeats continuously. The current can stay on all the time because the current switches direction in the coil, when the coil changes sides, to keep the rotor magnetic field consistent and torque consistent.

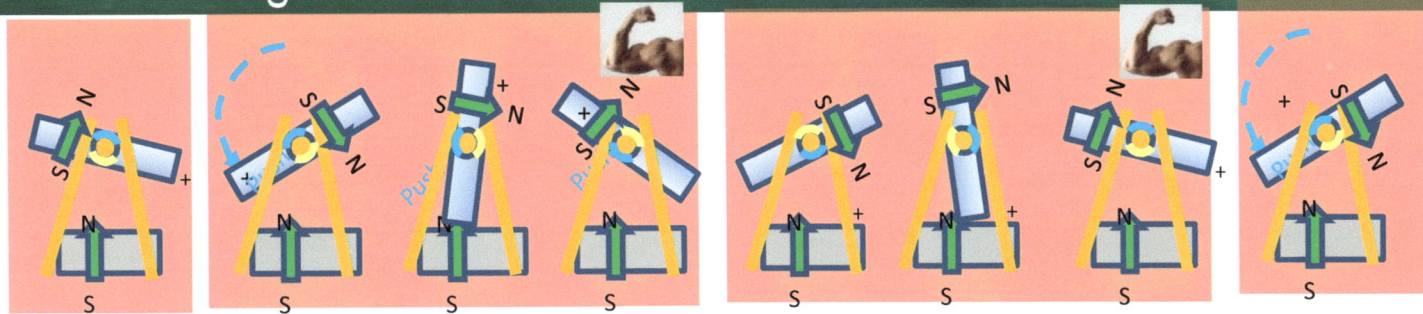

One cycle

Current →

ON

ON

Time →

OFF

**Current ON,
One direction,
Torque**

Get magnetic
field interaction

**Current ON,
Other direction through coil,
Torque**

**Coasting,
No torque**

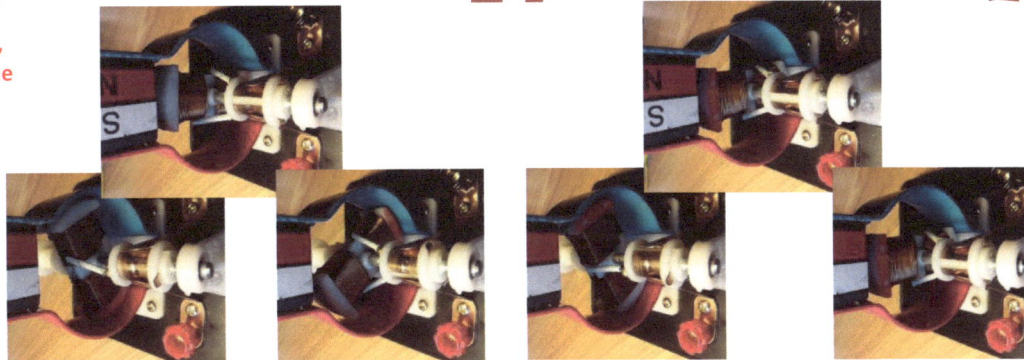

Current on: Current is mostly on all the time for a 2-pole motor. Half the time the current is flowing one way, and the other half the time the current is flowing the other way with the coil pointed in the other direction as well. The keeps the magnetic field of the coil always pointing to only one side.

The coil will keep spinning until the battery runs dry, or your finger stops the spin. This 2-pole coil has typical motor improvements (multi-pole commutator with more coils, iron core, tighter spacing) and so is on the way to being useful.

2-pole or more motors have power almost all the time.

How can we see currents and magnetic fields? Just run a current through the rotor coil and observe that a magnetic field is created using a compass. Because of the commutator, the rotor magnetic field is created in only one direction as the rotor spins.

"Rotors have their own magnetic field."

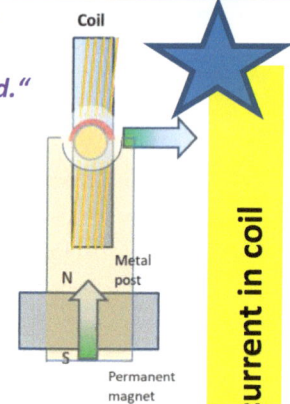

Coil
Metal post
N
S
Permanent magnet

1-pole hands-on learning of electro-magnet coil:

Experiment D.1: Show compass needle deflection when have current in coil.

- Show current only flowing for less than half of spin, using detected magnetic field

Let's demonstrate that electric currents in the rotor generate a magnetic field, using a compass to measure the field.

1-pole commutator: Only have current ON over half a turn, when rotor is sideways to the permanent magnet.

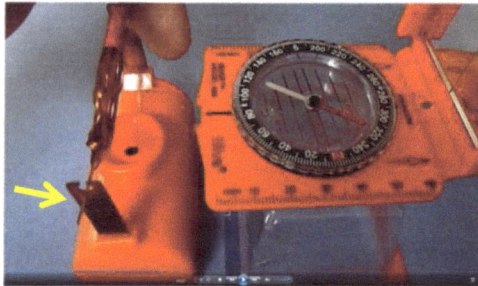

OFF Current
- Compass points along Earth's magnetic field (north)

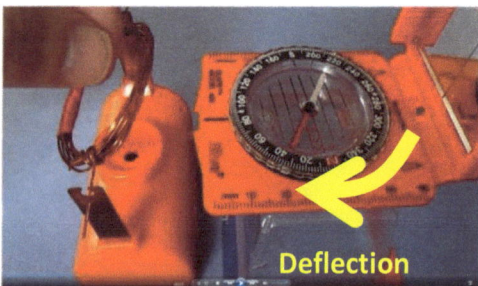

Deflection

ON Current
Result: Compass deflected along coils magnetic field

2-pole hands-on learning of electro-magnet coil:

Experiment D.2: Show compass needle deflection when have current in coil.

- Show current flowing for full spin, using detected magnetic field

No current in rotor (disconnect battery):
- Needles goes back to Earth's magnetic north.

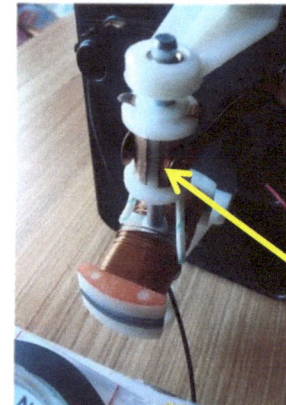

2-pole Commutator: ON all the time, switches current direction halfway through turn.

Exposed rotor and commutator

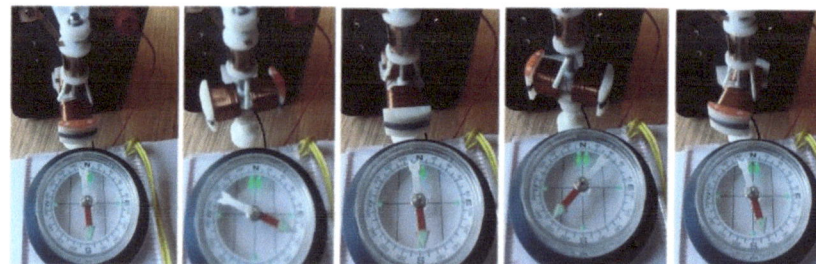

| 0 turn | 1/4 turn | 1/2 turn | 3/4 turn | full turn |

ON Current when re-connect battery, and needle deflection:
Shown is a full turn of the 2-pole commutator

Result: View the compass needle. The 2-pole rotor magnetic field is always pointed in the same general sideways direction because the commutator switches the current halfway through.

Experiments: Deflect compass needle using current in coil

You've created a magnetic field! Thank Mr. Ampere in year 1820 for showing this.
Apply current to the coil, and the coil deflects the compass needle. There must be a created magnetic field.

Observation 1: Consistent Spin Direction

"I am a creature of predictability."

The consistent spin direction of the rotor is determined by the product of the two magnetic fields: stator magnetic field direction, and rotor magnetic field direction from current direction and clockwise or counter-clockwise coil winding.

Experiment D.3: Observe that coil spins in consistent direction, for un-changed battery direction and coil alignment to posts.

Coil spins in a consistent direction, when keep the same current I direction and permanent magnet direction B.

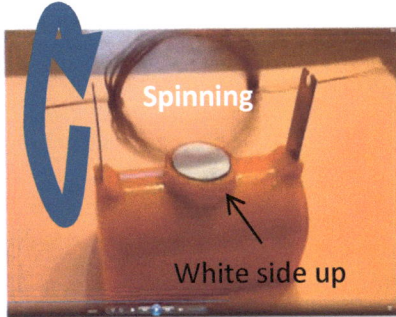

Spinning

White side up

Spinning 1-pole motor

Kick start the spin: Sometimes, the 1-pole coil needs a push (twist, torque) in the same direction to start turning. The kick start is necessary when the enamel insulator in the commutator is resting on the posts and stopping the initial current draw.

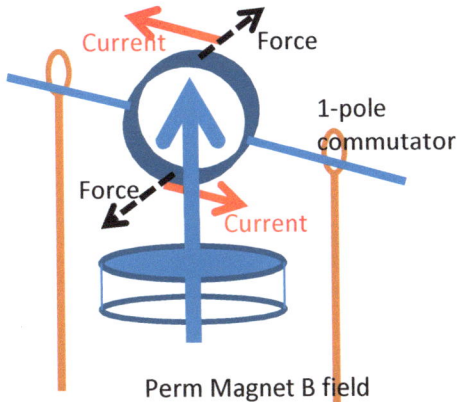

Current — Force
Force — Current
1-pole commutator
Perm Magnet B field

Force on Wire = Number Turns*Current*Magnetic Field B*Length

For a constant current direction, and magnetic field, the spin direction is predictable.

Observe that rotor spins in consistent direction, for un-changed battery direction and DC magnet stator field.

The permanent magnet can be inserted in different directions.

Coil spins in a consistent direction, when keep permanent magnet in same direction.

Spinning

Spinning 2-pole motor

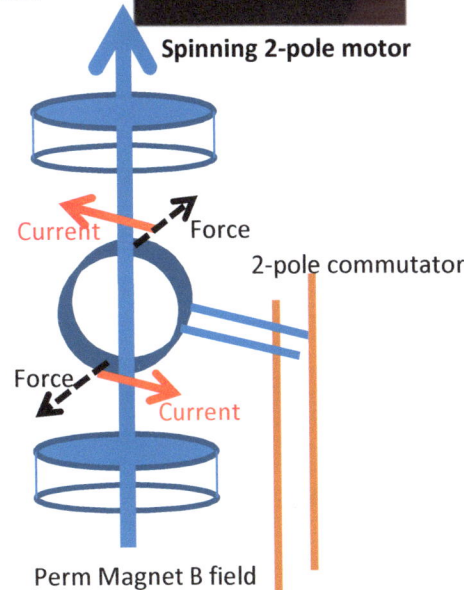

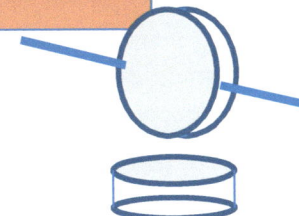

Current — Force
2-pole commutator
Force — Current
Perm Magnet B field

Equivalent to a permanent magnet pointing sideways, wanting to twist.

Spin in or out:
The drill bit always turns in the direction you want and expect.

One spin direction of engine works, and the other does not.
The starter motor better turn in the right direction to start your car.

Experiments: Stop and start spin, where spin in same direction.

The coil will spin in one consistent direction. The spin direction depends on the current direction and the North or South pole facing up on the permanent magnet.

Flip Permanent Magnet to Reverse Spin

Why would you flip the permanent magnet of the stator?
Flipping the permanent magnet shows that when the magnetic field changes direction, the torque flips and the spin is reversed. Change either the rotor magnetic field, or the stator magnetic field, but not both, to reverse the spin direction.
Remember being able to reverse direction of spin on a drill. The AC motor reverses direction by switching the current direction to either stator or rotor, but not both.

"I can spin in either direction, so I work in robots, cars, drills, garage door openers."

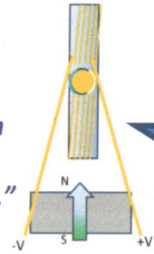

Experiment D.4: Flip permanent magnet

1. **Observe that coil spins in opposite direction, after flip the permanent 'stator' magnet.**

When flip the permanent magnetic field, by flipping the magnet, the coil spins in the opposite direction.

Before flip magnet

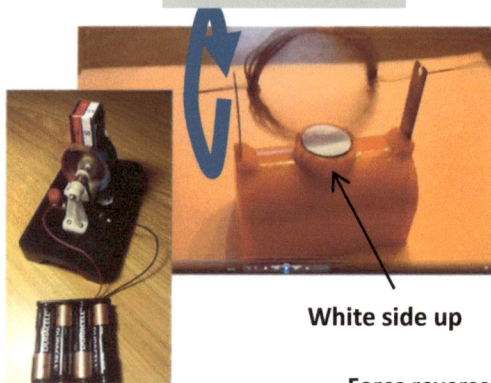

White side up

After flip magnet, reverse spin direction

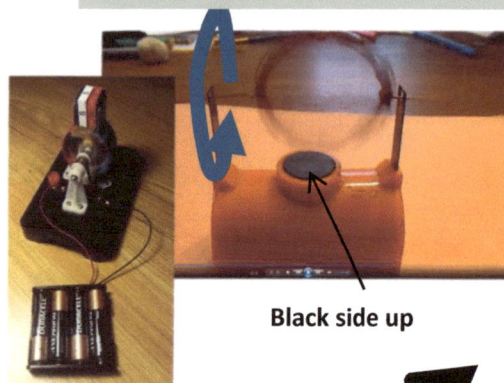

Black side up

Force reverses direction when flip magnet

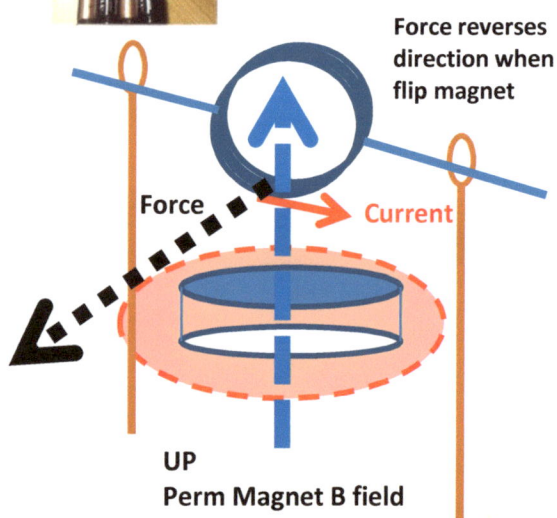

Force   Current

UP
Perm Magnet B field

Force on Wire = Current*Magnetic Field B*Length

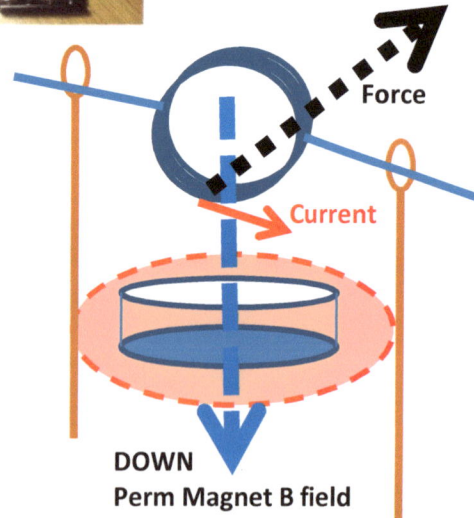

Force

Current

DOWN
Perm Magnet B field

Change direction of magnetic field, and force on wire reverses, and the coil will spin in reverse direction.

Electric drills use both spin directions:

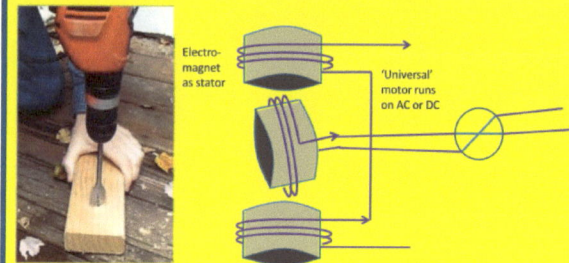

Flip one of the coil voltages on AC motor, and spin is reversed.

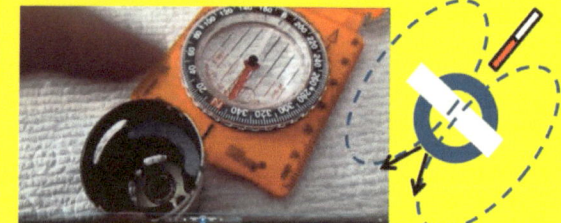

Field of permanent magnet
The fixed stator magnet is the field the rotor wants to align with.

If flip stator permanent magnetic field, the rotor would spin in the opposite direction.
But the motor direction is changed just by reversing the voltage, so no problem.

Force on Wire = Current*Magnetic Field B*Length

Change sign of magnetic field B of stator, so change sign (direction) of force.

Experiments: Flip stator magnet and spin goes in opposite direction.

Flipping the permanent magnet will reverse the spin. We need to reverse spin for printers and electric cars.

Good and Bad Commutators, Like Good and Bad Anything

"Without a commutator, I'll just rock from side to side."

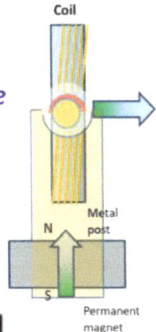

Here are two ideas that have current flowing all the time. One works, and the other doesn't. As a lesson in failure, here is a failed 1-pole commutator, with all the enamel stripped away, not just half. For success, here is also the path to 'goodness' using a multi-pole 2-pole commutator with the current switching direction.

Just a bare wire, no enamel at ends, no current switching

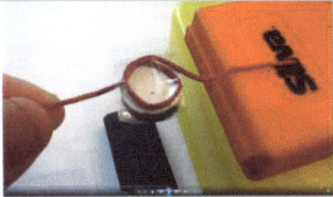

Coil stays parallel to permanent magnet.

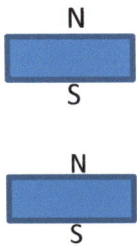

Coil flipped 180 degrees, but current in same direction.

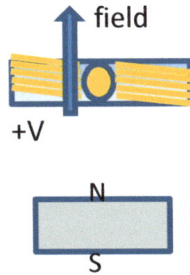

field

+V

N
S

N
S

N
S

N
S

Commutator, with current switching

Using commutator, coil flipped 180 degrees, but current is in same direction relative to the outside world.

Coil Edge 1

field   field

+V

Coil Edge 1

N
S

N
S

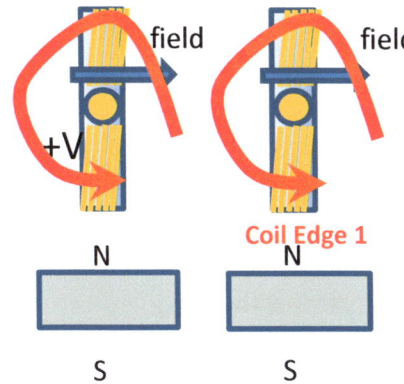

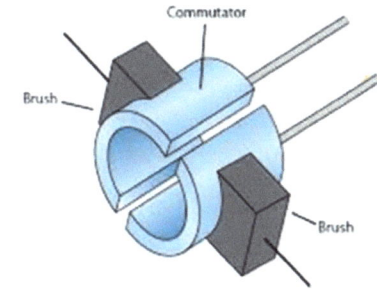

2. A split ring wrapping around the axle, the commutator makes physical contact with the brushes, which connect to opposite poles of a power source to deliver positive and negative charges to the commutator.

Bad: For a 1-pole motor, what happens if all the wire is stripped at the ends of rotor, and current flows all the time?

- The coil will not spin. The coil will snap to the North direction of the coil, parallel to the North direction of the permanent magnet. The coil should just oscillate about the aligned position. (note: coil has the potential to spin weakly if the wire skips and current is not flowing all the time)

With one pole, you don't want current flowing all the time. If current stays on, you just have two permanent magnets together, and that does not spin. That's a problem with a 1- pole commutator. The magnets would just align and stop.

Good: For a 2-pole or more-pole motor, what happens when the current in rotor is reversed on the other half of the turn?

- When the current reverses when the coil is reversed, then the motor never coasts, always has power, always has torque in the same direction, and we then have ourselves a 2-pole modern motor with a 'commutator', at the location where the coil meets the DC current power supply. This 2-pole commutator is part of a good motor, except for all the other improvement stuff, like lots of iron for increased magnetic fields and more torque.

When current switches, then you do want current flowing all the time, with multi-pol commutator. Then you've got spin, and torque in same direction. Rotor magnetic field stays sideways, pointed in same direction.
Have torque all the time, instead of coasting unpowered through half the cycle like a 1-pole motor.

You'll know a bad commutator by its low torque.

Electromagnet Rotor Magnetic Field,
It Makes the Motor Go 'Round, Q and A

These questions about the rotor are 'pivotal'.

The rotor magnetic field is created by the current when a voltage is applied. This rotor field is always kept in the same direction to keep constant torque, independent of the spin of the rotor hardware. This rotor field depends on the iron core of the electro-magnet to create a much larger rotor magnetic field to make a strong mechanical torque.

Question 1: Can we measure the magnetic field produced by the rotor coil?
- Yes, we can see the direction of the magnetic field using a compass. To actually measure the magnitude of the magnetic field, we would need a more advanced instrument, like a magnetometer.
- Magnetometers inserted into more advanced motors can be electronic chips, such as Hall sensors or Giant-Magneto-Resistive sensors.

Question 2: Which way should people understand the torque on the motor's rotor, from the point of view of currents or point of view of magnetic moments?
- Either way is fine. Both ways give the same final torque.
- The point of view of currents is easier to understand for a simple electro-magnet rotor, without an iron core. With an iron core, the effective current on its surface which creates its magnetic field needs to be added to the torque.
- The point of view of magnetic moments is easier to understand when the iron core of an electro-magnet rotor is present, which can increase the magnetic field by a factor 10.

Question 3: For a 2-pole rotor, what direction is the rotor magnetic field as the rotor is rotated 360 degrees?
- The rotor magnetic field stays pointed in the same direction relative to the observer, or relative to the stator. That is the purpose of the commutator. The rotor is always an electro-magnet that craftily keeps its magnetic field stationary to the side in the same direction, even as the coil is spinning. Thank the switch due to the commutators. This keeps the twisting torque always on in the same direction.

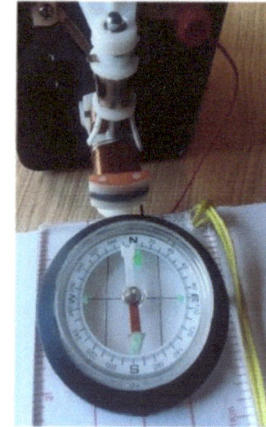

Rotor electro-magnet field is observed with a compass.

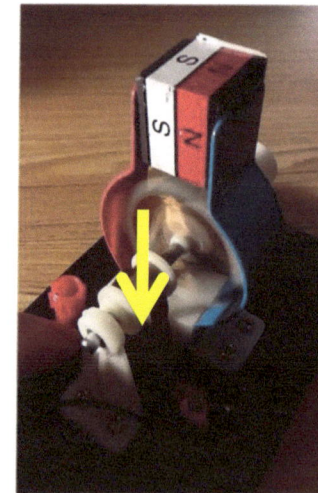

Rotor electro-magnet field (yellow arrow) is kept in one steady direction by the commutator, to keep the spin going.
The electro-magnet field is kept down, and the stator field is across, or sideways.

Let's show the electric current as a function of time, by inserting current and voltage meters in the power circuit for the electric motor.

Experiment E.1: Measure current versus time in the spinning coil, using an AC multi-meter or oscilloscope.

- Place the coil at an orientation such that the current is flowing and creating a torque, and the motor will start spinning. Or just flick the coil rotor, and the coil will keep spinning, although it might want to reverse direction if you flicked it in the direction opposite the direction of spin.

"My current flips on and off, to keep my magnetic field sideways."

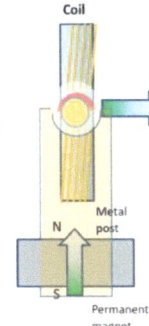

1-pole motor kit

Circuit: a coil and a battery

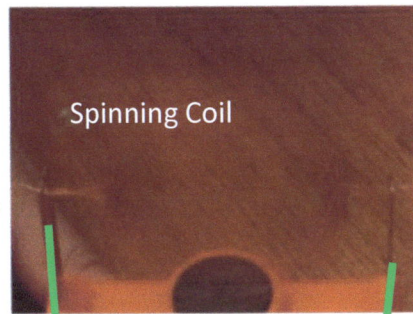

Spinning Coil

Magnet

Battery (1.5 Volts)

- Negative   + Positive

On/Off current in wire as coil rotates

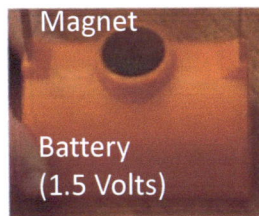

Measurement:
- If have oscilloscope, then can hook the oscilloscope up across a series resistor to measure current.
- If have a multi-meter, then note that measure a voltage when set voltage meter to DC or AC, which means the current is AC.
- See measurement setup in following pages.

If added meter

1-pole Current versus time

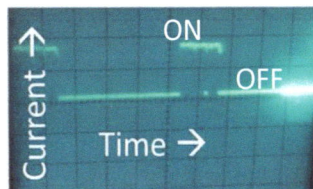

Current ↑ ON OFF Time →

One spin

2-pole motor kit

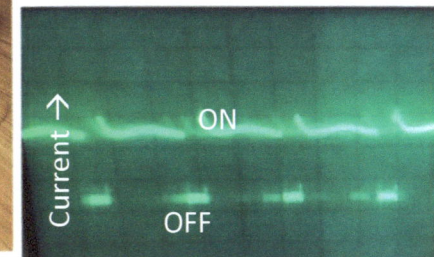

Current → ON OFF

One spin

Experiments: Observe on/off current as current powers the motor

Our toy 1-pole motor is the simplest circuit. Just connect a battery to the two posts that the coil rotor is riding on, and the coil will spin.

A 2-pole motor will draw current and keep steady torque as the rotor coil is rotating up and rotating away.

1-Pole Motor with Current Meter

You don't just have to see a coil spinning, and just guess at the power source. You can also look at the current passing through the coil using a standard multi-meter or oscilloscope. Here is a multi-meter hooked up.

"A simple multi-meter will measure average current. An oscilloscope will show the shape of the current."

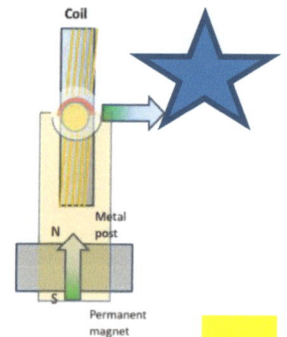

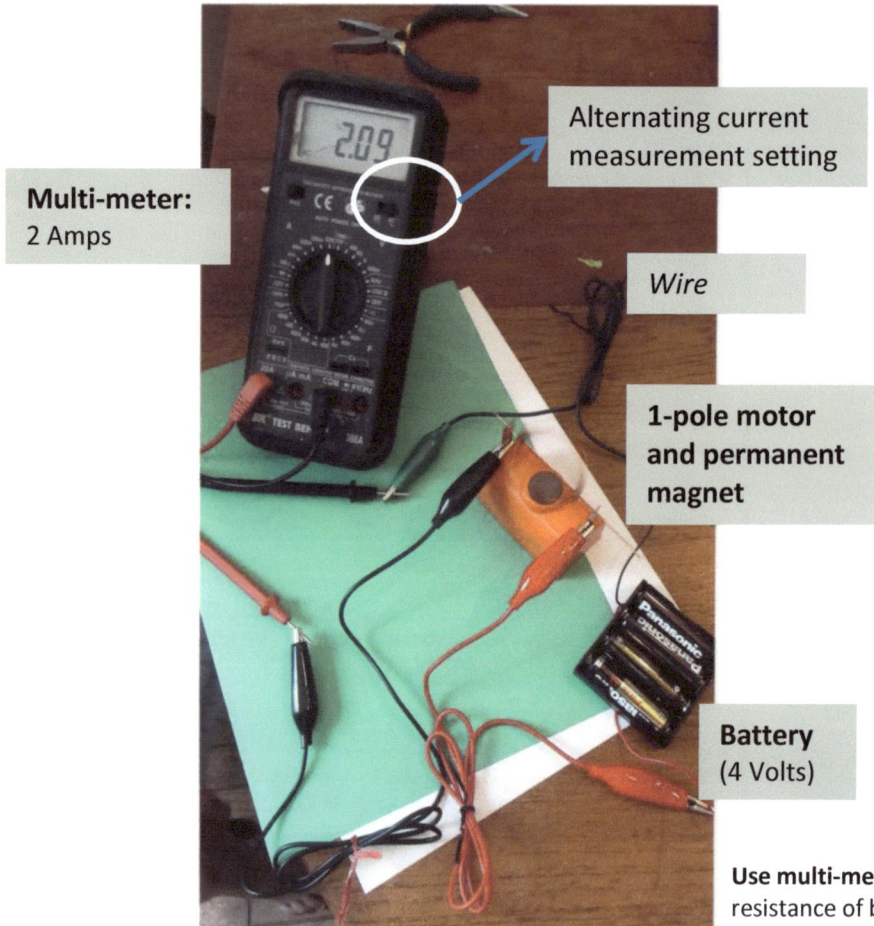

Coil

Metal post

N

S

Permanent magnet

Experiment E.2 with multi-meter: If have a multi-meter, then measure the AC current passing through coil by inserting the multi-meter in series with the current.

Multi-meter:
2 Amps

Alternating current measurement setting

Wire

1-pole motor and permanent magnet

Battery
(4 Volts)

1-pole motor with battery and current meter.

Multi-meter:
AC Current amplitude, the average. Multi-meters can measure current, voltage, resistance, and other things like capacitance and inductance.

Any current measuring instrument will work

Oscilloscope:
Current versus time, instant, using an older analog oscilloscope, not digital.

See the current yourself using a meter

Use multi-meter to get resistance value: Multi-meter records current, so can determine resistance of battery and coil using Ohms law:
- Resistance R = Voltage/Current = 4Volt/2Amp = 2 ohms

Option for more power:
- Double batteries to supercharge the 1-pole motor, using wires and alligator clips: Need a separate power boost to overcome the extra resistance of the multi-meter.

Experiments: Measure average AC current using meter

You can measure the current going through the coil by a current meter or by measuring the voltage across a resistor in series with the coil.

Motor Circuit with Multi-meter

A multi-meter has an input to measure current. The multi-meter can be put in the direct path of the current. That is because the multi-meter input is designed to have a very low resistance and not load the circuit.

If have a multi-meter, then measure the current passing through coil. The current is mostly AC for a 1-pole motor, with half off time, and the current is mostly DC for a 2 or more pole motor, with full on time.

"Pick your demo motor, and measure the voltage across a resistor to get current."

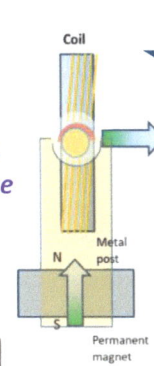

Coil

Metal post

N

S

Permanent magnet

On/Off current as coil rotates

Multi-meter

Multi-meter: 2 Amps

2.09

Alternating AC current measurement setting for 1-pole motor

1-pole demonstration motor

2-pole demonstration motor

+ Positive

Battery: 6 Volts using 4 AAA batteries

Add DC or AC current measurement using series multi-meter

Multi-meter

- Negative

Measure the current with multi-meter in series

Average Current:
- This meter is measuring AC current. Here we see 2 Amps going through the coil. That might sound like a lot, but the power is only 3 Watts with a 1.5 Volt battery, using the relationship that power equals voltage times current.

DC or AC setting:
- For a 1-pole motor, a multi-meter needs to be on alternating current (AC) setting. If on DC setting, then will get half or less of the current reading, because DC measures the average current, which is midway between 0 Amps and 2 Amps.
- For a 2-pole motor, the current is flowing out of the battery almost full time, so the multi-meter setting should be DC.

The current meter is directly measuring the current going around the loop, through the coil.

Experiments: Measure average AC current using voltage across series resistor

Motor Circuit with Oscilloscope, to Make Currents Visible

"Oscilloscopes are great. They can measure the number of poles of your motor."

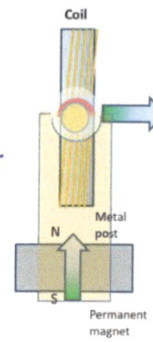

Coil

Metal post

N

S

Permanent magnet

Either meter, an oscilloscope or a multi-meter, can measure alternating voltage across the resistor. Here is an oscilloscope hooked up.

Experiment E.3 with oscilloscope: If have an oscilloscope, then measure the current versus time passing through coil by inserting the oscilloscope across or in parallel to a low resistance resistor.

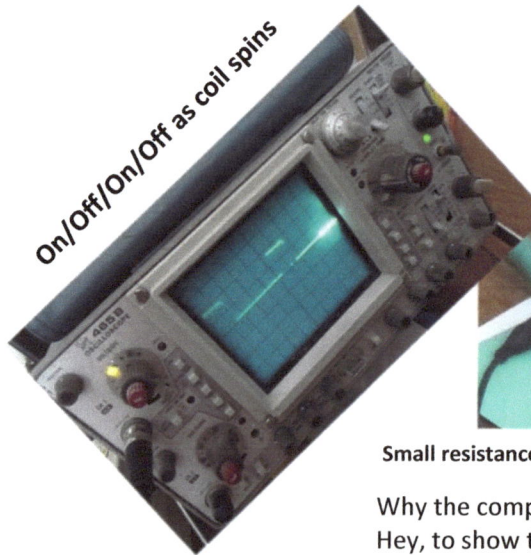

On/Off current in wire as coil rotates

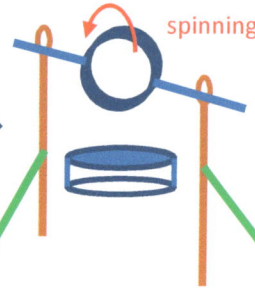

spinning

1-pole demonstration motor

2-pole demonstration motor

On/Off/On/Off as coil spins

Resistor

Small resistance in series to see voltage drop

+ Positive Volt

Battery: 6 Volts using 4 AAA batteries

- Negative Volt

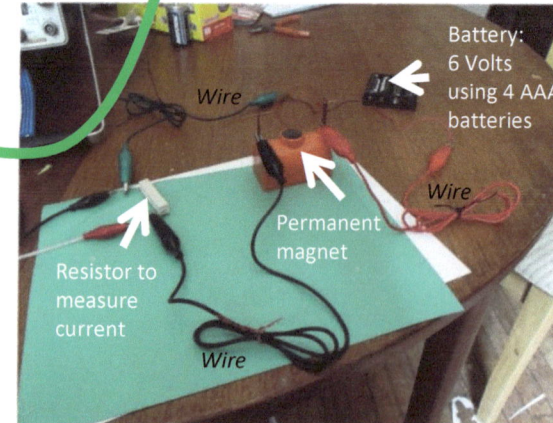

Why the complication to add a resistor?? Hey, to show the currents, we needed to add a resistor and pump up the battery voltage a little to handle the small power drop across the resistor.

Battery: 6 Volts using 4 AAA batteries

Wire

Wire

Permanent magnet

Resistor to measure current

Wire

The oscilloscope is measuring the voltage across a low resistance resistor, say 1 ohm resistor, in series with the coil. The battery power is beefed up to 6 volts, to allow a small voltage drop across the resistor, and still have current to power the spinning coil.

You can make this circuit too. The circuit is just a battery across a resistor and coil. For measurement, you could use a multi-meter across the resistor, on voltage setting, or an oscilloscope across the resistor.

Experiments: Measure voltage versus time using oscilloscope

1-pole Motor Circuit with Oscilloscope: Current versus Time

Oscilloscopes let you see current versus time, which is more visual wand shows the exact or instant current shape, instead of a multi-meter which only measures average current or power. Oscilloscopes offer a lot of information about spin frequency, current level, and duty cycle (fraction of time the current is on).

For the impractical 1-pole motor, the current is ON for less than half the time per full spin of the coil. There is 0.1 seconds per full spin, so the coil is spinning at 10 cycles per second, or 10 Hz.

Example uses of oscilloscope:
Oscilloscopes are common automotive garages and in engineering and science labs.

- Car spark plug pulses. You'll see them at car garages.

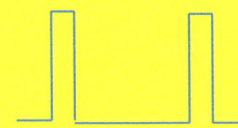

60 Hz

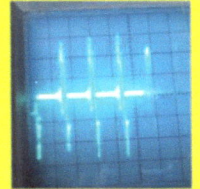

- Alternating voltage on wall electrical outlet.

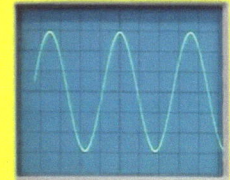

10 GHz

- Modulation of communication signals (AM/FM).

- Viewing radar pulses.

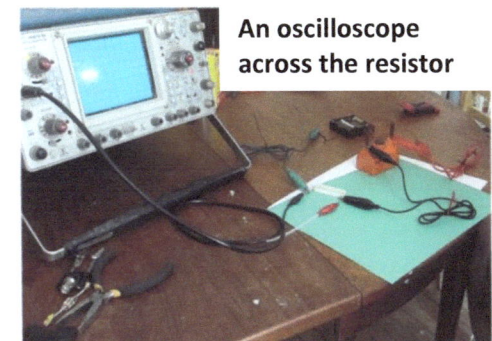

2 Volts per division (across 1 ohm resistor) = 2 Amps per division

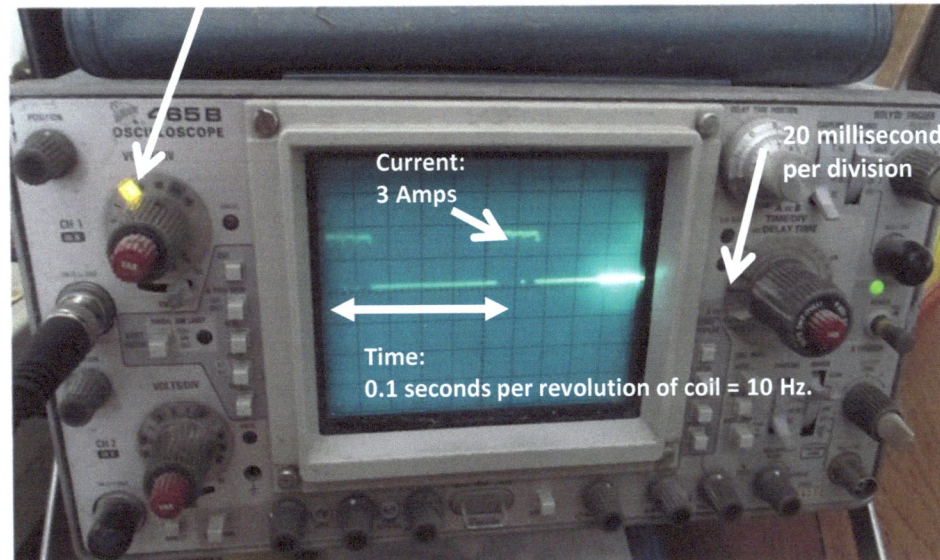

Current:
3 Amps

20 millisecond per division

Time:
0.1 seconds per revolution of coil = 10 Hz.

2 Volts per division (across 1 ohm resistor) = 2 Amps per division

An oscilloscope across the resistor

Measure the current with an oscilloscope across the small value resistor.

Make a simple circuit on your kitchen table.

Here we have instantaneous current versus time, which is the beauty of using an oscilloscope. This is the real stuff, the direct proof that the current is turning on and off, due to the enamel insulator on the wire.

Circuits to See Current in Rotor, Q and A

Electric currents make motors work. The commutator, through the brushes, lets the current flow only in whatever coil is sideways at the time and causes maximum torque.
An N pole rotor will have N/2 coils. Each coil gets current when facing one way, and the opposite current when facing the other way. That current reversal in synch with the coil reversal keeps the rotor magnetic field in a steady direction.

Question 1: How do we know the direction of the current in the rotor?

- We are able to measure the current, like anything involving practical things like motors.
- Currents have a direction. A DC current stays in one direction. An AC current keeps reversing directions, usually at a 60 Hz wall outlet power.

Question 2: What is the difference between a multi-meter and an oscilloscope?

- A multi-meter will report the average of current or voltage. An oscilloscope will show the actual time dependence of the current or voltage.
- A multi-meter just reports a number for current or voltage. An oscilloscope will show a picture of the time dependence.

Question 3: How is current measured?

- Typically current is measured by channeling some of the current through a very low resistance resistor and measuring a voltage drop across the resistor ('in-series' measurement).
- Alternatively, part of the current can be channeled though a coil in a magnetic field in a 'galvanometer', and the deflection of the coil due to the torque will indicate the current ('in-parallel' measurement).
- Alternatively, an iron ring and coil can be placed around the wire with the current, and the magnetic field from the current is measured. With an dc current, a Hall effect sensor is used. With an ac current, the induced voltage is measured, like a generator.

Measure the current through the torque on a coil (old way) or magnetic field sensor around a wire, or a slight voltage drop across a resistor.

Measure the current by measuring the magnetic field induced around the current.

Measure the current by measuring rotation of an electro-coil in a magnetic field.

All these forces on the coil go back to the forces on the charged particles inside the metal wire. If the electrons are moving because a battery voltage is applied, then the magnetic field exerts a force sideways to the wire. This is a motor. If the coil is forced to move by water or steam, then the electrons are physically moving sideways with the wire, as the wire moves, and the magnetic field exerts a force on the charges (electrons) along the wire. This is a generator.

Motor: Push current (those electrons) around using a voltage:

- Electron charge is moving in a magnetic field and a magnetic force is applied to any moving charges, which creates force on wire.
- The goal is to push current sequentially through coils when they are sideways to the stator magnetic field, to create twist.

Hanging wire loops around the magnet: first demonstration of force on a current.

What is observed:
- The liquid is conductive, although poorly.
- The steady current in the wire gets a sideways push due to the magnetic force, so the wire rotates around the center magnet.

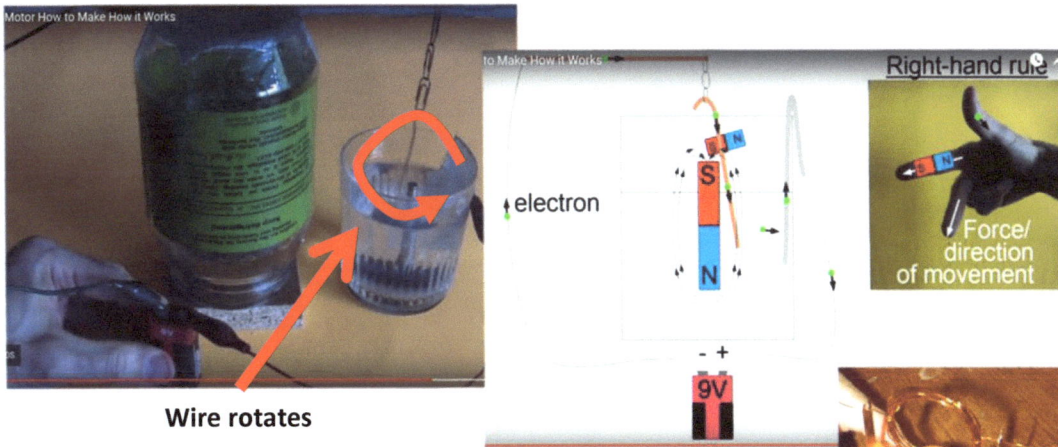

Wire rotates

Motor: Michael Faraday discovered that a current can cause a wire to rotate around a permanent magnet.
Queen (1821): 'Why should I care?'
Faraday: 'Someday you can tax it.'

The one pole motor could have been the first demo too, instead of a twizzling wire.

Generator: Move a wire mechanically through a magnetic field and electrons move with the wire and feel a force:

- Magnetic force on electrons because mechanically moving the electrons in a magnetic field, which creates a voltage.
- The goal is to rotate a metal coil in a huge magnetic field, to create current and voltage.

Magnet pushed into a coil and a voltage is generated: first demonstration of the voltage.

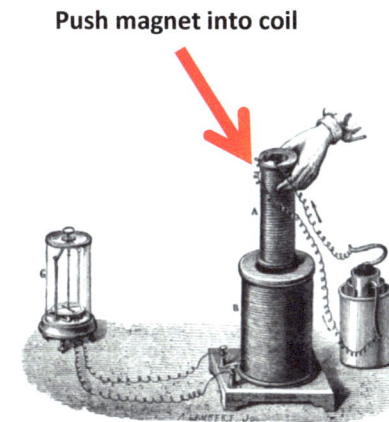

Push magnet into coil

Michael Faraday, with many discoveries in magnetic field effects.

Generator: Michael Faraday discovered a voltage from a magnet moving in a coil: when one electro-magnet is moved in and out of a second coil, a voltage is generated in second coil.

Electric charging from a shake flashlight, with a magnet sliding through a coil.

Here are the first experiments by Faraday showing a motor and a generator.

Fundamental Magnetic Force on Moving Charge (a Current)

Aaahhh, yes. Here is the beauty of fundamental forces, which then leads to current and voltages. Some people love getting everything to be expressed as a force on an electron, although maybe others don't.

"All roads lead to forces on electrons."

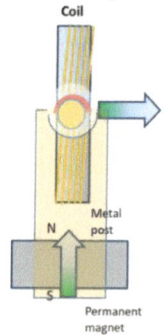

Coil

Metal post

N

S

Permanent magnet

Push those electrons around using a voltage for a motor, or move a wire with electrons in it for a generator: either way, the electrons are moving in a magnetic field and a magnetic force is applied to these electron charges.

Motor

Applied voltage

Electron velocity or current

Current

Field

Force

Force on electrons (and wire)

Metal wire

Permanent Magnet B field

Motion of wire from magnetic force

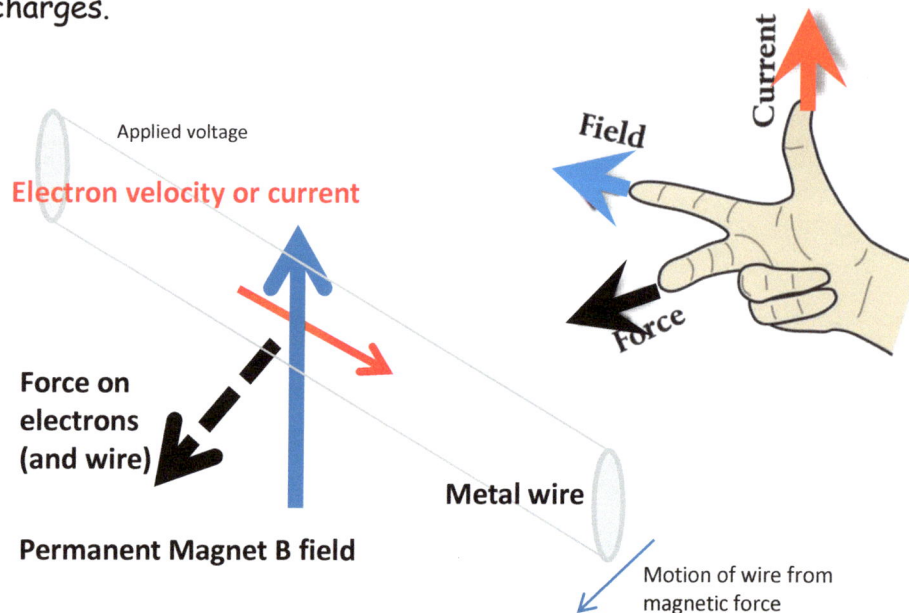

For a motor the magnetic force (black) is sideways from the wire due to a moving charge in a magnetic field

Electrons moving in a magnetic field have a force given by :

*Force per electron = electron charge * velocity of electron * magnetic field*

This force is perpendicular to the electron velocity and to the magnetic field.

Solar Wind:
Charged particles, called solar wind, from the Sun spin around magnetic field lines of Earth and head for the N and S poles.

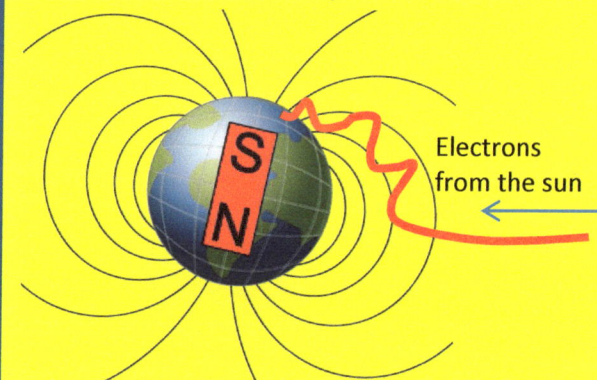

S

N

Electrons from the sun

Northern lights
Charged particles radiate light when their path is bent or accelerated in the Earth's magnetic fields.

Force on a moving electron in a magnetic field is proportional to velocity and magnetic field.

Reciprocity: Force for Motor, and Voltage for Generator

Conceptually, even a 1-pole coil can be a motor or a generator. For a motor, an electric current will push the rotor around. For a generator, the opposite also occurs where a spinning rotor in a magnetic field will generate a voltage.
Practically, this commutator for a 1-pole coil is just held down by gravity and is too difficult to spin mechanically to be a generator.

"Motors and Generators are understood from forces on electrons, same thing."

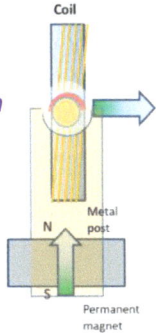
Coil / N / S / Metal post / Permanent magnet

Motor fundamental force:
force on electric current to make mechanical torque

When a current is in the wire, the force propels the coil around. This is a mechanical force and mechanical power.

We are directly converting electric power into the spinning mechanical power of the motor. The charges in the wire are now moving along the wire, but getting pushed by the magnetic field around the axis of the rotor.

There is beautiful symmetry here, where a motor can either use electricity to generate mechanical motion, or mechanical motion can generate electricity.

Motor: Run a current in the wire with a battery

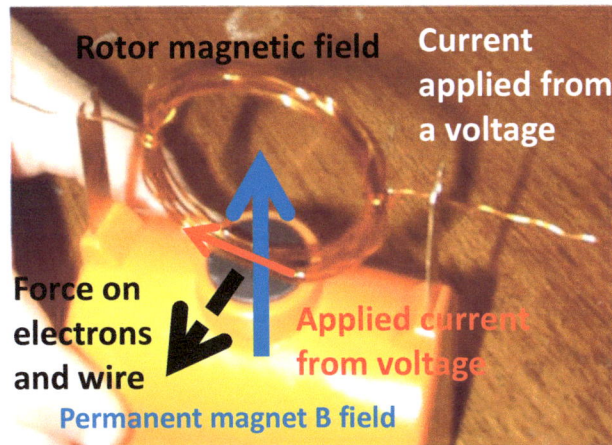

Rotor magnetic field
Current applied from a voltage
Force on electrons and wire
Applied current from voltage
Permanent magnet B field

Magnetic fields, currents, and forces for motor

Here, for a motor, the electrical current is along the wire. This current is caused by the voltage applied to the wire. When the electrons move, then experience a force from the permanent magnetic field. This force is sideways and is a torque which rotates the motor around.

Motor:
- Electric Current velocity is along wire
- Force is sideways to wire

*Force on Wire = Current * Magnetic Field * Length*

Generator fundamental force: moving wire causes force on electrons in wire to make voltage.

When a coil in a motor is forced to spin from a mechanical source, then we are not applying an electrical current to the motor to cause the rotor to spin, but instead are directly and mechanically spinning the motor. The charges in the wire are now moving sideways, or spinning around the shaft of the rotor. The permanent magnetic field then pushes these charges along the wire: this current is the electrical output of the generator.

Generator: Spin the wire mechanically

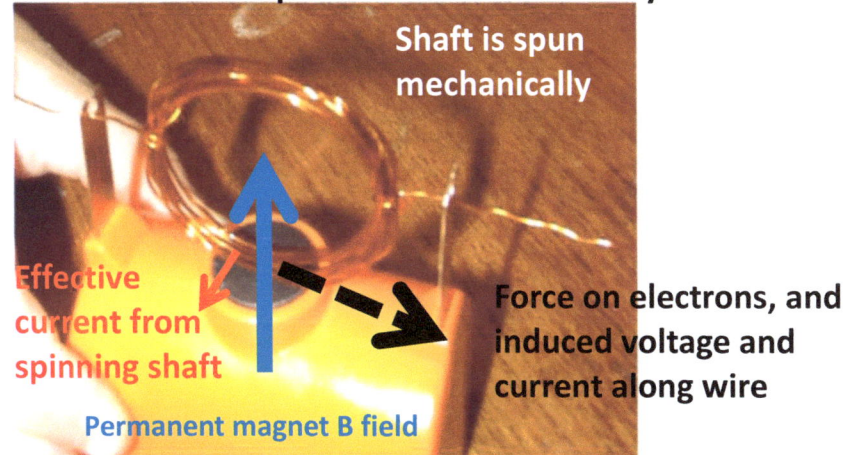

Shaft is spun mechanically
Effective current from spinning shaft
Force on electrons, and induced voltage and current along wire
Permanent magnet B field

Magnetic fields, currents, and forces for generator

Generator:
- Electric Current velocity is perpendicular (sideways) to wire
- Force is along the wire

*Voltage = Area * 2*π* frequency * Magnetic Field*

Currents in a magnetic field experience a force (turning the coil). For a motor, the current is from the applied voltage. For a generator, the current is due to physical movement of the coil wire.

Reciprocity: Motor Force and Generator Voltage

Here is the force on a current or rotor based on magnetic force on a single moving charge, allowing a motor. Here is also the voltage in a moving coil from magnetic force on a single charge, allowing a generator.

These explanations are for those who want to know how the math that shows that an electric current will push the rotor around, or that the opposite also occurs where a spinning rotor in a magnetic field will generate a voltage.

"Hey, some math, it happens, for motors and generators."

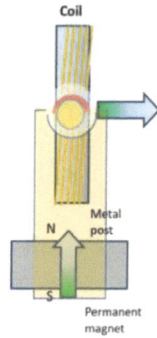

Extra credit for those physics types...

Motor: Get force sideways on wire, from an electric current in a magnetic field

◆ **Force = Current*Magnetic Field*length**

Derivation of force on a wire from force on single charge:

Force per charge: A moving charge in a magnetic field feels a force.
   Force per charge = (charge e)*(velocity of charge)*(magnetic field B)

Get the number of charges flowing in a certain length of wire to get the total force on the wire.
Number of charges:
   Number charges = (number charge density)*(Area wire)*length

Force per wire:
   Force total = Number of charges*Force per charge
   Force total = (number charge density)*(Area wire)*length*(charge e)*(velocity of charge)*B

Isolate the current terms to get a simpler equation for force.
Charge density:
   Charge density = (charge e)*(Number charge density)
   Current = (charge density)*(Area wire)*(velocity charge)

Replace the Current terms by simply Current in the Force total above:
   Force = Current*B*length

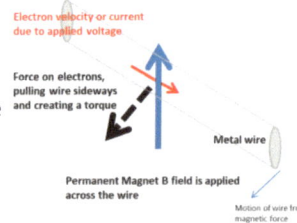

> This force on a current converts electrical to mechanical motion, and causes the rotor to spin:
> • The motion of electrons from a voltage in a magnetic field B causes sideways force on the wire.

Generator: Get voltage on wire, from spinning generator coil in a magnetic field

Voltage = 2*π*frequency *Area coil*Magnetic Field

Derivation of voltage around a coil from force on a single charge:

Equilibrium voltage when
   Magnetic force = Electrostatic force (charge e*E)

Where the electric field E is
   E = Voltage/length

Force per charge: Use magnetic force per charge.
   Force = (charge e)*(velocity wire)*B = (charge e)*Voltage/length

Solve for voltage:
   Voltage = (velocity wire)*B*length

For a rotor of a motor: Get voltage based on spin rate and coil length.
   Velocity wire = 2*π*frequency*radius
   length~2*radius

Insert velocity wire and length into voltage equation above.
   Voltage ~ 2*π*frequency*(2*radius^2)*B
   Voltage ~ 2*π*frequency *((Area coil)*2/π)*B

If did exact math,
   Voltage = 2*π*frequency *(Area coil)*B

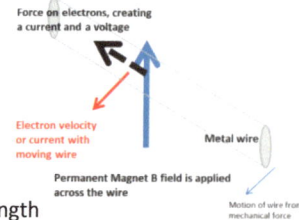

> *This validates Lens's law:*
> $$V = \frac{d\phi}{dt} = \omega\mathrm{NAB} = \omega\mathrm{LB}$$
> **Equation G.7**

> This voltage from a wire moving in a magnetic field converts mechanical motion to electrical power, from mechanically spinning the rotor:
> • Motion of wire is moving the electrons through a magnetic field B, which pushes the charge along the wire and causes a voltage.

Here are the actual equations for force for a motor and voltage estimates for a generator, from the previous page.

Why Coil Spins: Both Force and Magnetic Moment Understanding

Does the rotor spin because of forces on currents, or forces and torques on magnetic moments? Well, both. You pick your favorite explanation for why a motor spins.

"Motors can be understood from forces on magnet fields, or forces on electrons, same thing."

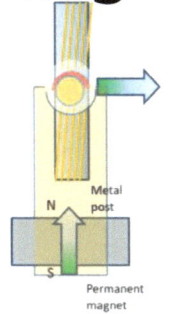

Interpretation 1 why the coil spins: Torque on magnetic moment

Here is one way to understand why coil spins using magnetic moments: Current in the coil makes the rotor magnetic field, or rotor magnetic moment, which then wants to align with permanent magnet.

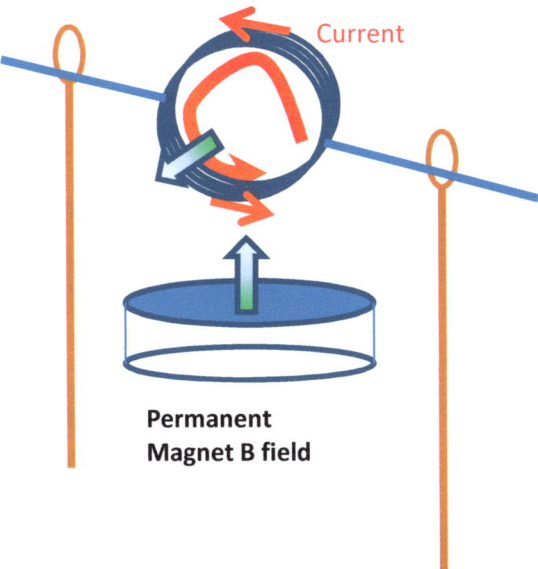

Current

Permanent Magnet B field

Maximum Torque = Coil Magnetic Moment * Magnetic Field B

Where
Coil Magnetic Moment = Area Coil * Number turns * Current

Magnetic fields rotate to align

Interpretation 2 why the coil spins: Force on electric currents

There is a second but equivalent way to understand why the coil spins using force on currents: The current in the coil is moving charges, and moving charges feel a force from a DC magnetic field. The force on the current then pushes on the wire and then pushes the coil around in a spin.

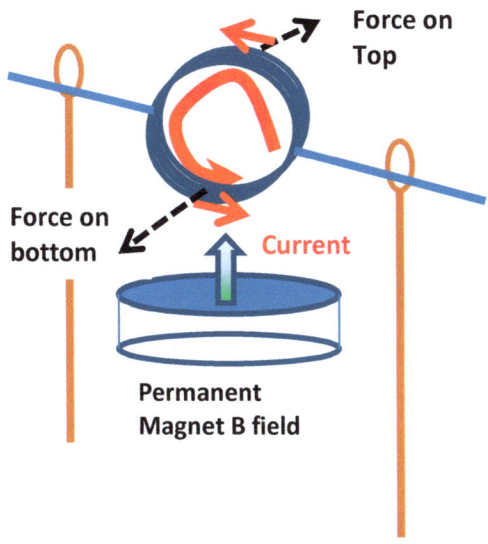

Force on Top

Force on bottom

Current

Permanent Magnet B field

Force on Wire = Current*Magnetic Field B*Length

Current in wire gets pushed sideways because of Permanent Magnet's field

Magnetic moment of rotor, inside stator field

...or...

Electric current through coils of rotor, inside stator field.

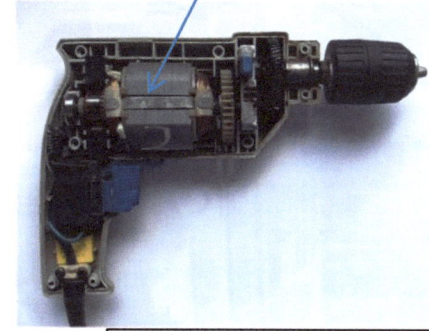

What about iron cores?
- The torque way, using magnetic moments, is better because this approach can easily account for iron cores, using coil magnetic moment.
- The force way needs to create an effective current flowing around the outside of the core, and then the math is the same.

Both ways of looking at the twist on the coil – force on currents or torque on magnetic moment – give the same value of the torque.

Spin: Both Force and Magnetic Moment Understanding

The graphical display of force on currents below can show that there are sideways forces on the top and bottom of the coil.
Or we can treat the whole coil as a magnetic moment and say the coil has less energy when it rotates to be parallel to the stator magnetic field.

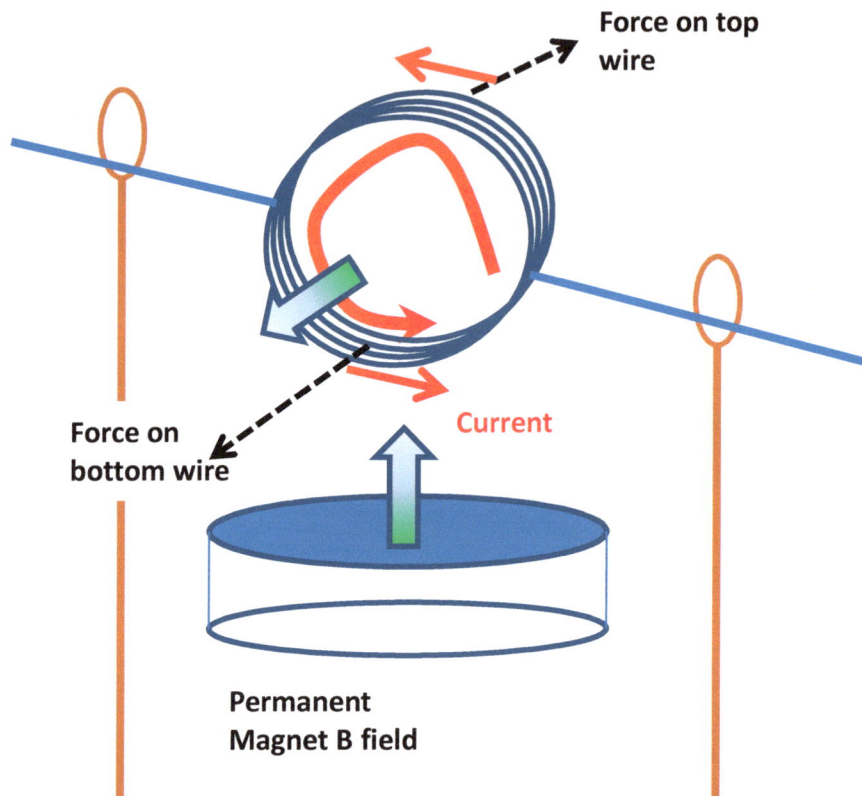

"Motors can be understood from forces on magnet fields, or forces on electrons, same thing."

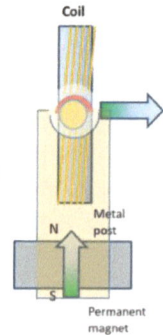

Force on top wire

Force on bottom wire

Current

Permanent Magnet B field

Both ways – magnetic alignment and force on current in wire – are the same and explain why the coil spins.
The current generates the magnetic field, so current or magnetic field interpretation are the same thing. Both ways of understanding will give the same torque equation.

Coil Magnetic Moment without core
=Area Coil * Number turns * Current

Coil Magnetic Moment with core
=k * Coil Magnetic Moment without core

Where k ~ 3 to 10.

Magnetic moment interpretation:
Torque = Coil Magnetic Moment * Magnetic Field B

Estimated value:
Torque = π*radius^2 * Number Turns * Current *Magnetic Field B

Force on current interpretation:
Force on Wire = Current*Magnetic Field B*Length
Torque = radius coil * 2*Force on Wire

Estimate value:
Torque = 4*radius^2 * Number Turns * Current * Magnetic Field B

Both ways of understanding the torque ways – force on currents or twist on magnetic moment – are the same

		parameter	equation	variables
○	1	Voltage V across resistor, as a function of current (Ohm's law):	$V=IR$	V: voltage I: electric current R: resistance
■	2	Magnetic moment m of a coil:	$m=NAI$	I: electric current A: area of single turn loop N: number of loops
■	3	Magnetic moment m of a bar magnet:	$m=MV$	M: magnetization of the material V: volume of the material
◆	4	Force on a magnetic moment m due to magnetic field gradient:	$F = m \cdot \dfrac{dB}{dr} = \|m\| \left\|\dfrac{dB}{dr}\right\| cos\theta$	F: force m : magnetic moment B: magnetic field r : distance θ: angle between magnetic moment and field gradient
◆	5	Force on a wire carrying an electric current due to an external magnetic field:	$F=ILB$	I: electric current L: length of wire B: magnetic field
◆	6	Torque on a magnetic moment in an external magnetic field, change in energy versus angle:	$\tau = m \times B = \|m\|\|B\|sin\theta$	For largest torque, you want large magnetic fields B of the coil and large permanent magnetization of the outer motor. τ: torque θ: angle between magnetic moment and external magnet field
◆	7	Voltage generated in coil by rotating in magnetic field (a generator): Voltage is change in flux versus time	$V = \dfrac{d\phi}{dt} = \omega NAB = \omega LB$	ϕ: flux, or magnetic field times area of coil. ω: frequency in radians per second. N: number loops A: area of a single turn loop. L: inductance of the coil
◆	8	Magnetic field in coil when current is applied:	$B = \dfrac{\mu_o NI}{r}$	Example: Current ~ 2 amps, N turns ~ 10, Radius ~ 0.015 meters →B = 1300e-6 T = 13 Gauss (Earth's magnetic field is 0.5 Gauss)
◆◆	9	Energy of a magnet in an external magnetic field	$Energy = U = -\|m\|\|B\|cos(\theta)$	B: external magnetic field m: magnetic moment of magnet θ: angle between magnetic moment m and external field B
◆◆	10	General expression of force, based on energy gradient versus distance	$F = \dfrac{dU}{dr}$	F: force U: Energy r : distance
◆◆	11	General expression of torque, based on energy gradient versus angle:	$\tau = \dfrac{dU}{d\theta}$	τ: torque U: Energy θ : angle between magnetic moment and external field

These are equations for magnetic fields from currents, and forces and torques from changes of energy.

Right Hand Rules for Magnetic Fields and Forces

Students all over the world are getting indoctrinated with curl rules and which finger goes where rules. They work. The main point is that magnetic fields wrap around a current, and that magnetic forces on a current push sideways to both the current and to any external magnetic field.

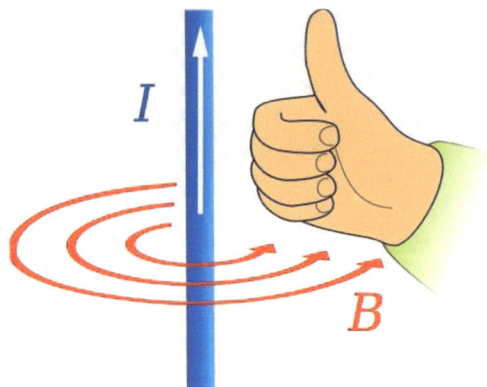

Curl Rule: Right Hand Rule for magnetic field

A straight line current causes a magnetic field to orbit around it.

An example is the current loop direction to create a magnetic field in the direction you want.   Just curl your right hand around the loop to confirm the magnetic field in the middle is in the direction that you want. Your thumb direction gives the current direction.

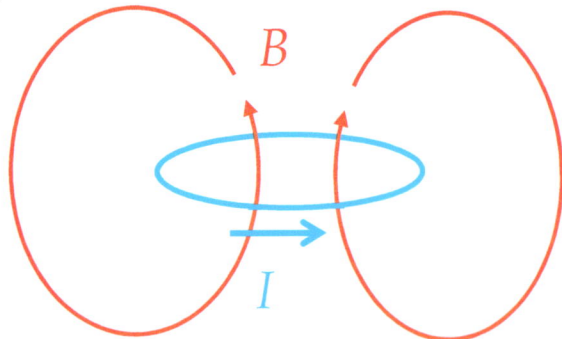

Curl Rule: Fleming's Right Hand Rule for direction of force

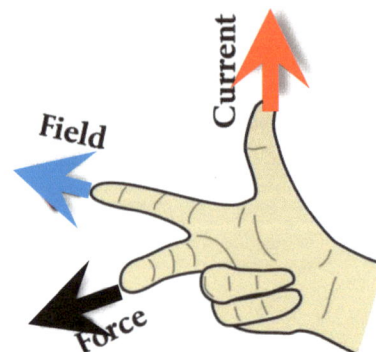

$$F = I \, L \, B \quad \text{(from equation 5)}$$

Or,
Force = Current * Length * Magnetic field

For a motor, the current is due to the applied voltage.

For a generator, the current is due to the sideways motion of the wire.

> **Why all these perpendicular directions?**
> Because the force on a moving charge in a magnetic field is always sideways to the velocity of the charge.

Some rules of thumb for getting directions of forces, torques, and magnetic fields

For the 1-pole motor, does the spin direction change when flip the coil? That is, does the current and rotor magnetic field change direction or stay the same?

Experiment H.1: flip coil between the two posts and observe that the spin direction does not change

Result: Same spin direction when switch posts on coil

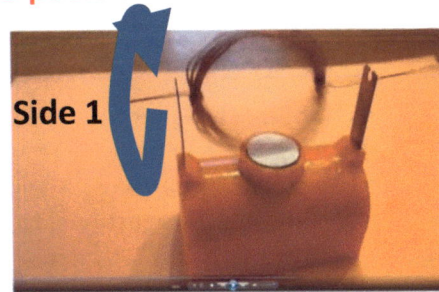

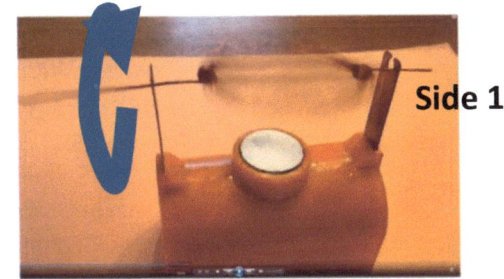

Side 1 | Side 1

Switch posts on coil: Same spin direction when reverse coil between the posts

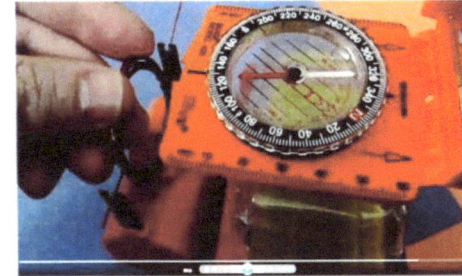

Same coil field when flip (reverse) coil between the posts.

- That is, the currents in the coil stay in the same direction. Two negatives: coil direction reverses and current relative to the wire reverses, so two negatives cancel.
- Coil still spins in the same direction because coil magnetic field is still in the same direction

Why?
Two negatives make the same field:
- Reverse sense of winding the wire (coil clockwise instead of coil counter-clockwise), AND
- Reverse + and – voltages applied to the particular wire ends.

Two reversals makes the same direction coil magnetic B field.

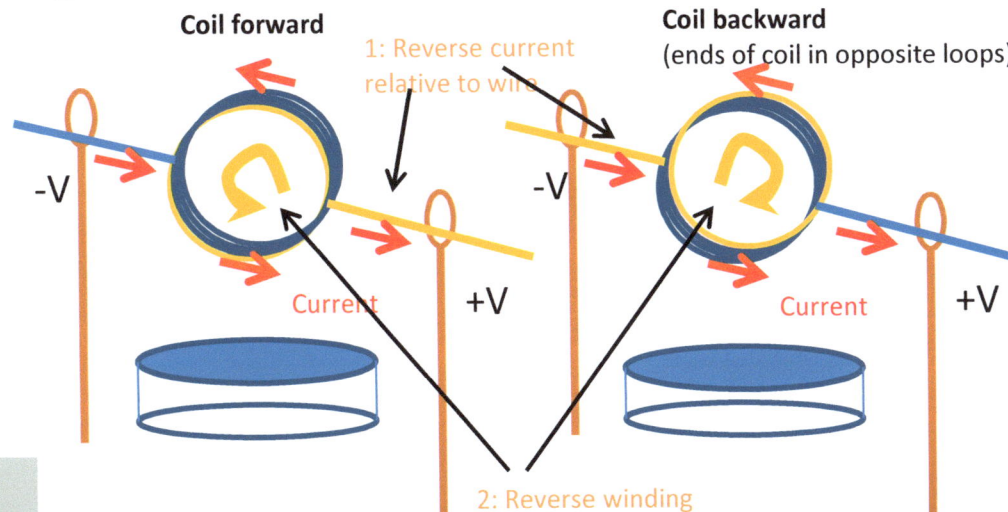

Coil forward | Coil backward (ends of coil in opposite loops)

1: Reverse current relative to wire

-V +V | -V +V

Current | Current

2: Reverse winding

To get the coil to spin in other direction, need to switch voltage (flip battery), or flip magnet (flip permanent B field), but not both together.

When flip coil between posts, the coil spins in the same direction. The current stays in same direction because windings are now facing the other direction

Experiments: Reverse coil and explain why spin direction is the same

Best Number of Turns for Coil

Let's demonstrate the best number of turns of wire in the coil to build the 1-pole commutator.

Winding more turns lets the wire resistance be larger than the battery's own internal resistance, and then most of the battery power is lost in the coil as we want, not in the battery.

"Generally, the more wire turns in the coil, the better, until reach the battery current limits."

Coil / Metal post / N / S / Permanent magnet

More turns and with iron core, in real rotor.

Not spinning

Question: How many turns should we make the coil?

Answer: We want enough turns to maximize the coil magnetic field. We want enough turns and wire length to get the coil resistance much higher than the battery internal resistance. More than that, and there is no benefit to more turns. When the wire resistance dominates the total resistance, the coil magnetic field does not change with extra turns. The magnetic field is proportional to the number of turns, but the current will be inversely proportional to the number of turns.

However, for this 1-pole design, a fat wire is needed mechanically to create the axles to rest the coil and spin. Using this fat wire, we can't get enough turns for the coil to have a wire resistance exceed the internal battery resistance.

There is an internal resistance in the battery. The coil magnetic field is proportional to the number of turns and its resistance is proportional to the number of turns. So, without internal resistance of the battery, there would be the same coil magnetic field no matter how many turns of the coil, because more turns means more field at same current, but the current drops off because the coil resistance is growing by the same amount.

In a nutshell, the number of turns of the coil should be enough to generate much more resistance than the internal resistance of the battery. Any number of turns above that threshold will produce the optimum torque.

Battery's internal resistance

Battery Size	Preferred Nominal Charging Current	Preferred Nominal Test Current	Series Resist. 118 In Ohms For Nominal Current
N	8.4 ma	12.2 ma	87
AAA	9.73 ma	14.2 ma	75
AA	14.6 ma	21.4 ma	50
C	47.8 ma	70 ma	15.27
D	63.4 ma	93.8 ma	11.53

https://www.rapidtables.com/calc/wire/wire-gauge-chart.html

Wire resistance per meter

Wire gauge calculator

Select gauge #:	14	
Or enter gauge #:	18	AWG
Select wire type:	Copper	
Resistivity:	1.72e-8	Ω·m

Calculate   Reset

Diameter in inches:	0.0403	in
Diameter in millimeters:	1.0237	mm
Cross sectional area in kilo circular mils:	1.6243	kcmil
Cross sectional area in square inches:	0.0013	in^2
Cross sectional area in square millimeters:	0.8230	mm^2
Resistance per 1000 feet*:	6.3697	Ω/kft
Resistance per 1000 meters*:	20.8980	Ω/km

diameter (meter)	Number turns	length wire (meter)
0.025	50.00	3.93

This wire length is still no where near enough length to get optimum magnetic field, with wire resistance greater than battery internal resistance.

Instead, the coil should use many turns of very thin wire to get the wire resistance above resistance internal of battery.

Resistance internal battery (~10 ohm)

Resistance wire (~3 ohm)

Resistance R of wire:

$$R = \frac{\rho l}{A}$$

Where ρ=resistivity of metal, l=length, A=cross sectional area of wire.

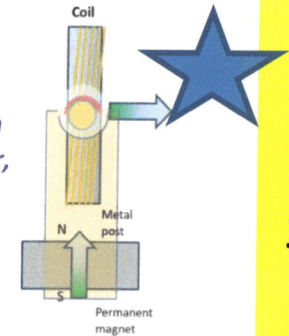

A battery has internal resistance which limits current. The coil's magnetic field, and torque, is largest when the resistance of the wire is much larger than the resistance of the battery.

Experiments: Use enough turns so coil resistance is larger than battery resistance

Index

Alternator, in cars (see generator)

Bullhorn, public speaking   48

Compass   10-11, 25-26, 38, 40-42, 68, 74-77, 91, 189

Cars
- Electric full and hybrid   122-135
- Gasoline engine   122-135

Circuits   215-220

Currents, electric
- Measured   208-210, 215-220
- Loops   64-65

Earth's magnetic field   10-11, 17

Electric motor parts   66-77
- Commutator
- Rotor
- Stator

Electro-magnet
- Lifting   35-38, 41-43
- Rotor   58-59, 61-62, 68-78, 79-82, 85-86, 96-97

Energy
- Force derivative   23, 54-55, 195-202, 203-206
- Torque derivative   54-55, 203-206

Equations for force, voltage   221-226, 227-228

Generators   148-165, 166-180, 221-226

Generator demonstrations
- Air-power   168
- Hand-turn power   156-157
- Steam power   166-173, 176-177

History
- Manual labor   7-9
- Spinning shaft, making motors   13-15
- Electricity applications   17-20
- Electric motors, everyday use   103-114

Imagine, no magnetic fields
- Motors, alternative to electric   115-121
- Generators, alternative   181-183

Iron core of electro-magnet   35-43

Magnet, permanent
- Alignment, flip to   53-54, 56-65
- Rare earth   26
- Toys   29-31, 44-46

Magnetic attraction   22-52

Magnetic repulsion   22-52

Magnetic memory   17-18, 21

Magnetic relay (switch)   44, 50-51

Motors
- AC   79-82
- DC   53-65, 66-78, 79-89
- Math   221-226, 227-228

Motor demonstrations
- 1-pole   90-102
- 2-pole   90-102

Music   48-49, 159-160

People / Scientists and Inventors:
- Ampere   39
- Edison   161
- Faraday   219
- Jedlik   84
- Kettering   144
- Oersted   39
- Sprague   144
- Tesla   162

Power Plants:
- Coal   166, 169, 170-172, 176-177
- Oil   166, 170, 176-177
- Nuclear   166, 170-171, 176
- Solar   150, 176-177, 181-182
- Water, hydroelectric   150, 173, 184-186
- Wind   166, 174-175, 176-176

Robots   145-147

Speaker coils   44, 48-49, 51

Back Cover

"I hope I spun you up, and you go and see magnets, electric motors, and generators everywhere."

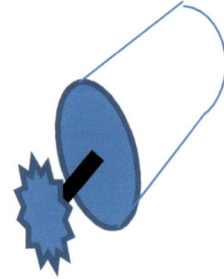

Court Rossman first got into showing kids the basics of motors when helping with Cub Scouts. He was impressed that there was a basic motor to demonstrate the concept, and that the kids could build this demo themselves. It is possible that the scout leaders themselves learn more from doing the activity.

This book relates the magnet and motor experiments to real uses and applications. It should not be hard for the reader to realize how important motors are to modern life styles. The concepts are all in the main part of the book, and the details or the more advanced material are placed in the appendices, so the main part of the book keeps a lighter flow of concepts.

Court is participating in the emphasis on hands-on learning. Science kits are readily available these days, and that is great. He just wants to help that trend, so the next generation has practical knowledge and creativity.

Court has also published 'Scout Pinewood Derby Cars and Real Cars', 'Water Bottle Rockets', and 'Gamut of Speedy Rockets'. Court has a life-long interest in physics and science, and has a Ph.D. in physics, partly around magnetic field effects.

Two audiences were in mind when writing this book. One is a young person who is interested in how things work, probably with the help of a parent. Another is two college students, probably history or business majors, discussing how things work over a beer. These two business majors will probably be the boss of the engineer young person when they all grow up.

Thanks to Brian Stevens for running a 1-pole motor experiment during cub scouts merit badge for engineering This scout meeting made me realize that a book could be written at a basic level to describe motors. Thanks to Peach Rossman for enjoying magnets and engines and letting me take pictures. Thanks to Doug Jansen for providing the analogy of a clunky 1 pedal bicycle. Thanks to James Kurdzo for describing low friction brushless motors, and how he uses them to power his electric bicycle. Thanks to the many motor and generator kits readily available to buy.

References:
http://www.animations.physics.unsw.edu.au/jw/electricmotors.html
http://spiff.rit.edu/classes/phys213/lectures/niagara/niagara.html
http://www.tva.gov/power/hydro.htm
http://users.owt.com/chubbard/gcdam/html/hydro.html
https://www.thoughtco.com/rare-earth-metals-2340169
47+ Magnet Jokes That Will Make You Laugh Out Loud (jokojokes.com)

www.ingramcontent.com/pod-product-compliance
Lightning Source LLC
Chambersburg PA
CBHW041059210326

41597CB00004B/137

* 9 7 8 0 5 7 8 3 9 4 3 9 8 *